Natural Active Compounds from Foods: Applications in Antioxidant and Bioactivity Evaluation

Natural Active Compounds from Foods: Applications in Antioxidant and Bioactivity Evaluation

Guest Editors

Chia-Hung Kuo
Hui-Min David Wang
Shu-Ling Hsieh

MDPI

Basel • Beijing • Wuhan • Barcelona • Belgrade • Novi Sad • Cluj • Manchester

Guest Editors
Chia-Hung Kuo
Department of Seafood Science
National Kaohsiung University of Science and Technology
Kaohsiung
Taiwan

Hui-Min David Wang
Graduate Institute of Biomedical Engineering
National Chung-Hsing University
Taichung
Taiwan

Shu-Ling Hsieh
Department of Seafood Science
National Kaohsiung University of Science and Technology
Kaohsiung
Taiwan

Editorial Office
MDPI AG
Grosspeteranlage 5
4052 Basel, Switzerland

This is a reprint of the Special Issue, published open access by the journal *International Journal of Molecular Sciences* (ISSN 1422-0067), freely accessible at: https://www.mdpi.com/journal/ijms/special_issues/6EFT85Z7K1.

For citation purposes, cite each article independently as indicated on the article page online and as indicated below:

Lastname, A.A.; Lastname, B.B. Article Title. *Journal Name* **Year**, *Volume Number*, Page Range.

ISBN 978-3-7258-7663-1 (Hbk)
ISBN 978-3-7258-7664-8 (PDF)
https://doi.org/10.3390/books978-3-7258-7664-8

Contents

About the Editors

Chia-Hung Kuo

Chia-Hung Kuo currently holds the positions of Professor and Chair of the Department of Seafood Science, and Director of the Center for Aquatic Products Inspection Service at the National Kaohsiung University of Science and Technology (NKUST), Taiwan. He earned his MS in Food Science and Technology from the National Taiwan University, and a Ph.D. in Chemical Engineering from the National Taiwan University of Science and Technology. Throughout his career, he has garnered several awards, including Special Outstanding Academia-Industry Collaboration Talent Award (2023), Special Outstanding Teaching Talent Award (2024-2025), Special Outstanding Counseling and Service Talent Award (2024) and Special Outstanding Research Talent Award (2018–2025) from NKUST, the second-largest university in Taiwan. He has authored over 120 international SCI papers, 19 books or book chapters, and holds 3 patents. His H-index is 37 with more than 4300 citations. Additionally, he serves as an Editorial Board Member for *Catalysts*, a Topical Advisory Panel Member for *International Journal of Molecular Sciences*, and acts as a Guest Editor for Special Issues in *Catalysts, Sustainability, International Journal of Molecular Sciences*, and *Molecules*. Recently, he has been recognized as one of the World's Top 2% Scientists by Stanford University from 2024 to 2025. His research focuses on process biochemistry, biocatalysis, food engineering, food analysis, extraction, oil and fat processing, and fermentation biotechnology.

Hui-Min David Wang

Hui-Min David Wang is currently a Professor at the Graduate Institute of Biomedical Engineering, National Chung Hsing University. To date, Professor Wang has published almost 300 SCI-indexed international journal articles and has contributed to three book chapters. His research interests span a wide range of interdisciplinary fields, including precision medicine, skin physiology, cancer therapy, drug discovery and development, micro- and nanotechnology, biomedical sensing, biochemical engineering, biomaterials, food chemistry, development of functional health ingredients, green and sustainable energy, and bioactive compound applications. His work demonstrates significant cross-disciplinary integration and impact. Beyond his academic research, Professor Wang actively serves on boards of directors and review committees for numerous domestic and international academic organizations, contributing to the advancement of scientific research and the promotion of industry–academia collaboration. His research outcomes have resulted in more than 20 granted patents, many of which have been successfully transferred to industry, facilitating the commercialization of academic innovations. Through these efforts, he has actively promoted cross-sector collaboration among industry, government, academia, research institutions, and medical centers, fostering international engagement and achieving mutually beneficial outcomes.

Shu-Ling Hsieh

Shu-Ling Hsieh, a Distinguished Professor at the Department of Seafood Science (National Kaohsiung University of Science and Technology), graduated from the Institute of Fisheries Science, National Taiwan University, Taiwan, in 2002. Throughout her career, she has garnered several prestigious recognitions, including being named among the World's Top 2% Scientists by Stanford University from 2022 to 2025, as well as receiving the Outstanding Research Award and the Excellence in Teaching Award multiple times from National Kaohsiung University of Science and Technology. She has research expertise in aquaculture, molecular biology, alimentology, and food chemistry, including food applications. She is currently focusing her research on: application of natural compounds in aquatic nutrition and immune enhancement; exploration of bioactive components from natural substances for health benefits; study on the sustainability of marine ecological environment about microplastics. She plays a crucial role as a reviewer in a wide range of internationally renowned journals. Additionally, she holds 18 patents, has edited 5 books, and published over 140 research articles. Her H-index is 40 with more than 4800 citations.

Preface

This Reprint showcases the diverse and growing importance of bioactive compounds in foods and their potential to promote health and prevent chronic diseases. The studies included herein delve into various aspects of these compounds, ranging from their extraction and analysis to their specific bioactivities and applications.

The Reprint begins by exploring the effect of high-dose vitamin C on tendon cell degeneration, providing insights into its potential as a treatment for tendinopathy. It then transitions to the fermentation of blackberry juice with lactic acid bacteria, highlighting how this process can enhance antioxidant properties and extend shelf life. Further, the chemopreventive effects of acetylated flavonoids in different cancer cell lines are evaluated, revealing promising strategies for anticancer applications. The subsequent chapters investigate the influence of natural variation and rootstock on the quality and volatile organic compounds of citrus, demonstrating the complex interplay between genetics and environmental factors. Moreover, the study examines the impact of antioxidant treatment as an adjuvant to standard therapy in patients with septic shock, examining the correlation between cytokine storm, oxidative stress, and therapeutic effects. Furthermore, the study explores the effect of *Hibiscus sabdariffa* L. on the metabolism of arachidonic acid in the isolated kidney of a rat model of metabolic syndrome, shedding light on its potential to improve renal function. The effects of L-cysteine and γ-aminobutyric acid treatment on the postharvest quality and antioxidant activity of loquat fruit during storage are also examined, providing valuable insights for the food industry. In addition, the study investigates the proteomic modulation in TGF-β-treated cholangiocytes induced by curcumin nanoparticles, highlighting the anti-inflammatory and hepato-protective properties of curcumin. The interaction between human hemoglobin and antipsychotics clozapine, ziprasidone, and sertindole is also explored. Finally, the production of high-molecular-weight exopolysaccharides from *Tuber borchii* is achieved by submerged fermentation, showcasing the potential of truffles as a source of valuable bioactive compounds.

Collectively, the articles in this Reprint represent a significant contribution to the field of natural active compounds in foods, providing valuable insights for researchers, food scientists, and healthcare professionals alike. It is our hope that this reprint will inspire further research and innovation in the quest to harness the power of natural compounds for the betterment of human health.

Chia-Hung Kuo, Hui-Min David Wang, and Shu-Ling Hsieh
Guest Editors

International Journal of *Molecular Sciences*

MDPI

Article

High-Molecular-Weight Exopolysaccharides Production from *Tuber borchii* Cultivated by Submerged Fermentation

Cheng-Chun Chen [1,†], Parushi Nargotra [2,†], Chia-Hung Kuo [2,3,*] and Yung-Chuan Liu [1,*]

1 Department of Chemical Engineering, National Chung Hsing University, Taichung 402, Taiwan
2 Department of Seafood Science, National Kaohsiung University of Science and Technology, Kaohsiung 811, Taiwan
3 Center for Aquatic Products Inspection Service, National Kaohsiung University of Science and Technology, Kaohsiung 811, Taiwan
* Correspondence: kuoch@nkust.edu.tw (C.-H.K.); ycliu@dragon.nchu.edu.tw (Y.-C.L.); Tel.: +886-7-3617141 (ext. 23646) (C.-H.K.); +886-4-22853769 (Y.-C.L.)
† These authors contributed equally to this work.

Abstract: Truffles are known worldwide for their peculiar taste, aroma, and nutritious properties, which increase their economic value. However, due to the challenges associated with the natural cultivation of truffles, including cost and time, submerged fermentation has turned out to be a potential alternative. Therefore, in the current study, the cultivation of *Tuber borchii* in submerged fermentation was executed to enhance the production of mycelial biomass, exopolysaccharides (EPSs), and intracellular polysaccharides (IPSs). The mycelial growth and EPS and IPS production was greatly impacted by the choice and concentration of the screened carbon and nitrogen sources. The results showed that sucrose (80 g/L) and yeast extract (20 g/L) yielded maximum mycelial biomass (5.38 ± 0.01 g/L), EPS (0.70 ± 0.02 g/L), and IPS (1.76 ± 0.01 g/L). The time course analysis of truffle growth revealed that the highest growth and EPS and IPS production was observed on the 28th day of the submerged fermentation. Molecular weight analysis performed by the gel permeation chromatography method revealed a high proportion of high-molecular-weight EPS when 20 g/L yeast extract was used as media and the NaOH extraction step was carried out. Moreover, structural analysis of the EPS using Fourier-transform infrared spectroscopy (FTIR) confirmed that the EPS was β-(1–3)-glucan, which is known for its biomedical properties, including anti-cancer and anti-microbial activities. To the best of our knowledge, this study represents the first FTIR analysis for the structural characterization of β-(1–3)-glucan (EPS) produced from *Tuber borchii* grown in submerged fermentation.

Keywords: *Tuber borchii*; submerged fermentation; exopolysaccharide; gel permeation chromatography; Fourier-transform infrared spectroscopy; β-(1–3)-glucan

1. Introduction

Truffles are ectomycorrhizal fungi that belong to the phylum Ascomycota, the order Pezizales, and the family Tuberaceae genus *Tuber* [1,2]. Truffles live in symbiotic relationships with the roots of certain trees, where they coexist and produce hypogenous fruiting bodies. True truffles are those that belong to the genus *Tuber*, while other fruiting bodies, such as from the genus *Melanogaster*, are called false truffles, truffle-like, etc. [2,3]. Truffles are globally known as the most economically valuable edible fungi due to their distinctive taste, aroma, and nutritional characteristics [1,4]. They are a great source of vitamins, minerals, essential amino acids, polysaccharides, phenolic compounds, essential fatty acids, sterols, sphingolipids, and androstanolone. Additionally, they are also an untapped source of medicinal compounds with anti-inflammatory, antioxidant, antibacterial, immune-suppressant, anti-mutagenic, anti-carcinogenic, antiviral, and hepato-protective properties. The presence of medicinally important molecules and biomedical properties shows their promising potential for application in pharmacology and medicine [5–8]. Moreover, they, like other fungi [9–12], also produce various enzymes, such as cellulase, xylanase, amylase, laccase, and so on, during their growth cycle to breakdown/degrade the complex organic matter present around the host plant into simple compounds [5].

Like many other species of truffles, *Tuber borchii* is highly appreciated for its flavor and sensorial qualities [1]. It is a whitish truffle that can be found throughout Europe thriving in both sub-alkaline and acidic soils in climates ranging from cold temperate to Mediterranean. The host trees and shrubs for its growth include oak, poplar, strawberry, and pine, among others [4]. Its ecological attributes and the gastronomic properties of its fruiting bodies contribute to its significance. It is regarded as the *Tuber* species with the largest ecological magnitude due to its ability to adapt to various environmental conditions. Therefore, it can be planted in an environment that is unfavorable for other species of truffles, thereby increasing the geographical area where truffles can be cultivated [13]. Moreover, unlike other *Tuber* species, its mycelium can easily by isolated and cultivated in vitro [14], which makes it a desired *Tuber* species on which to execute various biochemical analyses for in-depth study. *T. borchii* is also known for its ability to produce aroma due to the presence of various volatile organic compounds, which can be exploited for commercial production of truffle-flavored food with natural aroma [15]. Since *T. borchii* has not been used for the production of mycelial biomass, exopolysaccharides (EPSs), and intracellular polysaccharides (IPSs) in submerged fermentation, and due to its aforementioned properties, it was chosen in the current study.

The majority of truffles are obtained by either natural or semi-artificial cultivation. The natural cultivation is impacted by various environmental conditions, such as the temperature, soil type, and host tree. [16]. Additionally, harvesting the truffle fruiting body is a time-consuming and laborious process, with inconsistency in truffle quality, and typically takes 4 to 12 years [4]. Therefore, to address the growing global market demand and reduce the overexploitation of natural truffles, which is endangering the species, in-depth research is being carried out on the semi-artificial cultivation of truffle fruiting bodies. In this regard, submerged fermentation has emerged as a viable alternative for the effective production of bioactive mycelia. Submerged fermentation can help overcome the limitations of natural cultivation by offering a shorter production time and higher biomass amount [17]. Moreover, it results in the production of various EPSs, IPSs, enzymes, and other metabolites of commercial importance [4]. Extracellular amylase, cellulase, and xylanase production from *Tuber maculatum* mycelia has been reported by submerged fermentation with varying culture conditions [1,5,18].

The polysaccharides produced by truffles are known to have different biomedical properties, including immunomodulating, anti-tumor, hypoglycemic, and blood-lipid-lowering activities, and do not affect normal cells [8]. Truffles contain either α- or β-glucan, or mixed α-, β-glucan, which mainly includes (1→3)/(1→6)-β-D-glucans, (1→3)-α-D-glucans, and (1→6)-β-D-glucan. The β-D-glucans are reported to increase the production of macrophages, natural killer cells, T cells, and cytokines, and a few possess antioxidant activity. Moreover, they are also regarded as the utmost significant immunomodulating polysaccharides. Heteroglycans, which consist of sugars, such as mannose, fructose, galactose, and glucose, are also present in the truffle [19]. The molecular weight (MW) of polysaccharides has a critical impact on their biological function. The biological activity of polysaccharides with a high MW may generally be superior. The enhanced activity of high-MW polysaccharides is attributed to their effective binding affinity towards the carbohydrate receptors of immune cells. Polysaccharides with low MW have the potential to penetrate into immune cells and generate an immune response [20]. However, the antioxidant property of a polysaccharide is inversely associated with its MW, i.e., polysaccharides with lower molecular exhibit greater antioxidant activity, whereas high-MW polysaccharides are thought to have potent immunomodulatory and anti-tumor activities [20–22].

Therefore, considering the importance of polysaccharides production from truffles and the role of MW in determining the biomedical properties of a polysaccharide, *T. borchii* was cultivated in submerged fermentation in the current study. The effects of different medium compositions (carbon and nitrogen sources) and different culture conditions on the growth of truffle mycelia and polysaccharides production was evaluated to find the optimal growth conditions for enhanced truffle mycelia and polysaccharides production. The MW distribution and the structure of EPS were estimated by using Gel Permeation Chromatography (GPC) and Fourier-transform infrared spectroscopy (FTIR), respectively.

2. Results and Discussion

2.1. Effect of Different Carbon Sources on Mycelial Biomass and Polysaccharides Production

Living organisms utilize different carbon sources for energy, growth, and development. Therefore, in this study, four different sugars, i.e., two monosaccharides (glucose and fructose) and two disaccharides (sucrose and lactose), were screened to determine the best carbon source. It was found that the dry cell weight (DCW, 3.47 g/L) of truffle with sucrose as the carbon source was significantly higher than the other three carbon sources after 28 days of submerged fermentation (Figure 1). The DCW obtained in the presence of glucose, fructose, and lactose was 2.52 g/L, 2.14 g/L, and 0.67 g/L, respectively (Figure 1). The poor mycelial growth observed in the presence of lactose could possibly be due either to the inability of β-galactose (one of the monomers of lactose) to favor mycelial growth or to low β-galactosidase activity expression in truffle *T. borchii* to hydrolyze lactose into glucose and galactose before it can be used effectively as a carbon source [23]. A study reported galactose, along with lactose and maltose, to be unsuitable for the growth of a medicinal mushroom of genus *Phellinus*, as they had less impact on promoting mycelial growth [24].

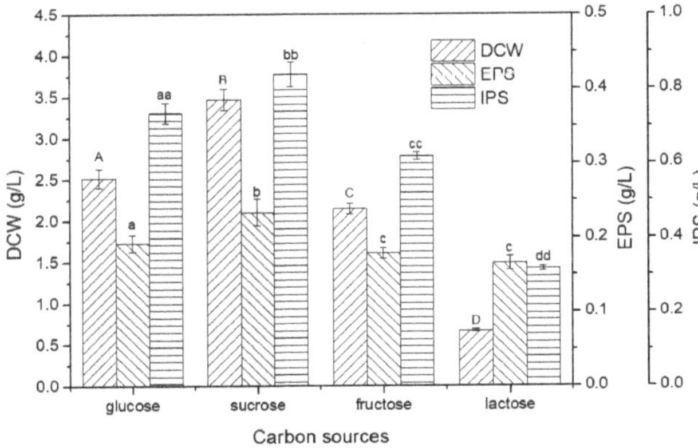

Figure 1. Effect of different carbon sources on mycelial growth (DCW: dry cell weight) and polysaccharides (EPS: exopolysaccharide; IPS: intracellular polysaccharide) production. Different letters above bars in the upper case indicate significant difference ($p < 0.05$) in DCW, different letters above bars in lower case indicate significant difference ($p < 0.05$) in EPS, and different letters above bars in double lower case indicate significant difference ($p < 0.05$) in IPS.

In terms of polysaccharides production, the presence of glucose, fructose, sucrose, and lactose resulted in the EPS amount of 0.192 ± 12 g/L, 0.178 ± 7 g/L, 0.233 ± 1.78 g/L, and 0.165 ± 9 g/L, respectively, after 28 days of submerged fermentation. The consumption of sucrose as a carbon source resulted in a maximum EPS production that was about 1.41 times higher than the lowest amount produced with lactose as a media component (0.165 ± 9 g/L). A similar pattern of results was obtained in the case of IPS production. The maximum IPS production was obtained in the presence of sucrose (0.838 ± 32.3 g/L), followed by glucose (0.735 ± 27.1 g/L) and fructose (0.619 ± 99.5 g/L). The presence of lactose in the medium resulted in the lowest IPS production (0.316 ± 6.2 g/L), which was about 2.648 times lower than that obtained from sucrose (Figure 1). The results indicated that the production of polysaccharides was proportional to the total amount of mycelia. Since the presence of sucrose resulted in maximum growth and EPS production, it was chosen as the carbon source for further studies. Research studies are available for the use of various carbon sources to produce mycelial biomass, EPS, and IPS from different mushrooms. A study reported the use of different carbon sources (glucose, sucrose, lactose, and fructose) to assess the impact of these sources on the mycelial dry weight and EPS and IPS production from *Pleurotus* sp. mushroom grown in submerged fermentation [25]. It was observed that the maximum mycelium dry weight was observed in the presence of glucose; however, the highest amount of EPS and IPS was obtained in the presence of sucrose [25]. In another study, although sucrose's presence as the carbon source resulted in increased biomass and IPS production from *Tuber sinense* in submerged fermentation, the maximum EPS was produced in the presence of lactose, even with poor cell growth [23]. The result of the present study correlates with those reported in the literature indicating that the choice and consumption of carbon sources vary among different mushroom varieties for mycelial growth and polysaccharides production.

When various carbon sources were used in the submerged cultivation, the substrate yields for both biomass and EPS were also estimated, which are listed in Table 1. As shown in the table, sucrose gave the highest yields for both biomass and EPS, whereas lactose displayed a very low biomass yield. The EPS yield for lactose, however, remained at a specific level of 0.41 (%), resulting in the high biomass-based EPS yield of 0.245 g/g. It is assumed that the lactose is not fit for *T. borchii* growth due to the aforementioned reasons, whereas the slow growth of cells did not appear to prevent the release of EPS from cells. This special high yield of EPS based on biomass could provide us a basis for further exploration when using lactose as a carbon source in *Tuber* cultivation.

Table 1. Effect of different carbon sources on the total yield of mycelial biomass and polysaccharides production from *T. borchii* submerged cultivation.

Carbon Source	Yx/s (%)	Y_{EPS}/s (%)	Y_{EPS}/x (g/g)
glucose	6.30	0.48	0.076
sucrose	8.68	0.59	0.068
fructose	5.36	0.45	0.084
lactose	1.67	0.41	0.245

Yx/s: biomass yield based on substrate, Y_{EPS}/s: EPS yield based on substrate, Y_{EPS}/x: EPS yield based on biomass.

2.2. *Effect of Sucrose Concentration*

The impact of varying sucrose concentrations (40 g/L, 50 g/L, 60 g/L, 80 g/L, and 100 g/L) on the amount of mycelial biomass and polysaccharides production was also investigated. The amount of DCW analyzed after 28 days of growth was found to be 3.47 g/L, 3.25 g/L, 3.37 g/L, 3.58 g/L, and 3.31 g/L at the concentration (g/L) of 40, 50, 60, 80, and 100, respectively (Figure 2). Although there was not much difference in the mycelium amount, however, the maximum mycelial growth was observed at the sucrose concentration of 80 g/L. With regard to EPS production, the maximum amount (0.547 ± 28.34 g/L) was obtained at the concentration of 80 g/L and decreased to 0.464 ± 31.51 g/L at the 100 g/L sucrose concentration (Figure 2). The EPS obtained at a 50 g/L and 60 g/L concentration was 0.386 ± 46.95 g/L and 0.429 ± 39.54 g/L, respectively. The EPS amount at 80 g/L sucrose was 2.196 times higher compared to the lowest (0.249 ± 4.14 g/L), which was obtained at 40 g/L.

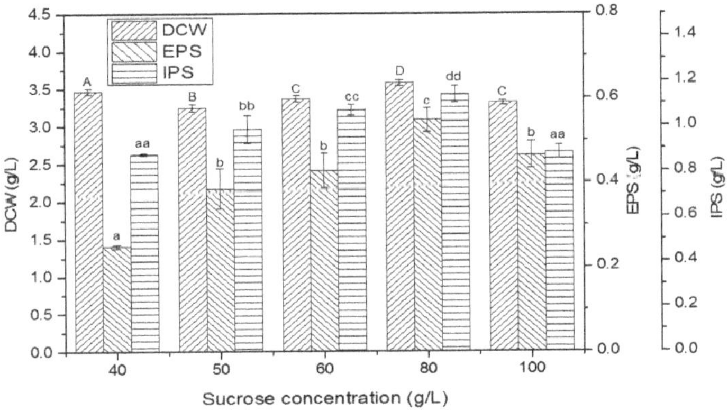

Figure 2. Effect of different sucrose concentrations on mycelial growth (DCW: dry cell weight) and polysaccharides (EPS: exopolysaccharide; IPS: intracellular polysaccharide) production. The letters above bars are the same indication as shown in Figure 1.

The amount of IPS at the sucrose concentrations (g/L) of 50, 60, and 100 were 0.987 ± 64.29 g/L, 1.071 ± 24.82 g/L, and 0.884 ± 29.82 g/L, respectively (Figure 2). Moreover, at the sucrose concentration of 80 g/L, the IPS production was up to 1.142 ± 36.16 g/L, which was 1.304 times higher than the lowest amount obtained at 40 g/L (0.875 ± 5.1 g/L). The results indicated that the polysaccharides production is significantly affected by the sucrose concentration. The decrease in mycelial biomass and polysaccharides production at 100 g/L sucrose could be attributed to the increased viscosity of the fermentation medium due to the increased amount of the carbon source, which may limit the nutritional and oxygen diffusion, represses mycelial growth, and consequently lead to low polysaccharides production. Moreover, the osmotic pressure brought on by the high sugar content may reduce polysaccharides production [18,23]. The results were in line with the reported study on polysaccharides production from *Grifola umbellate* in submerged fermentation. The cell density and polysaccharides production was enhanced, with an increased glucose concentration from 1% to 3%; however, at a further increase from 3% to 5% concentration, the cell density and polysaccharides production was inhibited [26]. Accordingly, sucrose at the concentration of 80 g/L was used as a carbon source in subsequent experiments.

2.3. Effect of Different Nitrogen Sources on Mycelial Biomass and Polysaccharides Production

Several different organic and inorganic nitrogen sources were screened for their effects on mycelial biomass and polysaccharides production. A lack of nitrogen hinders the growth of fungi and the production of polysaccharides, since it is a crucial component of proteins and nucleic acids. The organic sources of nitrogen used were yeast extract, peptone, and malt extract, while ammonium chloride (NH_4Cl) and diammonium phosphate ($(NH_4)_2HPO_4$) were inorganic nitrogen sources. The nitrogen sources were used at a concentration of 10 g/L, and sucrose (80 g/L) was used as a carbon source. After 28 days of submerged fermentation, the presence of yeast extract as a nitrogen source in the medium resulted in a DCW of 3.41 ± 0.04 g/L. Compared with yeast extract, the DCW decreased in the presence of peptone (1.40 ± 0.05 g/L), followed by malt extract (0.41 ± 0.01 g/L) and NH_4Cl (0.27 ± 0.04 g/L). However, when $(NH_4)_2HPO_4$ was used, a much lower final cell density (DCW), i.e., 0.08 ± 0.01 g/L, was obtained (Figure 3).

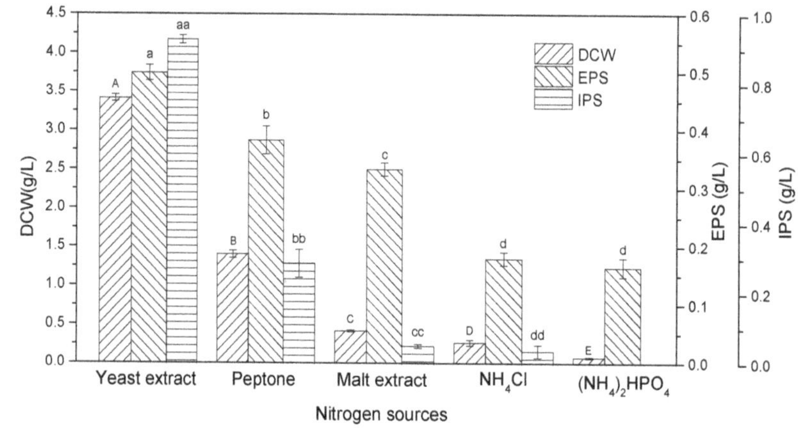

Figure 3. Effect of different nitrogen sources on mycelial growth (DCW: dry cell weight) and polysaccharides (EPS: exopolysaccharide; IPS: intracellular polysaccharide) production. The letters above bars are the same indication as shown in Figure 1.

The presence of yeast extract in the medium also favored EPS production, as the maximum EPS of 0.498 ± 13.24 g/L was achieved, which was about 3.45 times higher than the lowest amount obtained in the presence of $(NH_4)_2HPO_4$ (0.144± 16.15 g/L). EPS of 0.382 ± 23.98 g/L, 0.333 ± 11.38 g/L, and 0.179 ± 11.59 g/L was obtained when peptone, malt extract, and NH_4Cl were used as a nitrogen source, respectively. In terms of IPS production, a similar trend was found, with the highest amount of IPS (0.929 ± 12.36 g/L) in the presence of yeast extract as compared to peptone (0.286 ± 40.31 g/L), malt extract (0.051 ± 5.55 g/L), and NH_4Cl (0.036 ± 18.5 g/L). When $(NH_4)_2HPO_4$ was used as a nitrogen source, no IPS was detected due to the negligible mycelial growth. From these findings it was evident that the organic nitrogen may have promoted cell growth and polysaccharide formation by supporting the synthesis of essential amino acids since the cell density and polysaccharides production were low in the presence of inorganic nitrogen. Moreover, yeast extract is known to be a rich source of proteins, amino acids, and vitamins, which favors cell growth [27,28]. Contrary to our results, some reports suggest that different nitrogen sources may stimulate high mycelial growth and polysaccharides production from the same organisms under the same conditions. Wang and co-workers [28] observed that the maximum mycelial biomass (2.32 g/L), EPS (1.58 g/L), and IPS (29.1 mg/L) were obtained when yeast extract, malt extract, and peptone, respectively, were used as a nitrogen source for the cultivation of *Grifola frondosa* under submerged fermentation. Similarly, the mycelium dry weight (27.3 ± 2.1 g/L) was maximum when yeast extract was present in the medium, whereas meat extract produced the maximum EPS (24.1 ± 0.8 mg/mL) and IPS (5.23 ± 0.07 mg/mL) amount when *Pleurotus* sp. was grown in submerged fermentation using these nitrogen sources [25].

2.4. Effect of Yeast Extract Concentration

Since yeast extract promoted mycelia growth and polysaccharides production, it was used at varying concentrations (g/L), i.e., 10, 15, 20, 25, and 30, to further identify its optimal concentration for enhanced mycelial growth and polysaccharides production. The results indicated that cell growth increased with an increase in yeast extract concentration and was highest at 20 g/L with a DCW of 4.06 ± 0.11 g/L (Figure 4). A decrease in growth was observed upon further increasing the concentration, with the lowest DCW at 30 g/L yeast extract (1.27 ± 0.22 g/L). In contrast to mycelial growth, EPS production increased with the increasing concentration and was found to be maximum (0.905 ± 4.85 g/L) when 30 g/L yeast extract was used in the fermentation medium. However, as seen in Figure 4 and Table S1, for IPS production, the concentration of 20 g/L yeast extract gave a maximum IPS of 1.385 ± 41.03 g/L, and on further increasing the concentration, a negative effect on IPS production was observed. IPS production at 20 g/L yeast extract was 3.37 times greater than the lowest amount obtained at 30 g/L (0.411 ± 103.05 g/L). The high amount of IPS at 20 g/L could be attributed to the high mycelial growth obtained at the same concentration. Liu et al. [23] reported that the maximum dry cell weight (19.53 ± 0.13 g/L), EPS production titer (0.66 ± 0.06 g/L), and IPS production titer (0.90 ± 0.03 g/L) was obtained at a 25 g/L concentration of yeast extract from *Tuber sinense* in submerged fermentation. In another study, yeast extract as the organic nitrogen source and ammonium chloride as the inorganic nitrogen source resulted in the highest mycelium dry weight and EPS production from *Lentinus squarrosulus* [29]. From the findings of the current study, it was evident that the maximum mycelial biomass and IPS production was obtained at the 20 g/L yeast extract concentration; however, EPS production was highest with 30 g/L yeast extract. Therefore, in the subsequent experiments, the analysis of the molecular weight of the EPS produced with different yeast extract concentrations using the GPC technique was considered as the basis to ascertain the optimum yeast extract concentration.

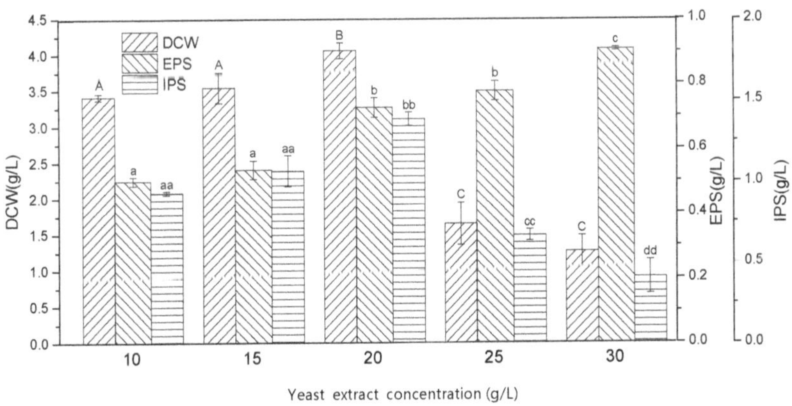

Figure 4. Effect of different yeast extract concentrations on mycelial growth (DCW: dry cell weight) and polysaccharides (EPS: exopolysaccharide; IPS: intracellular polysaccharide) production. The letters above bars are the same indication as shown in Figure 1.

2.5. Effect of Carbon/Nitrogen Ratio

The concentration of both carbon and nitrogen sources and their ratio in the medium is highly crucial for enhanced cell growth and polysaccharides (EPS and IPS) production. Therefore, the carbon/nitrogen (C/N) ratios using sucrose and yeast extract as the carbon and nitrogen sources, respectively, were determined. The effect of different C/N ratios on the mycelial growth and EPS and IPS production is shown in Table 2. When the C/N ratio was higher than 4, as in case I, the DCW obtained was almost 3.3–3.6 g/L, which was associated with higher IPS of about 0.880–1.150 g/L. In this case, the EPS level obtained was in the range of 0.250–0.550 g/L. However, in case II, when the C/N ratio was lower than 4, a lower DCW (1.3–1.7 g/L) was obtained, which was associated with a lower IPS amount (0.400–0.700 g/L). Nevertheless, in this case, higher EPS (0.770–0.900 g/L) was achieved. This indicated that whereas the production of IPS may be positively correlated with the quantity of cell biomass, EPS production appears to be adversely correlated with cell biomass. When the C/N ratio and sucrose concentration were 4 and 80 g/L (case II), respectively, it was possible to achieve both high cell biomass (4.06 ± 0.11 g/L) and a greater IPS level (1.385 ± 41.03 g/L). Additionally, in this case, the EPS obtained was 0.725 ± 30.17 g/L. This suggested that not only the C/N ratios, but also the carbon source levels are critical for both biomass and polysaccharides (EPS and IPS) production. Similar to our study, the C/N ratio was important for the biomass and polysaccharides production from mushroom *Pleurotus* sp. A C/N ratio of 80:1 yielded enhanced mycelial biomass and IPS and EPS production of 20.1 g/L, 11.89 ± 1.15 mg/mL, and 3.38 ± 1.21 mg/mL, respectively [25].

Table 2. Effect of different C/N ratios on mycelial biomass and polysaccharides production when sucrose and yeast extract were used for *T. borchii* submerged cultivation.

C/N Ratio (*w/w*)	Dry Cell Weight (g/L)	EPS (g/L)	IPS (g/L)
Case I			
4 (40:10)	3.47 ± 0.04	0.249 ± 0.41	0.875 ± 0.51
5 (50/10)	3.25 ± 0.05	0.386 ± 4.70	0.987 ± 6.13
6 (60/10)	3.37 ± 0.04	0.429 ± 3.95	1.071 ± 2.48
8 (80/10)	3.58 ± 0.04	0.547 ± 2.83	1.142 ± 3.62
10 (100/10)	3.31 ± 0.03	0.464 ± 3.15	0.884 ± 2.98
Case II			
8 (80/10)	3.41 ± 0.04	0.498 ± 13.24	0.929 ± 12.36
5.3 (80/15)	3.54 ± 0.21	0.603 ± 28.14	1.064 ± 96
4 (80/20)	4.06 ± 0.11	0.725 ± 30.17	1.385 ± 41.03
3.2 (80/25)	1.65 ± 0.29	0.776 ± 29.72	0.667 ± 34.38
2.66 (80/30)	1.27 ± 0.22	0.905 ± 4.85	0.411 ± 103.05

2.6. Truffle Growth and Biomass/Polysaccharides Production

Generally, in a batch culture, the classical growth curve of all the fungi, including truffles, include different phases, i.e., lag phase, log phase, and stationary phase. These phases are different from one another on the basis of the amount of cell biomass, metabolite production, and nutrient depletion. The lag phase is characterized by the fact that not all of the cells from the inoculum begin to proliferate at the same time, followed by the log phase or exponential phase, wherein cells utilize nutrients and grow at an exponential rate. The stationary phase is represented by the exhaustion of nutrients and no further increase in the biomass amount, as cell growth slows down, and the number of living cells drops off exponentially. In order to identify the optimal time of incubation to achieve maximum growth and polysaccharides production, it is crucial to understand and monitor the time profile of the growth. Therefore, truffle was grown in submerged fermentation using the selected carbon and nitrogen sources in the optimized concentration, and the growth and polysaccharides production (EPS and IPS) was monitored for 35 days. Figure 5a shows the growth curve of truffle *T. borchii*, where the initial stage was from the 0th to 7th day, when the inoculum was in the adaptation period, and little growth was observed. Exponential growth was seen from the 11th to 28th day, as on the 28th day, the maximum mycelia biomass (5.38 ± 0.01 g/L), EPS (0.70 ± 0.02 g/L), and IPS (1.76 ± 0.01 g/L) were obtained (Figure 5a,b). A decline in the biomass growth and polysaccharides production was observed after the 28th day of submerged fermentation. Therefore, in the subsequent experiments, submerged fermentation was carried out for 28 days to obtain the maximum biomass and polysaccharides production.

2.7. Molecular Weight Determination of Exopolysaccharides

The molecular weight of a polysaccharide is an important parameter for its biological activity. It is generally believed that higher-MW polysaccharides typically exhibit improved biological and anti-cancer activity owing to the ability of polysaccharides with MWs > 90 kDa to form a triple-helix structure. Furthermore, polysaccharides with MW > 200 kDa are reported to be good immunogens with immunomodulating activity [30,31]. Therefore, in the present study, the MW of the EPS produced by the truffle after 28 days of fermentation was analyzed using the GPC method, using pullulan as the standard. The molecular weight of EPS produced at varying yeast extract concentrations on different days of submerged fermentation and extracted by various extraction methods, including water extraction, alkali extraction, and acid extraction, was estimated.

The MW distribution of the polysaccharides was divided into high molecular weight (280 kDa > MW > 100 kDa), medium molecular weight (100 kDa > MW > 10 kDa), and low molecular weight (10 kDa > MW > 800 Da) within the calibration curve.

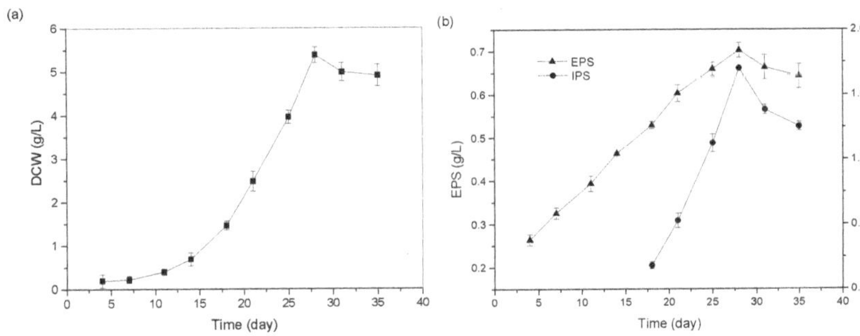

Figure 5. Time profile of *T. borchii* (**a**) enhanced dry cell weight; (**b**) polysaccharides production under submerged fermentation.

2.7.1. Molecular Weight of EPS Produced at Different Concentrations of Yeast Extract

The results from the previous experiments showed that the maximum mycelial biomass was obtained at 20 g/L yeast extract; however, the best EPS amount was obtained at the highest concentration of yeast extract (30 g/L). Since the molecular weight of the polysaccharide also affects its functional activity, the molecular weight of EPS produced at different yeast extract concentrations was analyzed. The MW and molecular weight distribution of EPS extracted from water, NaOH, and HCl extraction were compared. It was found that the EPS extracted by using water, NaOH, and HCl at the yeast concentrations 15 g/L and 20 g/L had the maximum average MW. Figure 6a shows that the EPS from water extract was mostly distributed in the small-MW range and slightly in the large-molecular-weight range. The NaOH extraction of EPS at a 20 g/L yeast extract concentration contained EPS of higher and medium MW (Figure 6b). More than 70% of the EPS was in the small MW range when the concentration of yeast extract was 25 g/L. However, when HCl was used for the extraction, 90% of the EPS was in the small MW range, with an average MW of only about 5 kDa (Figure 6c). It was evident from the water and NaOH extraction of EPS that at a yeast concentration lower than 15 g/L and higher than 20 g/L, small-MW EPS was more prominent; therefore, only 2 yeast extract concentrations, i.e., 15 g/L and 20 g/L, were used for HCl extraction. The high ratio of small-molecular-weight EPS even at 15 and 20 g/L yeast extract could be attributed to the hydrolysis of EPS by HCL. From these findings it was concluded that a 20 g/L yeast concentration is the optimal concentration for obtaining high mycelial biomass and high-molecular-weight EPS.

2.7.2. Molecular Weight of EPS Produced on Different Days of Submerged Fermentation

The maximum mycelial growth and polysaccharides production was achieved on the 28th day of submerged fermentation, after which both were decreased. However, since the polysaccharides production was different on different days of fermentation, there could be a possibility of variation in the MW of EPS produced during these days. Therefore, the relationship between EPS production and the MW of EPS produced on different days of fermentation was investigated. It was observed that the average MW of polysaccharides was gradually increased with fermentation time. On the 0th day of fermentation, the average MW of EPS was 47,349 Da, which increased to 92,445 Da on the 28th day, with the highest average MW achieved throughout the course of fermentation. The average MW remained stable to about 92 kDa, even after the 28th day of fermentation (Figure 7a). The EPS produced from day 0 to day 14 was mostly in the range of small MWs, with EPS of about 0.22 g/L on day 14. A shift from the small-MW range to the medium range was observed for EPS produced between the 18th to 21st day, and about 0.28 g/L EPS was recorded on the 21st day. However, the ratio of high-molecular-weight EPS increased after the 21st day and reached the average MW of 92 kDa on the 28th day of fermentation. A decrease in small- and medium-molecular-weight EPS was observed on the 28th day. The amount of high-MW EPS increased from about 0.053 g/L on day 0 to 0.32 g/L on the 28th day (Figure 7b). Similar to our study, Yeh et al. [32] also reported the increase in the ratio of high-MW EPS produced from *Hypocreales* sp. NCHU01 under submerged fermentation with the increase in the cultivation time. The average MW of EPS produced from edible fungus *Stropharia rugosoannulata* was reported to be 2.1×10^4 D [33]. The high-MW EPS produced using the optimal concentration of the carbon and nitrogen source under submerged fermentation may possess multiple biomedical applications in the health care sector.

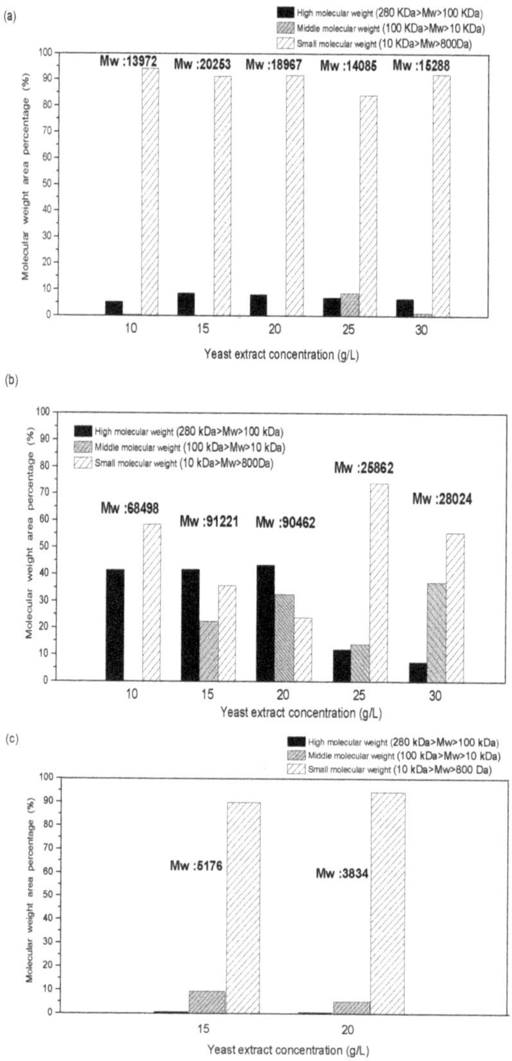

Figure 6. Molecular weight distribution of (**a**) water-extracted, (**b**) NaOH-extracted, and (**c**) HCl-extracted EPS produced at different concentrations of yeast extract.

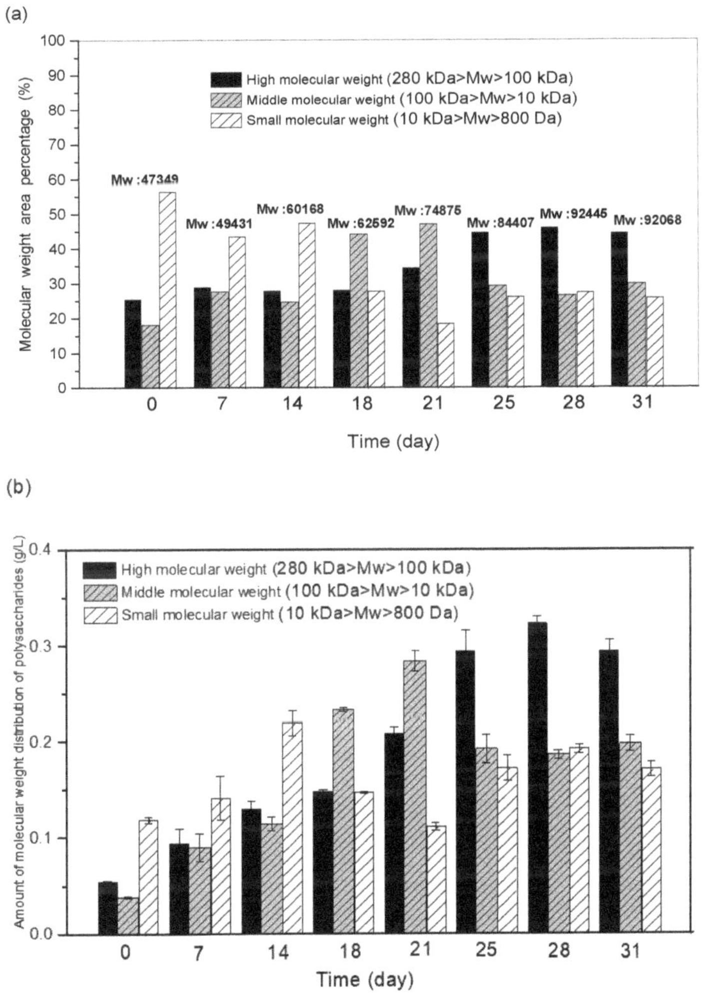

Figure 7. (**a**) Average molecular weight of EPS produced; (**b**) Molecular weight distribution of EPS produced during different days of submerged fermentation.

2.8. FTIR Polysaccharide Analysis

The FTIR spectroscopy technique was used to analyze the structural characterization of EPS produced by the truffle owing to its sensitivity to the position and anomeric arrangement of glycosidic bonds. Figure 8 represents the FTIR spectra of EPS produced by *T. borchii* at the sucrose and yeast extract concentrations of 80 and 20 g/L, respectively, extracted with NaOH and collected on the 28th day of cultivation by submerged fermentation. The broad and intense peak observed at 3372 cm^{-1} represented the stretching vibration of a hydroxyl group (O–H), suggesting the presence of a molecule with several free hydroxyl groups. The existence of the methylene group in the EPS was confirmed by the absorption peak at 2935 cm^{-1}, which corresponded to the stretching vibration of C–H bonds [34]. The peak at 1648 cm^{-1} was assigned to the bound-water bending vibration, as the polysaccharides possess a significant affinity for water and are rapidly hydrated in the solid form [35]. The -CH_3 stretching vibration was denoted by the presence of a peak at 1400 cm^{-1}. The characteristic absorption peaks at 1114 and 881 cm^{-1} were attributed to (1→3) glycosidic linkages and a β-configuration in the anomeric region, which confirmed that the EPS was β-(1–3)-glucan [34,36]. The β-glucans are recognized to have antimicrobial, anti-cancer, and glucose-lowering properties. They are the strong immune response modifiers due to their ability to trigger innate and adaptive immune responses by interacting with a variety of immune cell receptors [6,36]. The production of β-glucans as EPS may pave the way for its biomedical application.

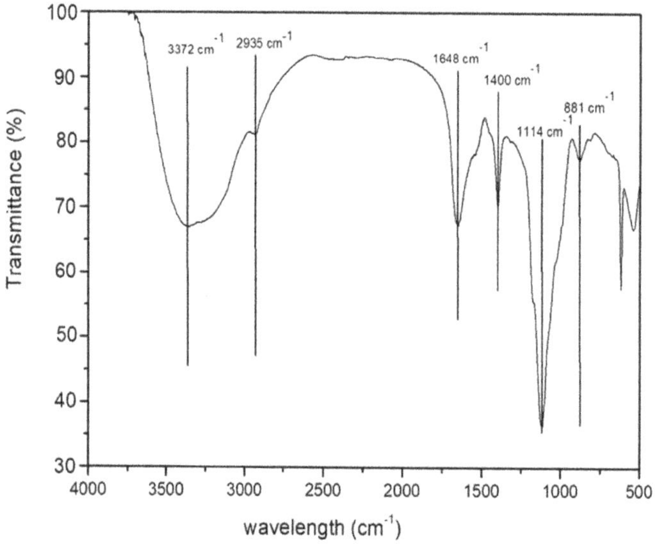

Figure 8. FTIR analysis of exopolysaccharides (EPSs) produced from *T. borchii* cultivated by submerged fermentation.

3. Materials and Methods

3.1. Strain and Chemicals

The truffle strain used in the present study was *Tuber borchii* (ATCC® MYA1019™), which was purchased from the American Type Culture Collection (ATCC, Manassas, VA, USA). It was identified using molecular PCR (T100, Bio-rad, Hercules, CA, USA) analysis, and the strain identification report is added as Supplementary materials (Table S2). The mycelium was maintained on a potato-dextrose-agar (PDA) plate. The PDA plate was inoculated with mycelium and incubated at 25 °C for 28 days for growth. The biomass was then grown in 500 mL Erlenmeyer flasks containing 100 mL of seed media with the following composition: 0.5% glucose, 0.1% yeast extract, 0.1% peptone, 0.02% KH_2PO_4, 0.02% $MgSO_4{\cdot}7H_2O$ [26]. The seed media was inoculated with four pieces of cutter square (20 mm × 20 mm) mycelial biomass previously grown on a PDA plate. The cutter square was homogenized (V100, Osterizer, Mexico) prior to the seeding and selected from the outer part of the plate culture. The inoculated media was incubated on a rotary shaker set at 100 rpm and 25 °C for 21 days. Chemicals, including glucose, magnesium sulphate ($MgSO_4{\cdot}7H_2O$), potassium dihydrogen phosphate (KH_2PO_4), and pullulan standard, were purchased from Showa Denko K.K, Tokyo, Japan. Bacto™ yeast extract and Vitamin B_1 were acquired from Becton, Dickinson and Company, Sparks, MD, USA and Merck Ltd., Taipei, Taiwan, respectively. PDA was procured from HiMedia, Bombay, India, HCL and ethanol from Echo Chemical Co. Ltd., Miaoli, Taiwan, and NaOH from Showa, Saitama, Japan. All the chemicals used in the current study were of analytical grade.

3.2. Flask Cultivation

A shake-flask culture was performed in a 500 mL Erlenmeyer flask containing 100 mL of the basal fermentation medium. The medium consisted of the following components: glucose, 40 g/L; yeast extract, 10 g/L; KH_2PO_4, 1 g/L; $MgSO_4 \cdot 7H_2O$, 1 g/L; vitamin B_1 hydrochloride, 0.15 g/L. The pH was adjusted to 7 by the addition of either 0.1 N HCl or 1 N NaOH. Media were sterilized at 121 °C for 20 min, and the carbon source was autoclaved separately. A total of 90 mL medium in a 500 mL shake flask was inoculated with 10 mL seed broth. The flasks were incubated on a rotary shaker at 100 rpm and 25 °C for 28 days.

3.3. Effect of Different Carbon, Nitrogen Sources, and Carbon/Nitrogen Ratio

Various carbon and nitrogen sources were screened for mycelial growth and polysaccharides production. The carbon sources used in the study were glucose, sucrose, fructose, and lactose. The effect of the carbon source was studied by using either of the carbon sources in basal fermentation medium at 40 g/L concentration and keeping other growth parameters and conditions constant. Similarly, the effect of various organic substrates (yeast extract, peptone, and malt extract) and inorganic salts (ammonium chloride (NH_4Cl) and diammonium phosphate ($(NH_4)_2HPO_4$)) as nitrogen sources was also evaluated by using either of the organic and inorganic nitrogen sources in the basal fermentation medium, as per the concentration mentioned above in Section 3.2. Furthermore, varying concentrations of the best carbon (40 g/L, 50 g/L, 60 g/L, 80 g/L, and 100 g/L) and nitrogen sources (10 g/L, 15 g/L, 20 g/L, 25 g/L, and 30 g/L), which resulted in the maximum growth and polysaccharides production (EPS and IPS), were used in order to find the optimal concentration of the respective source. Additionally, the effect of C/N ratio on mycelial biomass and polysaccharide (EPS and IPS) production was also evaluated. Two cases were studied. In case I, the carbon source concentration was varied, and the nitrogen source was used at a constant concentration with the following C/N ratios: 4 (40/10), 5 (50/10), 6 (60/10), 8 (80/10), and 10 (100/10). In case II, the concentration of the carbon source was maintained constant whereas the nitrogen source concentration was changed, resulting in the following ratios; 8 (80/10), 5.3 (80/15), 4 (80/20), 3.2 (80/25), and 2.66 (80/30).

3.4. Assays

3.4.1. Cell Biomass

The cell biomass content was expressed as dry cell weight (DCW). Each time, three flasks were taken, and the broth was filtered to obtain the DCW. The filter paper used was Advantec No.1 with a pore size of 6 μm and a diameter of 90 mm (Toyo Roshi Kaisha Ltd., Tokyo, Japan). The cells were then washed three times with distilled water and dried at 60 °C to a constant weight.

3.4.2. Exopolysaccharides Extraction and Analysis

The EPS amount was quantified by removing the mycelial biomass in the broth through filtration. To the filtered broth, 95% (*v*/*v*) ethanol (four times the volume of the broth) was added and kept at −20 °C overnight in order to precipitate out the crude EPS. The ethanol and broth mixture was then subjected to centrifugation (Megafuge 16R, Thermo Scientific, Taichung, Taiwan) at 10,000× *g* for 20 min. The supernatant was discarded and to the pellet, 75% ethanol (8 mL) was added, followed by centrifugation at 10,000× *g* for 20 min. The supernatant was removed, and the pellet was kept at 60 °C for 20 min to remove water and ethanol. For EPS extraction using water, HCl, or NaOH, the pellet was resuspended in deionized (DI) water, 1 N HCl, or 1 N NaOH, respectively, and kept at 85 °C and 100 rpm for 1 h. The total sugar concentration of the crude EPS was assayed using the phenol-sulfuric acid method, and glucose was used as a standard [37].

3.4.3. Intracellular Polysaccharides Analysis

For the determination of IPS, the mycelial biomass obtained after filtration was subjected to freeze drying. A total of 100 mg of powder was mixed with 10 mL of distilled water, and the mixture was kept at 121 °C for 20 min. The process was repeated twice, followed by centrifugation at 8000 rpm for 10 min. The supernatant was collected and subjected to the same ethanol precipitation and washing process to obtain the crude IPS, as described in Section 3.4.2. The total sugar was estimated using the phenol-sulfuric acid method [37].

3.4.4. Molecular Weight Analysis

The molecular weight of the crude EPS was determined on the basis of a calibration curve built by the pullulan standard (5.9–1600 kDa). For estimation, a gel permeation chromatography (Analytical Scientific Instruments, Richmond, CA, USA), together with a RI-2000 detector (Showa Denko K.K, Tokyo, Japan), was employed. Three columns, viz., KB-802.5, KB-804, and KB-805 (0.8 cm × 30 cm, Showa Denko K.K, Tokyo, Japan), were connected in a series. The flow rate was kept at 1.0 mL/min with the column temperature at 35 °C, and the injection volume was 20 mL. DI-water was utilized as a mobile phase [38]. The average molecular weight of the EPS extracted using the water extraction, alkali extraction, and acid extraction methods at different yeast extract concentrations was also analyzed. Additionally, the molecular weight of the polysaccharides produced at different fermentation time intervals was estimated to determine the average MW of the polysaccharides produced by truffle. The area percentage of the EPS with high MW was estimated using the following equation:

$$\text{Molecular weight area } (\%) = \frac{\text{Area of EPS with MW higher than 300 kDa}}{\text{Area of total EPS}} \times 100$$

The molecular weight distribution tendency of the EPS, along with the fermentation time, from small MW converted to medium MW and then to large MW was estimated by multiplying the polysaccharide amount (g/L) with the MW area percentage (%), and the relative amount of each polysaccharide was obtained.

3.4.5. Fourier-Transform Infrared Spectroscopy Analysis

The crude EPS produced by *T. borchii* at the sucrose and yeast extract concentration of 80 and 20 g/L, respectively, extracted with NaOH and collected on the 28th day of cultivation by submerged fermentation, was subjected to FTIR (FT-720; Horiba, Ltd., Tokyo, Japan) for the analysis of the structure and different functional groups in the EPS. The sample and potassium bromide (KBr) were mixed in the ratio 1:100, ground into powder by agate mortar, and then pressed into a pellet. The FTIR analysis conditions set were as follows: scanning 32 times per measurement; resolution set at 2 cm^{-1}; and scanning wavelength range 500~4000 cm^{-1}.

3.5. Statistical Analysis

All the experiments were conducted in triplicates. Several flasks were run simultaneously, and daily sampling was done in triplicates by taking three flasks every time. A mean ± standard deviation was used to express each result. The statistical significance was assessed by the multiple comparisons Tukey post-hoc analysis of variance (ANOVA) using Origin Pro ver.9.0 (Origin Lab Corporation, Northampton, MA, USA). The differences of the results were considered statistically significant at p-values < 0.05.

4. Conclusions

The current study presents the first ever report on the cultivation of *T. borchii* by submerged fermentation for the production of polysaccharides of biological importance. Various carbon and nitrogen sources were screened and their concentration was optimized in order to achieve the maximum amount of mycelial biomass and polysaccharides (EPS and

IPS). The most suitable carbon and nitrogen sources were found to be sucrose (80 g/L) and yeast extract (20 g/L), respectively, which yielded maximum mycelial growth and enhanced EPS and IPS production. The molecular weight of the EPS analyzed by the GPC method revealed that with the increase in concentration of yeast extract to 20 g/L, the molecular weight distribution shifted to the high-molecular-weight range. The highest average molecular weight of the EPS was 92 kDa, which was achieved in the EPS extracted using NaOH at a yeast extract concentration of 20 g/L on the 28th day of submerged fermentation. The results of this study demonstrated the production of high-molecular-weight EPS, which was found to be β-(1–3)-glucan after structural characterization. Moreover, the EPS (β-(1–3)-glucan) produced in this study can be screened for various functional properties, such as anti-cancer, antioxidant, immunomodulation, and glucose lowering, to determine its potential role in the biomedical field.

Supplementary Materials: The following supporting information can be downloaded at: https://www.mdpi.com/article/10.3390/ijms24054875/s1.

Author Contributions: Conceptualization, Y.-C.L.; methodology, C.-C.C.; validation, P.N.; formal analysis, C.-C.C.; investigation, C.-C.C.; writing—original draft preparation, P.N., C.-H.K. and Y.-C.L.; writing—review and editing, P.N., C.-H.K. and Y.-C.L.; supervision, Y.-C.L. All authors have read and agreed to the published version of the manuscript.

Funding: This work was supported by research-funding grants provided by the National Science and Technology Council of Taiwan, R.O.C. (Grant no. NSTC-111-2221-E-005-040 and NSTC-111-2221-E-992-005-MY3) and was also supported in part by the Ministry of Education, Taiwan, R.O.C. under the ATU plan.

Institutional Review Board Statement: Not applicable.

Informed Consent Statement: Not applicable.

Data Availability Statement: Data are contained within the article.

Acknowledgments: Author P.N. thankfully acknowledges the National Science and Technology Council of Taiwan, R.O.C., for providing the Post-Doctoral fellowship.

Conflicts of Interest: The authors declare no conflict of interest.

References

1. Bedade, D.K.; Singhal, R.S.; Turunen, O.; Deska, J.; Shamekh, S. Biochemical Characterization of Extracellular Cellulase from *Tuber maculatum* Mycelium Produced Under Submerged Fermentation. *Appl. Biochem. Biotechnol.* **2017**, *181*, 772–783. [CrossRef] [PubMed]
2. Tang, Y.-J.; Liu, R.-S.; Li, H.-M. Current Progress on Truffle Submerged Fermentation: A Promising Alternative to Its Fruiting Bodies. *Appl. Microbiol. Biotechnol.* **2015**, *99*, 2041–2053. [CrossRef] [PubMed]
3. Leonardi, M.; Iotti, M.; Pacioni, G.; Hall, I.R.; Zambonelli, A. Truffles: Biodiversity, Ecological Significances, and Biotechnological Applications. In *Industrially Important Fungi for Sustainable Development: Volume 1: Biodiversity and Ecological Perspectives*; Fungal, Biology; Abdel-Azeem, A.M., Yadav, A.N., Yadav, N., Usmani, Z., Eds.; Springer: Cham, Switzerland, 2021; pp. 107–146, ISBN 978-3-030-67561-5.
4. Shah, N.; Marathe, S.J.; Croce, D.; Ciardi, M.; Longo, V.; Juilus, A.; Shamekh, S. An Investigation of the Antioxidant Potential and Bioaccumulated Minerals in *Tuber borchii* and *Tuber maculatum* Mycelia Obtained by Submerged Fermentation. *Arch. Microbiol.* **2021**, *204*, 64. [CrossRef]
5. Bedade, D.; Berezina, O.; Singhal, R.; Deska, J.; Shamekh, S. Extracellular Xylanase Production from a New Xylanase Producer *Tuber maculatum* Mycelium under Submerged Fermentation and Its Characterization. *Biocatal. Agric. Biotechnol.* **2017**, *11*, 288–293. [CrossRef]
6. Cateni, F.; Gargano, M.L.; Procida, G.; Venturella, G.; Cirlincione, F.; Ferraro, V. Mycochemicals in Wild and Cultivated Mushrooms: Nutrition and Health. *Phytochem. Rev.* **2022**, *21*, 339–383. [CrossRef]
7. Yeh, C.-W.; Kan, S.-C.; Lin, C.-C.; Shieh, C.-J.; Liu, Y.-C. Polyhydroxylated Steroids and Triterpenoids from an Entophytic Fungus, *Hypocreales* Sp. NCHU01 Isolated from *Tuber magnatum*. *J. Taiwan Inst. Chem. Eng.* **2016**, *64*, 22–30. [CrossRef]
8. Zhang, Z.; Wang, J.; Liu, L.; Shun, Q.; Shi, W.; Liu, X.; Wang, F. The Optimum Conditions and Mechanism for Increasing Exo-Polysaccharide Production of *Truffles melanosporum* by Dioscorea Saponins. *LWT* **2019**, *107*, 331–339. [CrossRef]

9. Sharma, V.; Tsai, M.-L.; Chen, C.-W.; Sun, P.-P.; Patel, A.K.; Singhania, R.R.; Nargotra, P.; Dong, C.-D. Deep Eutectic Solvents as Promising Pretreatment Agents for Sustainable Lignocellulosic Biorefineries: A Review. *Bioresour. Technol.* **2022**, *360*, 127631. [CrossRef]
10. Sharma, V.; Bhat, B.; Gupta, M.; Vaid, S.; Sharma, S.; Nargotra, P.; Singh, S.; Bajaj, B.K. Role of Systematic Biology in Biorefining of Lignocellulosic Residues for Biofuels and Chemicals Production. In *Sustainable Biotechnology—Enzymatic Resources of Renewable Energy*; Singh, O.V., Chandel, A.K., Eds.; Springer: Cham, Switzerland, 2018; pp. 5–55. ISBN 978-3-319-95480-6.
11. Sharma, V.; Tsai, M.-L.; Nargotra, P.; Chen, C.-W.; Sun, P.-P.; Singhania, R.R.; Patel, A.K.; Dong, C.-D. Journey of Lignin from a Roadblock to Bridge for Lignocellulose Biorefineries: A Comprehensive Review. *Sci. Total Environ.* **2022**, *861*, 160560. [CrossRef]
12. Weng, Z.-H.; Nargotra, P.; Kuo, C.-H.; Liu, Y.-C. Immobilization of Recombinant Endoglucanase (CelA) from *Clostridium thermocellum* on Modified Regenerated Cellulose Membrane. *Catalysts* **2022**, *12*, 1356. [CrossRef]
13. Lancellotti, E.; Iotti, M.; Zambonelli, A.; Franceschini, A. Characterization of *Tuber borchii* and *Arbutus unedo* Mycorrhizas. *Mycorrhiza* **2014**, *24*, 481–486. [CrossRef] [PubMed]
14. Belfiori, B.; Riccioni, C.; Paolocci, F.; Rubini, A. Characterization of the Reproductive Mode and Life Cycle of the Whitish Truffle, *T. borchii*. *Mycorrhiza* **2016**, *26*, 515–527. [CrossRef] [PubMed]
15. Zambonelli, A.; Iotti, M.; Puliga, F.; Hall, I.R. Enhancing White Truffle (*Tuber magnatum* Picco and *T. borchii* Vittad.) Cultivation Through Biotechnology Innovation. In *Advances in Plant Breeding Strategies: Vegetable Crops: Volume 10: Leaves, Flowerheads, Green Pods, Mushrooms and Truffles*; Al-Khayri, J.M., Jain, S.M., Johnson, D.V., Eds.; Springer: Cham, Switzerland, 2021; pp. 505–532. ISBN 978-3-030-66969-0. [CrossRef]
16. Mustafa, A.M.; Angeloni, S.; Nzekoue, F.K.; Abouelenein, D.; Sagratini, G.; Caprioli, G.; Torregiani, E. An Overview on Truffle Aroma and Main Volatile Compounds. *Molecules* **2020**, *25*, 5948. [CrossRef]
17. Sharma, V.; Tsai, M.-L.; Nargotra, P.; Chen, C.-W.; Kuo, C.-H.; Sun, P.-P.; Dong, C.-D. Agro-Industrial Food Waste as a Low-Cost Substrate for Sustainable Production of Industrial Enzymes: A Critical Review. *Catalysts* **2022**, *12*, 1373. [CrossRef]
18. Bedade, D.; Deska, J.; Bankar, S.; Bejar, S.; Singhal, R.; Shamekh, S. Fermentative Production of Extracellular Amylase from Novel Amylase Producer, *Tuber maculatum* Mycelium, and Its Characterization. *Prep. Biochem. Biotechnol.* **2018**, *48*, 549–555. [CrossRef] [PubMed]
19. Maity, P.; Sen, I.K.; Chakraborty, I.; Mondal, S.; Bar, H.; Bhanja, S.K.; Mandal, S.; Maity, G.N. Biologically Active Polysaccharide from Edible Mushrooms: A Review. *Int. J. Biol. Macromol.* **2021**, *172*, 408–417. [CrossRef]
20. Gong, P.; Wang, S.; Liu, M.; Chen, F.; Yang, W.; Chang, X.; Liu, N.; Zhao, Y.; Wang, J.; Chen, X. Extraction Methods, Chemical Characterizations and Biological Activities of Mushroom Polysaccharides: A Mini-Review. *Carbohydr. Res.* **2020**, *494*, 108037. [CrossRef] [PubMed]
21. Li, X.; Zhu, Z.; Ye, L.; Kang, Z.; Zhang, X.; Huang, Y.; Zhang, B.; Zou, Y. Comparison of the Partial Structure and Antioxidant Activity of Polysaccharides from Two Species of Chinese Truffles. *Molecules* **2020**, *25*, 4345. [CrossRef] [PubMed]
22. Guan, T.; Wei, X.; Xu, P.; Chen, K.; Zou, Y.; Chen, M.; Zhu, Z. Comparison of Structural and Antioxidant Activity of Polysaccharide Extracted from Truffles. *J. Food Sci.* **2022**, *87*, 2999–3012. [CrossRef] [PubMed]
23. Tang, Y.-J.; Zhu, L.-L.; Li, D.-S.; Mi, Z.-Y.; Li, H.-M. Significance of Inoculation Density and Carbon Source on the Mycelial Growth and Tuber Polysaccharides Production by Submerged Fermentation of Chinese Truffle *Tuber sinense*. *Process Biochem.* **2008**, *43*, 576–586. [CrossRef]
24. Hur, H.; Imtiaj, A.; Lee, M.W.; Lee, T.-S. Suitable Conditions for Mycelial Growth of *Phellinus* Spp. *Mycobiology* **2008**, *36*, 152–156. [CrossRef]
25. Ahmad, R.; Clementine, I.N.; Er, J.S.; Mustaffa, S.N.A.B.S.; Mohtar, W.A.A.Q.W. Incorporation of Palm Oil for the Enhancement of Biomass and Polysaccharide Production through Submerged Fermentation from Locally Isolated *Pleurotus* sp. Mycelium. *Int. J. Adv. Res.* **2018**, *6*, 223–232. [CrossRef]
26. Huang, H.-C.; Liu, Y.-C. Enhancement of Polysaccharide Production by Optimization of Culture Conditions in Shake Flask Submerged Cultivation of *Grifola umbellata*. *J. Chin. Inst. Chem. Eng.* **2008**, *39*, 307–311. [CrossRef]
27. Sun, T.; Wang, R.; Sun, D.; Li, S.; Xu, H.; Qiu, Y.; Lei, P.; Sun, L.; Xu, X.; Zhu, Y. High-Efficiency Production of *Tremella aurantialba* Polysaccharide through Basidiospore Fermentation. *Bioresour. Technol.* **2020**, *318*, 124268. [CrossRef] [PubMed]
28. Wang, Y.C.; Lin, F.Y.; Hsu, T.H. Effects of Nitrogen from Different Sources on Mycelial Biomass and Polysaccharide Production and Pellet Morphology in Submerged Cultures of *Grifola frondosa*. *Bioresources* **2021**, *16*, 2937–2952. [CrossRef]
29. Ahmad, R.; Alias, S.M.U.; Hamid, A.A.; Yusoff, W.M.W.; Ismail, E.; Daud, F. Production and Antiproliferative Activity of Various Crude Extract from *Lentinus squarrosulus* Mycelium. *Sch. Acad. J. Biosci.* **2015**, *3*, 285–377. Available online: https://oarep.usim.edu.my/jspui/handle/123456789/13960 (accessed on 24 February 2023).
30. Leong, Y.K.; Yang, F.-C.; Chang, J.-S. Extraction of Polysaccharides from Edible Mushrooms: Emerging Technologies and Recent Advances. *Carbohydr. Polym.* **2021**, *251*, 117006. [CrossRef]
31. Park, H.G.; Shim, Y.Y.; Choi, S.O.; Park, W.M. New method development for nanoparticle extraction of water-soluble β-(1→3)-D-glucan from edible mushrooms, *Sparassis crispa* and *Phellinus linteus*. *J. Agric. Food Chem.* **2009**, *57*, 2147–2154. [CrossRef] [PubMed]
32. Yeh, C.-W.; Zang, C.-Z.; Lin, C.-C.; Kan, S.-C.; Chang, W.-F.; Shieh, C.-J.; Liu, Y.-C. Quantitative and Morphologic Analysis on Exopolysaccharide and Biomass Production from a Truffle Endophytic Fungus *Hypocreales* Sp. NCHU01. *J. Taiwan Inst. Chem. Eng.* **2014**, *45*, 108–114. [CrossRef]

33. Hao, H.; Cui, C.; Xing, Y.; Jia, X.; Ma, B.; Kang, W.; Li, T.; Gao, M.; Xu, C. Sulfation of the Extracellular Polysaccharide from the Edible Fungus *Stropharia rugosoannulata* with Its Antioxidant Activity. *J. Future Foods* **2023**, *3*, 37–42. [CrossRef]
34. Ahmad Usuldin, S.R.; Mahmud, N.; Ilham, Z.; Khairul Ikram, N.K.; Ahmad, R.; Wan-Mohtar, W.A.A.Q.I. In-Depth Spectral Characterization of Antioxidative (1,3)-β-D-Glucan from the Mycelium of an Identified Tiger Milk Mushroom *Lignosus rhinocerus* Strain ABI in a Stirred-Tank Bioreactor. *Biocatal. Agric. Biotechnol.* **2020**, *23*, 101455. [CrossRef]
35. Zhao, W.; Chai, D.-D.; Li, H.-M.; Chen, T.; Tang, Y.-J. Significance of Metal Ion Supplementation in the Fermentation Medium on the Structure and Anti-Tumor Activity of *Tuber* Polysaccharides Produced by Submerged Culture of *Tuber melanosporum*. *Process Biochem.* **2014**, *49*, 2030–2038. [CrossRef]
36. Orlandelli, R.C.; Corradi da Silva, M.d.L.; Vasconcelos, A.F.D.; Almeida, I.V.; Vicentini, V.E.P.; Prieto, A.; Hernandez, M.D.D.; Azevedo, J.L.; Pamphile, J.A. β-(1→3,1→6)-d-Glucans Produced by *Diaporthe* Sp. Endophytes: Purification, Chemical Characterization and Antiproliferative Activity against MCF-7 and HepG2-C3A Cells. *Int. J. Biol. Macromol.* **2017**, *94*, 431–437. [CrossRef] [PubMed]
37. DuBois, M.; Gilles, K.A.; Hamilton, J.K.; Rebers, P.A.; Smith, F. Colorimetric Method for Determination of Sugars and Related Substances. *Anal. Chem.* **1956**, *28*, 350–356. [CrossRef]
38. Rochas, C.; Domard, A.; Rinaudo, M. Aqueous GPC of Electrolytes and Polyelectrolytes. *Eur. Polym. J.* **1980**, *16*, 135–140. [CrossRef]

International Journal of
Molecular Sciences

MDPI

Article

Human Hemoglobin and Antipsychotics Clozapine, Ziprasidone and Sertindole: Friends or Foes?

Lena Platanić Arizanović [1], Nikola Gligorijević [2], Ilija Cvijetić [1], Aleksandar Mijatović [3], Maja Krstić Ristivojević [1], Simeon Minić [1], Aleksandra Nikolić Kokić [4], Čedo Miljević [5] and Milan Nikolić [1,*]

1 University of Belgrade-Faculty of Chemistry, Studentski trg 12-16, 11000 Belgrade, Serbia; ilija@chem.bg.ac.rs (I.C.); krstic_maja@chem.bg.ac.rs (M.K.R.); sminic@chem.bg.ac.rs (S.M.)
2 Institute of Chemistry, Technology, and Metallurgy, National Institute of the Republic of Serbia, University of Belgrade, Njegoševa 12, 11000 Belgrade, Serbia; nikola.gligorijevic@ihtm.bg.ac.rs
3 Faculty of Mining and Geology, University of Belgrade, Đušina 7, 11000 Belgrade, Serbia; aleksandar.mijatovic@rgf.bg.ac.rs
4 Institute for Biological Research "Siniša Stanković", National Institute of the Republic of Serbia, University of Belgrade, Bulevar despota Stefana 142, 11000 Belgrade, Serbia; san@ibiss.bg.ac.rs
5 Institute of Mental Health, Palmotićeva 37, 11000 Belgrade, Serbia; cedo.miljevic35@gmail.com
* Correspondence: mnikolic@chem.bg.ac.rs

Abstract: Packed with hemoglobin, an essential protein for oxygen transport, human erythrocytes are a suitable model system for testing the pleiotropic effects of lipophilic drugs. Our study investigated the interaction between antipsychotic drugs clozapine, ziprasidone, sertindole, and human hemoglobin under simulated physiological conditions. Analysis of protein fluorescence quenching at different temperatures and data obtained from the van't Hoff diagram and molecular docking indicate that the interactions are static and that the tetrameric human hemoglobin has one binding site for all drugs in the central cavity near αβ interfaces and is dominantly mediated through hydrophobic forces. The association constants were lower-moderate strength (~10^4 M^{-1}), the highest observed for clozapine (2.2 × 10^4 M^{-1} at 25 °C). The clozapine binding showed "friendly" effects: increased α-helical content, a higher melting point, and protein protection from free radical-mediated oxidation. On the other hand, bound ziprasidone and sertindole had a slightly pro-oxidative effect, increasing ferrihemoglobin content, a possible "foe". Since the interaction of proteins with drugs plays a vital role in their pharmacokinetic and pharmacodynamic properties, the physiological significance of the obtained findings is briefly discussed.

Keywords: antipsychotics; human hemoglobin; clozapine; ziprasidone; sertindole; binding; interactions

1. Introduction

Human hemoglobin (HHb), the most abundant blood protein, is an essential respiratory protein in vertebrates' erythrocytes (red blood cells). Structurally, it's a tetramer of two α- and two β-polypeptide globin chains, each bound to a redox iron heme in the crevices at the subunit exterior. The α-subunit contains 141 amino acid residues whereas the β-subunit comprises 146 amino acid residues, with seven and eight helical regions, respectively. Each heme group has a porphyrin ring and a ferrous atom capable of reversibly binding one oxygen molecule [1].

Antipsychotic medications are the mainstay for treating the symptoms of schizophrenia and other psychotic disorders, altering brain chemistry to help reduce psychotic symptoms like hallucinations, delusions, and disordered thinking [2]. Three second-generation (atypical) antipsychotics, acting as serotonin-dopamine antagonists, are particularly interesting due to their unique pharmacological properties or controversies associated with clinical use. Clozapine (Figure 1A) is the "gold-standard" antipsychotic for treatment-resistant schizophrenia. Despite its robust efficacy, there is evidence that clozapine is underutilized, partly due to concerns about its potential adverse effects [3]. Although it may be a slightly less efficacious antipsychotic drug, the main advantage of ziprasidone (Figure 1B) is its low propensity to induce weight gain and associated adverse effects [4,5]. Sertindole (Figure 1C) was withdrawn from the market due to concerns about its association with the prolongation of QT intervals, severe cardiac arrhythmias, and sudden cardiac death associated with its use. After re-evaluating its risks and benefits, it is reintroduced into the European market [6]. Still, the US Food and Drug Administration has not re-approved sertindole.

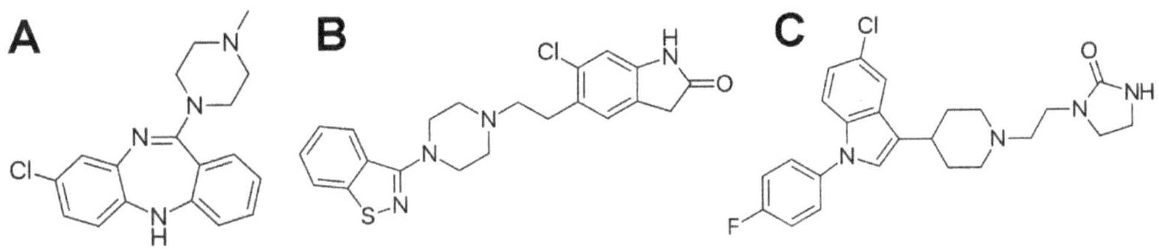

Figure 1. The molecular structure of the tested drugs used in the treatment of schizophrenia. A tricyclic dibenzodiazepine clozapine (3-chloro-6-(4-methylpiperazin-1-yl)-11*H*-benzo[b][1,4]benzodiazepine) (**A**). A benzothiazolylpiperazine derivative ziprasidone (5-[2-[4-(1,2-benzothiazol-3-yl)piperazin-1-yl]ethyl]-6-chloro-1,3-dihydroindol-2-one) (**B**). A phenylindole derivative sertindole (1-[2-[4-[5-chloro-1-(4-fluorophenyl)indol-3-yl]piperidin-1-yl]ethyl]imidazolidin-2-one) (**C**).

Various factors limit the effectiveness of antipsychotic drugs. Except for the most obvious accompanying adverse effects, the pleiotropic effects of drugs include interactions with biomolecules that are not the intended target of their action. Thus, knowledge of the binding characteristics of therapeutically important molecules is critical for understanding their availability in the requested tissues and their impact on the structure, activity, and function of the macromolecule partner(s). The erythrocyte compartment of the blood is often dismissed as an insignificant consideration in pharmacokinetics. However, many drugs accumulate significantly in human erythrocytes (e.g., [7]), cells that constitute a unique drug delivery system [8]. Mature erythrocytes act as "hemoglobin sacs", making this transport protein the first target for binding xenobiotics. Indeed, since the pioneering works of Nobel laureate Max Perutz, numerous studies have examined the binding of different exogenous substances to the HHb, including some drugs (e.g., [9]) and food antioxidants [10]. Therefore, this study aims to investigate for the first time the in vitro binding of antipsychotics clozapine, ziprasidone, and sertindole to (commercial and isolated) human hemoglobin (HHb) by various spectroscopic methods backed by molecular docking analysis. The influence of drug binding on hemoglobin's structure and thermal and oxidative stability was also determined. Findings are significant in assessing the pleiotropic effects of the tested antipsychotics ("friends or foes") on this miraculously vivid red human protein.

2. Results and Discussion

In the present study, we first explored the binding of the atypical antipsychotics clozapine, ziprasidone, and sertindole to HHb under simulated physiological conditions via steady-state fluorescence measurements combined with in-silico docking studies. Poor solubility of drugs in aqueous buffer solutions disables the use of isothermal titration calorimetry for biomolecular interaction analysis. Then, we studied the consequences of ligand binding, evaluating its influence on the α-helical structure and thermal stability of the protein by standard spectroscopic methods. Interestingly, the authors used commercial HHb preparations in most published papers to test the protein binding potential of various xenobiotics [10]. Therefore, the purchased HHb preparation (Section 3.1) was utilized to better compare our findings with the literature, i.e., the interaction of the HHb–various drugs. The exception was examining the oxidative stability of this erythrocyte hemoprotein in the presence of antipsychotics, for which freshly isolated, therefore completely native, oxy-HHb molecules were the starting material.

2.1. Fluorescence Quenching of Human Hemoglobin by Atypical Antipsychotics

Quenching studies of protein fluorescence by ligands is a convenient means for exploring ligand-protein interactions. The hemoglobin tetramer is assembled from two αβ dimers. Each dimer contains three tryptophan (Trp) residues: α-Trp14 and β-Trp15 are outside the subunit interface, while the β-Trp37 residue, located at the $\alpha_1\beta_1$ interface, has been regarded as the primary source of HHb fluorescence emission [11]. Hence, a study of the change in intrinsic fluorescence of HHb in the presence of the antipsychotics may give information about the local environment of the Trp moiety during interaction. Moreover, there are also five tyrosine (Tyr) residues in each αβ dimer: α-Tyr24, α-Tyr42, α-Tyr140, β-Tyr34, and β-Tyr144 [12].

Figures 2A, 3A and 4A present spectra showing the regular quenching of HHb fluorescence in the presence of increasing concentrations of the examined drugs. The working concentration of protein was 4 μM to prevent the quenching of Trp fluorescence by the neighboring heme group [13]. Human hemoglobin exhibited a strong fluorescence emission with a peak at 344 nm on excitation at 280 nm, and all ligands had comparatively low fluorescence compared with the protein. On the other hand, ziprasidone (Figure 3A) and sertindole (Figure 4A) showed strong fluorescence with a peak at ~400 nm, resulting in negative peaks in these areas of the collective emission spectra when sample mixtures were excited at 280 nm. In the case of clozapine and ziprasidone mixtures with HHb, no shift of the peak maximum was observed, while a blue shift of the peak maximum by 8 nm was observed with sertindole. The results suggest that the antipsychotics interact with HHb and the fluorescence quenching was due to specific complex formation, and that, since the β-37Trp is primarily responsible for the intrinsic fluorescence of HHb, changes in protein fluorescence intensity could indicate the presence of a binding site for antipsychotics near β-37 Trp residue [14].

The Stern-Volmer plots of the HHb–antipsychotic systems at different temperatures were essentially linear for drug concentrations up to 28 μM (Figures 2B, 3B and 4B), indicating that only one quenching type (static or dynamic) occurs. In static quenching, the complex formation occurs in the ground state. In contrast, dynamic quenching results from complex formation between the ligand and the protein in the excited state of the fluorophore. It was observed that K_{SV} values were on the order of 10^3 M^{-1} (Table 1), meaning there was a decrease in HHb fluorescence in the presence of the antipsychotic ligands. Based on Table 1 data and the variations of the K_{SV} values as a function of temperature, it was possible to characterize the preferential quenching mechanism for all three drugs as static since the values of the Stern-Volmer constants decreased with increasing temperature, thus forming a nonfluorescent supramolecular complex in the ground state [15].

The calculated value of K_q for the three antipsychotics binding is of the order of 10^{11} L mol^{-1} s^{-1} (Table 1), exceeding the value of 2×10^{10} L mol^{-1} s^{-1} acknowledged as the maximum K_q in the dynamic quenching of proteins [16], a further unequivocal proof that the described fluorescence quenching of HHb is not controlled by diffusion and does not originate from dynamic quenching. In all recent studies [14,17–20], the binding of the tested drugs to the commercial human hemoglobin molecules was characterized as static.

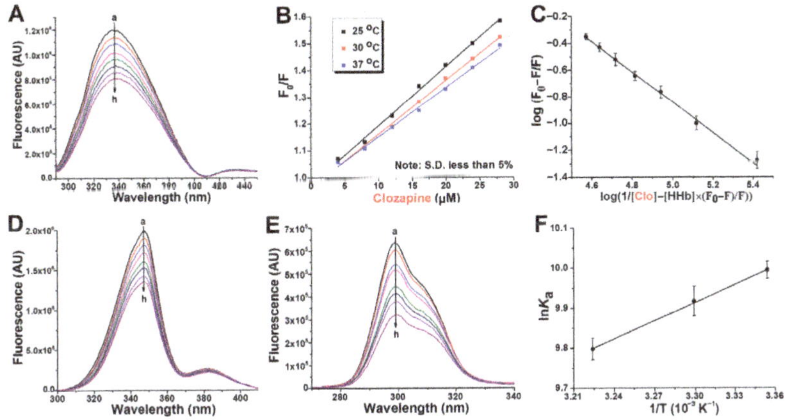

Figure 2. Binding of clozapine (Clo) to commercial human hemoglobin (HHb). Representative emission spectra (excitation at 280 nm) of HHb (4 μM) in the presence of different concentrations of Clo (0, 4, 8, 12, 16, 20, 24, and 28 μM, for curves **a** to **h**, respectively), at 37 °C (**A**). The Stern-Volmer plot of the protein fluorescence quenching at three temperatures (**B**). The plot of the fluorescence quenching data for the estimation of the binding constant and approximation of the number of binding places at 37 °C (**C**). Representative synchronous fluorescence spectra of HHb (4 μM) with (**D**) Δλ of 60 nm (Trp) and with (**E**) Δλ of 60 nm (Tyr) in the presence of increasing Clo concentration (0–28 μM for curves **a** to **h**, respectively). van't Hoff plot of the binding of Clo with 4 μM HHb for the thermodynamic analysis (**F**). Each data point indicates the average of three determinations for all plots, where error bars indicate the standard deviation.

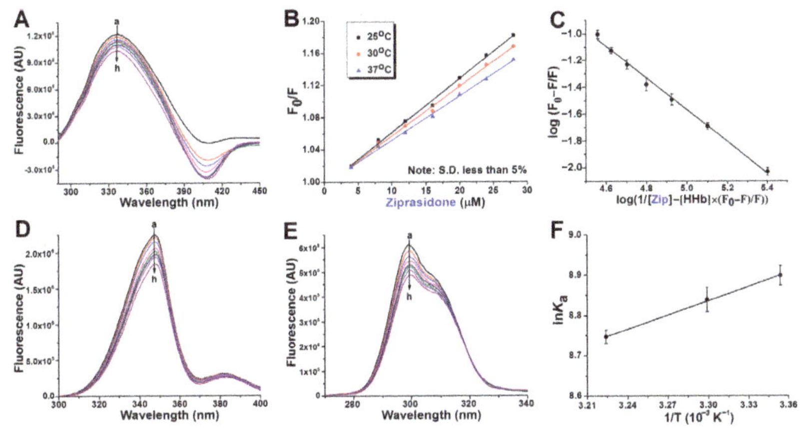

Figure 3. Binding of ziprasidone (Zip) to commercial human hemoglobin (HHb). Representative emission spectra (excitation at 280 nm) of HHb (4 μM) in the presence of different concentrations of Zip (0, 4, 8, 12, 16, 20, 24, and 28 μM, for curves **a** to **h**, respectively), at 37 °C (**A**). The Stern-Volmer plot of the protein fluorescence quenching at three temperatures (**B**). The plot of the fluorescence quenching data for the estimation of the binding constant and approximation of the number of binding places at 37 °C (**C**). Representative synchronous fluorescence spectra of HHb (4 μM) with (**D**) Δλ of 60 nm (Trp) and with (**E**) Δλ of 60 nm (Tyr) in the presence of increasing Zip concentration (0–28 μM for curves **a** to **h**, respectively). (**F**) van't Hoff plot of the Zip binding with 4 μM HHb for the thermodynamic analysis. Each data point indicates the average of three determinations for all plots, where error bars indicate the standard deviation.

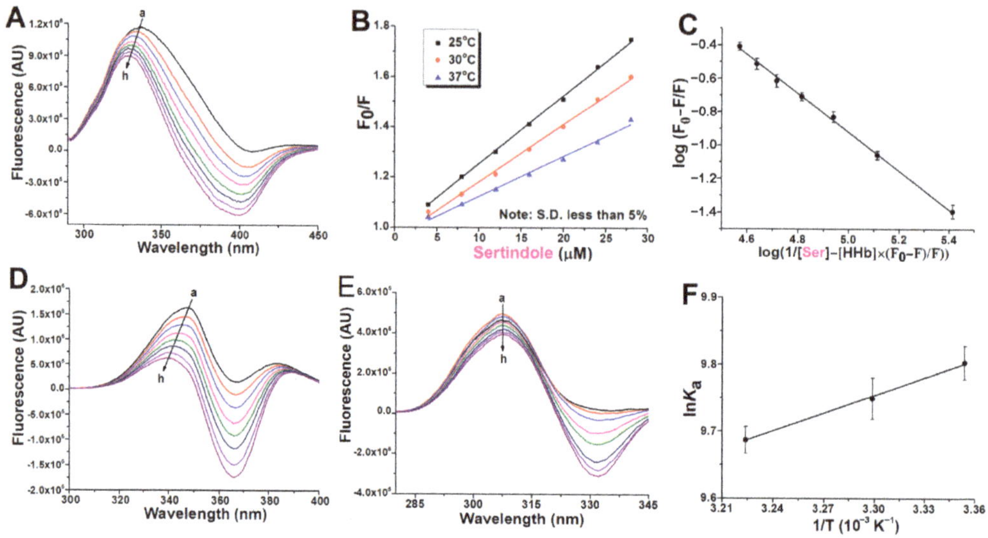

Figure 4. Binding of sertindole (Ser) to human hemoglobin (HHb). Representative emission spectra (excitation at 280 nm) of HHb (4 μM) in the presence of different concentrations of Ser (0, 4, 8, 12, 16, 20, 24, and 28 μM, for curves **a** to **h**, respectively) at 37 °C (**A**). The Stern-Volmer plot of the protein fluorescence quenching at three temperatures (**B**). The plot of the fluorescence quenching data for the estimation of the binding constant and approximation of the number of binding places at 37 °C (**C**). Representative synchronous fluorescence spectra of HHb (4 μM) with (**D**) Δλ of 60 nm (Trp) and with (E) Δλ of 60 nm (Tyr) in the presence of increasing Ser concentration (0–28 μM for curves **a** to **h**, respectively). (**F**) van't Hoff plot of the binding of Ser with 4 μM HHb for the thermodynamic analysis. Each data point indicates the average of three determinations for all plots, where error bars indicate the standard deviation.

Table 1. Quenching parameters of the interaction of atypical antipsychotics with human hemoglobin at three different temperatures.

Ligand	T (K)	K_{SV} (M^{-1})	R^2	K_q (M^{-1} s^{-1})
Clozapine	298.15	2.22×10^3	0.9958	2.22×10^{11}
	303.15	2.01×10^3	0.9957	2.01×10^{11}
	310.15	1.84×10^3	0.9952	1.84×10^{11}
Ziprasidone	298.15	0.67×10^3	0.9956	0.67×10^{11}
	303.15	0.62×10^3	0.9957	0.62×10^{11}
	310.15	0.55×10^3	0.9970	0.55×10^{11}
Sertindole	298.15	2.71×10^3	0.9981	2.71×10^{11}
	303.15	2.29×10^3	0.9957	2.29×10^{11}
	310.15	1.16×10^3	0.9911	1.16×10^{11}

K_{sv}, the Stern-Volmer (SV) quenching constant; R^2, linear correlated coefficient of SV plots; K_q, the bimolecular quenching constant. The standard deviation for all parameters was less than 5% of the mean.

2.2. Binding Constants Determination in Human Hemoglobin and Antipsychotic Systems

Parameters that characterize the interaction between a protein and its ligands are its binding strength and the number of ligand binding sites on the protein. Treating Equation (6) as the equation for a straight line, the value for the binding constant "K_a" and the number of binding sites "n" for the binding of clozapine, ziprasidone, and sertindole with HHb within the studied concentration range of the antipsychotics were obtained from the intercept and slope of the double log plot, respectively (Figures 2C, 3C and 4C).

The decreasing trend of K_a with increasing temperature for all ligands, although the differences are relatively minor (Table 2) and follow the dependence of K_{SV} on temperature, is consistent with the static quenching mechanism mentioned above. Most ligands are bound reversibly and display moderate affinities for the protein partner, e.g., binding constants are 1–15 × 10^4 M^{-1} [10,21]. Therefore, the K_a values reported in this paper indicate that the binding between clozapine (1.3 × 10^4 M^{-1} at 37 °C) or sertindole (1.6 × 10^4 M^{-1} at 37 °C) and HHb was lower-moderate and that these medications can be stored and carried by the hemoglobin in the human body, given the high concentration of this transport protein in the red blood cells of approximately 330 mg/mL [22]. In comparison with other drugs, these association constants are comparable to those of the anti-mitotic chemotherapy agent docetaxel (3.2 × 10^4 M^{-1} at 37 °C) [17], the synthetic analgesic tramadol (1.2 × 10^4 M^{-1} at 37 °C) [18], and the antimalarial chloroquine (1.6 × 10^4 M^{-1} at 37 °C) [19], but lower than those of commercial hemoglobin and the nonsteroidal anti-inflammatory medication diclofenac (5.0 × 10^5 M^{-1} at 37 °C) [14]. The estimated association constant for ziprasidone binding to HHb was found to be consistently lower, of the order of magnitude only 10^3 M^{-1} (Table 2), similar to the value of the analgesic and antipyretic drug acetaminophen (6.4 × 10^3 M^{-1} at 37 °C), obtained by calorimetry with isolated protein [23], but still higher than that recorded for promethazine (2.1 × 10^2 M^{-1} at 37 °C) and adiphenine (3.0 × 10^2 M^{-1} at 37 °C) amphiphilic drugs [20].

Table 2. Binding and thermodynamic parameters of human hemoglobin-antipsychotic complexes at three different temperatures.

Ligand	T (K)	$K_a \times 10^4$ (M^{-1})	n	ΔG (kJ mol^{-1})	ΔS (J mol^{-1} K^{-1})	ΔH (kJ mol^{-1})
Clozapine	298.15	2.19	1.12	−12.2	+41.0	−12.5
	303.15	2.03	1.03	−12.4		
	310.15	1.80	0.97	−12.7		
Ziprasidone	298.15	0.74	1.16	−12.3	+41.2	−9.8
	303.15	0.69	1.08	−12.5		
	310.15	0.63	1.06	−12.8		
Sertindole	298.15	1.81	1.10	−16.9	+56.8	−7.4
	303.15	1.71	1.06	−17.2		
	310.15	1.61	1.03	−17.6		

K_a, the binding (association) constant; n, the number of binding sites on the protein; ΔH and ΔS are the enthalpy and entropy changes, respectively; ΔG, the free energy change. The standard deviation for all parameters was less than 5% of the mean, and the linearly correlated coefficients of plots for K_a and n determination (Figures 2C, 3C and 4C) were higher than 0.99.

The value of n is nearly equal to 1 in all three systems and temperatures (Table 2), suggesting one binding site for these antipsychotics around the Trp residue (s) of the protein, identical to all the studies of interactions of human hemoglobin with drugs mentioned in the previous paragraph.

2.3. Synchronous Fluorescence Measurements in Human Hemoglobin and Antipsychotic Systems

Synchronous fluorescence spectroscopy gives information about the molecular environment near a protein fluorophore (Trp and Tyr). It involves simultaneous scanning of the excitation and emission monochromators while maintaining a constant wavelength interval ($\Delta\lambda$) between them. The emission maximum shift reflects the polarity changes around the fluorophore molecule [24]. The synchronous HHb fluorescence spectra with various amounts of antipsychotics are shown in Figures 2D, 3D and 4D ($\Delta\lambda$ stabilized at 60 nm, giving characteristic information of Trp residues) and Figures 2E, 3E and 4E ($\Delta\lambda$ of 15 nm for Tyr residues). The fluorescence intensity of HHb regularly decreased with the addition of all antipsychotics, further demonstrating the occurrence of fluorescence quenching in the binding process.

Expectedly, the contribution of Trp fluorophores (the signal intensity) to the internal fluorescence of HHb was greater. However, the contribution of fluorophores to protein signal quenching (the relative decrease of fluorescence intensity) in the presence of antipsychotics was different. The clozapine molecules affect the fluorescence intensity of Tyr more than that of Trp (Figure 2E vs. Figure 2D). In contrast, there is a greater decrease in the signal intensity of Trp residues as compared to Tyr with sertindole (Figure 4D vs. Figure 4E). Finally, it seems the fluorescence quenching of HHb comparably originated from both Trp and Tyr residues in the case of ziprasidone binding (Figure 3D vs. Figure 3E). The described results indicate that the binding site of these drugs on the hemoglobin molecule could be different.

It can also be seen that the maximum emission wavelength of Tyr residues in the HHb complex with clozapine (Figure 2E), ziprasidone (Figure 3E), and sertindole (Figure 4E) does not have a significant shift over the investigated concentration range, which indicates that these drugs binding does not affect the microenvironment of Tyr residues in HHb. Likewise, the environment of the Trp fluorophores appears to remain the same after clozapine (Figure 2D) and ziprasidone (Figure 3D) bind to HHb. In contrast, a blue shift (from 347 to 340 nm) of Trp residues was observed in the HHb–sertindole complex (Figure 4D), indicating the hydrophobicity of (at least one) Trp residue increased and this residue was moved to a more hydrophobic environment. Therefore, it seems the conformation of the HHb molecule was changed to some extent upon sertindole binding.

2.4. Thermodynamic Study of Antipsychotics Binding to Human Hemoglobin

The sign of thermodynamic parameters such as Gibbs free energy (ΔG), enthalpy change (ΔH), and entropy change (ΔS) involved in protein-ligand binding is helpful for the determination of the type of interaction involved between HHb and the antipsychotics analyzed. Therefore, the quenching experiments have been done at three different temperatures to determine the binding thermodynamic parameters. From the slope (the enthalpy) and intercept (the entropy) of the linear van't Hoff plots (ln K_a vs. 1/T; Figures 2F, 3F and 4F) and Equations (7) and (8), these parameters have been calculated and listed in Table 2.

In all cases, the negative sign of the ΔG value indicates the spontaneity of the molecular recognition process at all temperatures [25]. The values of ΔH were all negative (-7.4 kJ mol^{-1} for sertindole, -9.8 kJ mol^{-1} for ziprasidone, and -12.5 kJ mol^{-1} for clozapine), and ΔS were all positive (+41.0 J mol^{-1} K^{-1} for clozapine, +41.2 J mol^{-1} K^{-1} for ziprasidone, and +56.8 J mol^{-1} K^{-1} for sertindole) for all three HHb–ligand systems. The negative values of enthalpy changes (ΔH) and positive values of entropy contribution (TΔS) strongly suggest that it is an entropy-driven spontaneous and exothermic binding process. Moreover, the positive entropy changes also indicate that the randomness around the HHb-antipsychotic complexes increases, again evidence of hydrophobic interaction [23]. The negative enthalpy changes may also support the existence of hydrogen bonds in the complexation [25]. Indeed, these assumptions were supported by the docking study results (next section) and the high lipophilicity of these drugs (log P > 3; https://go.drugbank.com/ (accessed on 11 April 2023).

2.5. Docking Results

Unlike human serum albumin, the abundant transport protein for non-polar drugs through the bloodstream, which has at least two structurally well-defined major ligand binding sites [26], only the solvent-accessible central cavity of HHb contains functionally important binding sites for several classes of allosteric effectors that facilitate the lowering of oxygen affinity [9], within which various xenobiotics are also bound. According to HHb binding sites, the central cavity can be classified into two subsites: I, close to the corner of α_2 and β_1 subunits, may be suitable to bind electro-negative drugs, and II, close to the α_1 subunit, tends to bind electropositive drugs [27].

The molecular docking technique is frequently used to understand the interactions between proteins and ligands. The results presented in Figure 5 provide insights into the binding interactions of three antipsychotic drugs with the central binding site of HHb. The docking studies revealed that all three drugs bind within the HHb cavity, with clozapine and sertindole occupying similar binding pockets on the $\alpha_1\beta_1$ interface, while ziprasidone was preferentially placed near the α_2 and β_2 subunits (Figure 5A). The mean binding affinities of ziprasidone and sertindole were similar (-8.61 vs. -8.57 kcal mol^{-1}), while clozapine exhibited a somewhat lower binding affinity (-7.85 kcal mol^{-1}). Unsurprisingly, molecular docking has given the opposite result compared to fluorimetry since docking affinities do not include the entropic contributions; hence, they align with the binding enthalpies obtained from the fluorescence experiments.

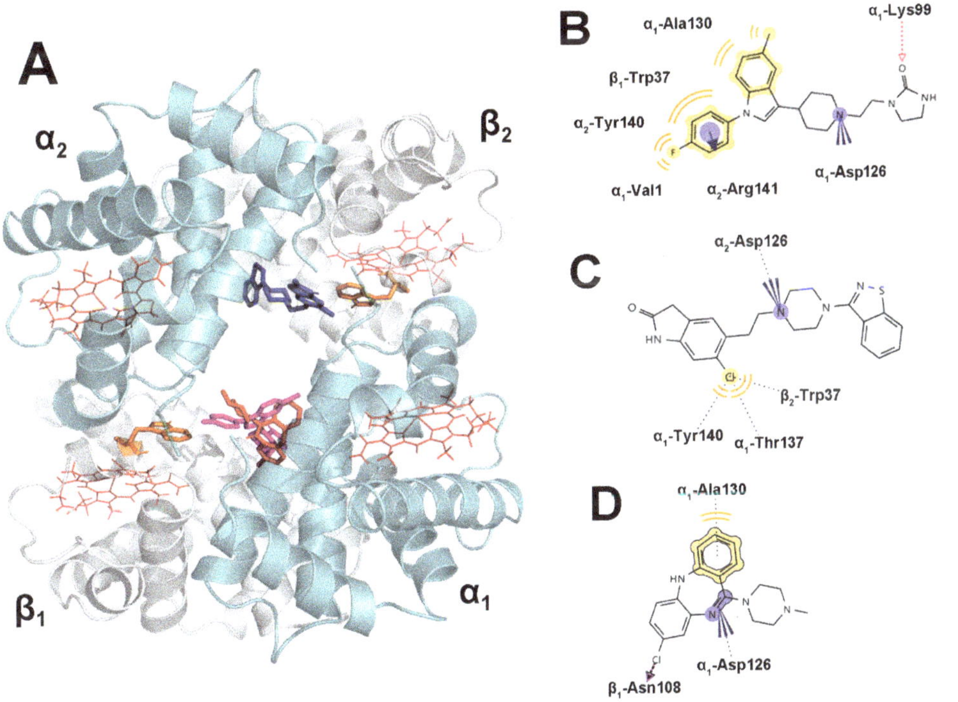

Figure 5. Schematic representation of human hemoglobin (HHb)-antipsychotic docked structures. The lowest-energy binding poses of ziprasidone (blue), clozapine (red), and sertindole (magenta) within the central cavity of HHb (PDB: 2D60); Trp37 residues on β1 and β2 subunits are colored oranges; the image was rendered in PyMOL (**A**). The 2D ligand-interaction diagrams for HHb with sertindole (**B**), ziprasidone (**C**), or clozapine (**D**); images were prepared in LigandScout 4.4.

Further analysis of the binding interactions revealed that sertindole forms several favorable hydrophobic interactions with specific amino acid residues such as α_1-Ala130, α_2-Tyr140, α_1-Val1, and β_1-Trp37 (Figure 5B). Additionally, the fluorobenzene ring of sertindole interacts with the positively charged α_2-Arg141 side chain. Also, the nitrogen atom from the piperidine ring forms electrostatic interactions with α_1-Asp126, while the carbonyl O atom forms hydrogen bond interactions with α_1-Lys99. The coplanarity between the fluorobenzene ring of the ligand and β_1-Trp37 indole ring, along with the relatively short distance between their centroids (5.7 Å), suggests the probability for π-π interactions that would complement hydrophobic interactions between these two moieties. These findings agree with the observed blue shift in the Trp synchronous HHb fluorescence spectrum upon sertindole binding (Figure 4D). On the other hand, ziprasidone preferably interacts with β_2-Trp37, α_1-Thr137, and α_1-Tyr140 through the chlorine substituent on the dihydrooxoindol ring (Figure 5C). It also forms favorable electrostatic interactions between the piperazine nitrogen atom and the carboxylic group of α_2-Asp126. In contrast, clozapine interacts with α_1-Ala130, β_1-Asn108, and α_1-Asp126 via hydrogen bonding, hydrophobic, and electrostatic interactions (Figure 5D).

Our findings, in agreement with thermodynamic and synchronous results, provide valuable insights into the binding mechanisms of sertindole, ziprasidone, and clozapine drugs within the central cavity of the HHb tetramer. Namely, the docking study confirmed that the hydrophobic effect followed by hydrogen bonds or electrostatic interactions is essential in the binding between HHb and the antipsychotics examined. The drugs docetaxel [17] and diclofenac [14] have also been suggested to bind to the central cavity of HHb, close to subunit α_1, and the possible binding site of acetaminophen is near β-Trp37 residue [23].

2.6. Circular Dichroism Spectra Analysis

To ascertain the possible effect of the antipsychotics binding on the α-helical secondary structure of HHb, far-UV CD spectroscopy was used. CD measurements in the presence of a fixed concentration of antipsychotics examined (a molar protein to drug ratio of 1:4) are shown in Figure 6A. Consistent with the literature, the CD spectrum of HHb in an aqueous buffer has two characteristic peaks of strong negative ellipticity near 208 nm and 222 nm, ascribed to a typical α-helical protein structure [28]. The negative peak near 208 nm is due to α-helix's π-π^* transition, while the peak near 222 nm is attributed to n-π^* transitions for both the α-helix and random coil [29].

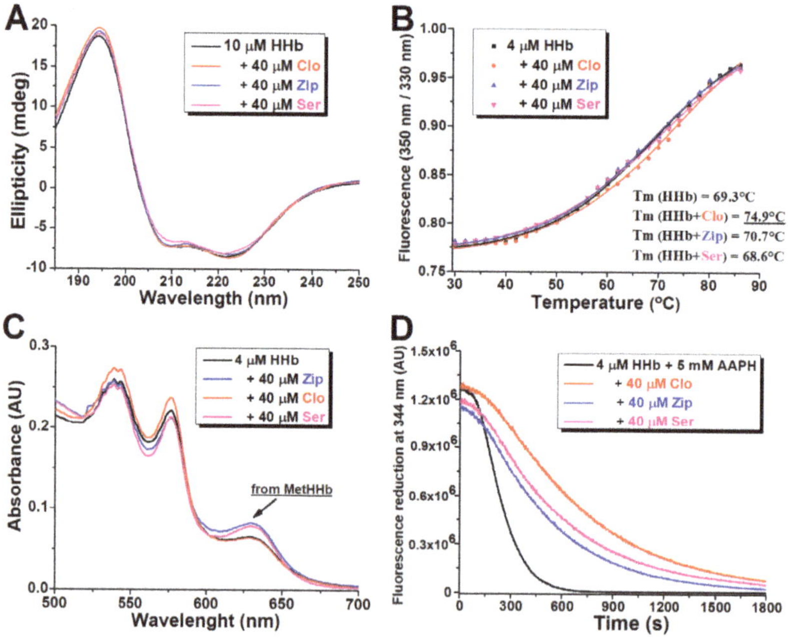

Figure 6. Binding effects of antipsychotics clozapine (Clo), ziprasidone (Zip), and sertindole (Ser) on the structure and oxidative stability of human hemoglobin (HHb). Representative far UV-CD spectra of commercial HHb in the absence and presence of tested drugs (**A**). Representative melting curves of commercial HHb without or with antipsychotic addition, with the specified mean values of the obtained protein melting points (Tm) (**B**); the standard deviations were less than 5% of the means. Representative visible absorption spectra of isolated HHb, alone and with the ligands (**C**). Representative changes in the isolated HHb emission intensity induced by 2,2′-azobis(2-amidinopropane) dihydrochloride (AAPH), without and in the presence of tested drugs (**D**).

The secondary structure of commercial HHb was found to contain 48.2% α-helical regions. The CD spectrum of the HHb showed only slight changes upon adding all three ligands, indicating minor effects of clozapine, ziprasidone, and sertindole binding on the secondary structure of the protein. Still, as shown in Table 3, the α-helical content of HHb borderline significantly ($p = 0.048$) increased to 52.2% in complex with clozapine, while it remained virtually unchanged with ziprasidone and sertindole. The increase in α-helical content suggested a more folded secondary structure and, therefore, a more compact conformation of the HHb molecule in the presence of bound clozapine.

Table 3. The effects of human hemoglobin (HHb)–antipsychotic complexes formation on selected structural-related protein parameters.

Sample & Monitored Parameter	α-Helices Content (%)	metHHb Content (%)	Protection from Oxidation (AU)
Control HHb (4 μM)	48.2 ± 0.7	3.96 ± 0.20	1 (reference)
+clozapine (10 μM)	52.2 ± 0.5 [a]	3.86 ± 0.23	118 ± 4.1
+ziprasidone (10 μM)	50.2 ± 0.6	5.32 ± 0.27 [c]	78 ± 3.2
+sertindole (10 μM)	49.9 ± 1.0	5.14 ± 0.42 [b]	60 ± 2.8

Significantly different from the control values: [a]—$p < 0.05$; [b]—$p < 0.01$; [c]—$p < 0.001$. MetHHb: oxidized, ferrihemoglobin; AU: arbitrary units.

Among other examined drugs, docetaxel, chloroquine, and tramadol were all found to decrease α-helical content in HHb, from 45.3 to 42.8% at the molar ratio of HHb to the ligand of 1:12 [17], from 35.6 to 31.0% at the molar ration 1:6 [19], and from 35.4 to 33.7% at molar ratio of 1:2 [18], respectively. Only diclofenac binding causes the gain of α-helix stability, increasing its content from an initial 41.5% to 49.6% in a 20-fold ligand molar excess [14]. Observed differences in the underestimated content of the α-helices in pure HHb can be mainly attributed to the diversity of used protein preparations. Indeed, in freshly isolated HHb, the calculated α-helix content is found to be ~76%, and it decreased after acetaminophen drug binding [23].

2.7. Effect of Atypical Antipsychotics on the Thermal Stability of Human Hemoglobin

Thermal-shift assays are a valuable complement to testing the overall protein stability *per se* and the biochemical consequences of drug actions. The thermal behavior of HHb was studied in the absence and presence of the examined drugs by measuring the unfolding transitions using fluorimetry at a temperature range of 30–85 °C. The relative concentration of protein in each antipsychotic was fixed at a molar ratio of 1:10. The melting temperature of proteins, Tm (midpoint transition temperature), was calculated from thermal denaturation curves (Figure 6B).

The obtained Tm for intact commercial HHb at experiment conditions was 69.3 °C, comparable with the CD results [30]. In the presence of clozapine, Tm was increased to 74.9 °C while remaining effectively unchanged in HHb complexes with ziprasidone and sertindole. Results confirmed the antipsychotic–protein complexing and the stabilizing effect of clozapine binding (CD study results) to higher levels of HHb structure and that ziprasidone and sertindole essentially do not affect (at least) the protein part of its tetrameric structure.

2.8. Effect of Atypical Antipsychotics on Spontaneous Human Hemoglobin Oxidation

Hemoglobin spontaneously oxidizes to methemoglobin (metHb) along with the concomitant production of superoxide radicals by a mechanism associated with the distal side of heme, which facilitates the nucleophilic displacement of bound oxygen by the histidine residue [31]. Visible absorption spectroscopy is an effective tool for examining structural changes in colored proteins and their complexation with ligands. Two regions on the normal hemoglobin spectrum can be considered in this part of the electromagnetic spectrum. The first region is the Soret (or B) band, related to heme iron-porphyrin complex absorption at 415 nm. The second region is the Q band, which is associated with the oxy- and deoxy-forms of heme at 550–600 nm. The formation of metHHb is followed by the appearance of a new, small absorption peak with a maximum of around 630 nm [32].

Figure 6C shows the deviation in the Vis absorption spectra of isolated HHb after prolonged incubation with antipsychotics. The calculated metHHb values are presented in Table 3. It was found that both ziprasidone and sertindole in ten-fold molar excess over protein slightly but significantly ($p < 0.01$) increased the amount of met-HHb formed under the applied conditions. This may be because (i) both drugs carry one acid hydrogen atom in their structure, which may be involved in the oxidation process through proton donation and thus enhance the rate of oxidation, or (ii) ziprasidone and sertindole modify the structure of HHb in such a way that it leads to an augmentation of the oxidation rate, or both. The addition of clozapine to the reaction mixture neither significantly promotes nor prevents the formation of oxidized HHb, judging by the identity of the absorption HHb profile in the wavelength range where the met form absorbs and the slightly lower oxidized protein content. The protective effect of clozapine on HHb autoxidation, in comparison with sertindole and especially ziprasidone, may be the consequence of (i) demonstrated (additional) stabilization of the protein's secondary structure, which slows down the initiation of the autocatalytic stage of the met-HHb formation process, or (ii) antioxidant, including radical scavenging activity of the drug itself [33,34].

Hemoglobin is a polyfunctional molecule involved in functions beyond oxygen transport, such as catalytic activity, nitric oxide metabolism, metabolic reprogramming, redox balance, and pH regulation [35]. Therefore, any substantial change in the active center of the protein (iron protoporphyrin IX prosthetic group) could affect the overall body homeostasis. Considering the low concentrations of ziprasidone (~100 ng/mL) [36] and sertindole (~10 ng/mL) [37] in the plasma of patients on chronic therapy and their relatively low binding constants to highly abundant HHb molecules in vitro that demonstrated a pro-oxidative effect and should not have any significant physiological consequences.

2.9. *Effect of Atypical Antipsychotics on Oxidant-Assisted Human Hemoglobin Oxidation*

Oxidative damage induced by the overproduction and/or accumulation of reactive oxygen species (ROS) can devastate the structure and activity of proteins. Along with by-products of oxygen metabolism, environmental stressors and some xenobiotics (i.e., antineoplastic drugs) greatly increase ROS production [38]. A hydrophilic free radical initiator, AAPH, is a suitable reagent tool for the evaluation of the oxidative stability of bioactive molecules [39]. The ability of antipsychotics to protect the structural elements of isolated HHb from oxidation was examined by monitoring the reduction of the intrinsic fluorescence of the protein induced by AAPH over time. The obtained spectra are shown in Figure 6D, and the calculated protective effect of each drug under the applied experimental setup (corresponding to the difference in areas under the curves in the presence and absence of the drug, expressed in arbitrary units) is depicted as a part of Table 3.

It was obvious that pure HHb is prone to oxidation, considering the fast and substantial decrease of its intrinsic fluorescence intensity, and that the addition of all three antipsychotics in a 10-fold molar excess slowed down this process (Figure 6D). It was found that clozapine delays the AAPH-induced oxidation of HHb molecules the most, followed by ziprasidone and sertindole (Table 3). Therefore, all drugs showed solid antioxidant properties and good protective activity against HHb oxidation.

The biggest increase in the oxidative stability of HHb in the complex with clozapine is not surprising. Its antioxidant properties are well known in various in vitro models [40], and it has shown similar protective effects on plasma proteins fibrinogen [41] and alpha-2-macroglobulin [42]. The suggested therapeutic plasma clozapine concentration (300 to 700 ng/mL) [43] is substantially higher than ziprasidone and sertindole. Therefore, clozapine may actually reduce the oxidation of all these proteins in vivo. All drugs have adverse effects, which can even be potentially life-threatening. There are blood dyscrasias and agranulocytosis/granulocytopenia with clozapine use [44]. Our results point only to the benefits of clozapine complexed with human hemoglobin, without whose evolution life as we know it would not exist.

3. Materials and Methods

3.1. Materials

Human hemoglobin (lyophilized powder), clozapine, ziprasidone hydrochloride hydrate (purity $\geq$ 97%), and sertindole (purity $\geq$ 97.5%) were purchased from Sigma-Aldrich Chemie GmbH (Taufkirchen, Germany) and used without further purification. Unless otherwise stated, HHb solutions were prepared in phosphate-buffered saline (PBS) at pH 7.4, and the protein concentration was determined using Drabkin's cyanmethemoglobin method [45]. Atypical antipsychotics were dissolved in 99.9% dimethyl sulfoxide (DMSO) to make 4 mM stock solutions. For all experiments, the final concentration of DMSO in various protein-ligand mixtures did not exceed 0.5% (*v/v*). All other chemicals were of analytical reagent grade, and Milli-Q water (Millipore, Molsheim, France) was used in the experiments.

Venous blood was drawn by venipuncture from a healthy volunteer, collected into a tube containing heparin as an anticoagulant, and immediately centrifuged at 2500 *g* for 10 min at 4 °C. Plasma and buffys coat were discarded. The erythrocytes were washed three times with isotonic (0.9%) saline. Packed erythrocytes were lysed in 20 volumes of 10 mM phosphate buffer, pH 7.4 (PB) at 4 °C overnight. Then the hemolysate was centrifuged at 12,000 *g* for 1 h at 4 °C to remove the membrane part. The supernatant hemolysate was collected, and aliquots were applied to size-exclusion chromatography through a Sephadex G-100 column (45 × 1.0 cm), preequilibrated, and eluted with PB. The concentration of obtained HHb was determined from its Soret absorbance at 415 nm (ε = 125,000 M^{-1} cm^{-1}) [45]. The purity of the isolated HHb preparations has been checked by 15% SDS-PAGE after staining with Coomassie blue.

3.2. Methods

3.2.1. Fluorescence Measurements

All fluorescence data were obtained on a FluoroMax®-4 spectrofluorometer (HORIBA Scientific, Kyoto, Japan) under thermostated conditions in a 1.0 cm quartz cell, with the width of the excitation and emission slits both adjusted at 3 or 5 nm. The appropriate blanks, corresponding to the various drugs in buffer concentrations, were subtracted to correct the fluorescence background.

The binding of antipsychotics to HHb was studied by the fluorescence quenching titration method using the intrinsic protein fluorescence probe at constant protein (4 μM) and various ligand concentrations (0 to 28 μM). The steady-state fluorescence spectra were measured at 25, 30, and 37 °C (298.15, 303.15, and 310.15 K, respectively). The excitation wavelength was set at 280 nm, and the emission spectra were read at 290 to 450 nm.

The synchronous fluorescence spectra of the HHb–antipsychotic complexes were recorded at two different scanning intervals: $\Delta\lambda$ of 15 nm (tyrosine residue excitation) and $\Delta\lambda$ of 60 nm (tryptophan residue excitation), where $\Delta\lambda = \Delta\lambda_{EM} - \Delta\lambda_{EX}$ [24].

The thermal stability of 4 μM commercial human hemoglobin was recorded in the absence and presence of 40 μM of each antipsychotic in the temperature range from 30 to 85 °C. The temperature increment was 2 °C/min, and the equilibration time was adjusted to 1 min. Emission spectra were recorded from 315 to 365 nm, with the excitation wavelength set to 280 nm. The melting curve was presented as the F_{350}/F_{330} ratio change with temperature, where F is the emission intensity at a corresponding wavelength, and fitted into a sigmoidal function where the inflection point represents the protein melting point (T_m) [46].

The stability of proteins towards oxidation was investigated by studying the impact of the strong oxidant 2,2′-azobis(2-amidinopropane) dihydrochloride (AAPH) (final concentration of 5 mM) on the intrinsic purified hemoglobin (4 μM) fluorescence in the absence and presence of drugs at fixed concentrations (40 μM). The fluorescence was monitored during a time course of 30 min at the HHb emission peak of 344 nm after excitation at 280 nm. The protective effect (PE) of each antipsychotic on oxidant-induced reduction of protein fluorescence, expressed in arbitrary units (AU), was calculated using the equation [47]:

$$\text{PE (AU)} \left[\times 10^{6}\right] = \text{AUC}_{\text{HHb+AAP}} - \text{AUC}_{\text{HHb}} \quad (1)$$

where $\text{AUC}_{\text{HHb+AAP}}$ represents the area under the curve obtained for the human hemoglobin/ atypical antipsychotic mixture, and AUC_{HHb} represents the area under the curve obtained for the protein alone.

3.2.2. Circular Dichroism (CD) Measurements

The CD measurements were carried out on a Jasco J-815 spectropolarimeter (Jasco, Tokyo, Japan) in the far-ultraviolet region (185–250 nm) at 37 °C, with two scans averaged for each CD spectrum, using a quartz cell with 0.01 cm path length and 50 nm/min scan speed. The concentration of HHb in PB was 10 µM, and the concentration of drug ligands was 40 µM. The obtained spectra were corrected by subtracting the appropriate spectra from the ligands alone.

The obtained data, presented in units of ellipticity (mdeg), were converted to mean residue ellipticity (MRE) using the following formula:

$$[\text{MRE}] = \frac{\theta \times \text{Mm}}{10 \times l \times C \times r} \quad (2)$$

where θ is ellipticity in mdeg, Mm is the molar mass of HHb (64.5 kDa), l is the length of the cuvette, C is the protein molar concentration in the experiment, and r is the number of amino acid residues in the protein (574).

The α-helical hemoglobin content (%) was calculated from these data using the CONTIN algorithm and SP29 database in the CDPro software package.

3.2.3. Visible Absorbance (Vis) Measurements

The absorption spectra of purified human hemoglobin (4 µM) in PB with and without the antipsychotics (40 µM) have been recorded on a UV-1800 spectrophotometer (Shimadzu, Kyoto, Japan) using a 1 cm path length cuvette over a wavelength range from 500 to 700 nm after 6 hrs incubation at 37 °C. The metHHb content (in % of total protein concentration) in samples was determined according to the relation [32]:

$$\text{metHHb} = 7.0 \times A_{577} + 76.8 \times A_{630} - 13.8 \times A_{560} \quad (3)$$

based on the millimolar extinction coefficients of the tetrameric hemoglobin (oxy, met, and hemichrome) species and the absorbances at the specified wavelengths and pH 7.4.

3.2.4. Statistical Analysis

All experiments were performed in triplicate, and the mean values, standard deviations, and all statistical data were processed using the OriginPro software (OriginLab Corporation, Northampton, MA, USA). Results were analyzed by one-way ANOVA ($p < 0.05$ was considered significant).

3.2.5. Molecular Docking

The initial structure of HHb was obtained from PDB, code 2D60 [48]. Two cocrystallized ligand molecules were removed, and the protein was ionized to pH 7.40 using PROPKA [49]. The protein molecule was then embedded in a 10 Å sphere of TIP3P water molecules and underwent 10 ps of conjugate gradient minimization to eliminate steric bumps while keeping the protein backbone fixed to maintain the experimental structure. The CHARMM force field was employed for this calculation, and NAMD 2.14 [50] was used with Vega ZZ 3.2.0 as the GUI [51].

The 3D structures of clozapine, sertindole, and ziprasidone were retrieved from PubChem, and their final structures were optimized at the PM7 semiempirical level of theory [52] in MOPAC 2016 [53]. The COSMO implicit solvation model of water was utilized by adding the EPS = 78.4 keyword, and the convergence criterion was increased 100 times using the PRECISE keyword.

Molecular docking of three drugs into the HHb was conducted in AutoDock Vina 1.1 [54]. Blind docking was performed by setting the entire protein structure as the receptor site. The exhaustiveness was increased to 200, and 20 binding modes were stored for each ligand.

3.3. Theory and Calculations

The fluorescence intensity of a compound decreases (fluorescence quenching) from various molecular interactions, namely, excited state reactions, molecular rearrangements, energy transfer, ground state complex formation, and collisional quenching. To determine the type of quenching, quenching experiments (Section 3.2.1) were performed at three different temperatures (25, 30, and 37 °C), and the data were analyzed by the classical Stern-Volmer equation [15]:

$$\frac{F_0}{F} = 1 + K_q \times \tau_0[Q] = 1 + K_{sv} \times [Q] \tag{4}$$

where F_0 and F are protein emission fluorescence at 344 nm without and with the addition of ligand, respectively, K_q is the quenching rate constant τ_0 is the average lifetime of the protein in the absence of ligand, which is of the order of 10^{-8} [55]. Further, [Q] is the quencher (ligand) concentration, and K_{SV} is the Stern-Volmer dynamic quenching constant, which is obtained from the slope of the linear F_0/F vs. [Q] plot.

Fluorescence intensities were corrected for the inner-filter effect (the absorption of the exciting light and reabsorption) according to the relationship [15]:

$$F_c = F_0 \times 10^{(AEX+AEM)/2} \tag{5}$$

where F_0 is measured fluorescence, F_c is corrected fluorescence, and A_{EX} and A_{EM} are the absorbances of the quencher at excitation and pick emission wavelengths (344 nm), respectively.

When ligand molecules bind independently to a set of equivalent sites on a macromolecule, the equilibrium between free and bound molecules is given by the following equation [56]:

$$\log \frac{F_0 - F}{F} = -n \times \log \frac{1}{[L] - [P] \times \frac{F_0 - F}{F_0}} + n \times \log K_a \tag{6}$$

where F_0 and F are the fluorescence intensities before and after the addition of ligand (an antipsychotic drug), [P] and [L] are the total protein (HHb) and the total ligand concentration, respectively. From the plot of log $((F_0 - F)/F)$ vs. $\log(1/([L] - [P]) \times (F_0 - F)/F))$, the binding (association) constant K_a and the number of binding sites n under a specified set of experimental conditions can be evaluated.

The binding forces contributing to interactions of small organic compounds with proteins frequently include van der Waals interactions, hydrophobic effects, electrostatic interactions, and hydrogen bonds [57]. As the main evidence used to determine the binding mode, the thermodynamic parameters: enthalpy change (ΔH), entropy change (ΔS), and Gibbs (free) energy change (ΔG), were calculated from the van't Hoff equations:

$$\ln K_a = -\frac{\Delta H}{R} \times \frac{1}{T} + \frac{\Delta S}{R} \tag{7}$$

and the relation:

$$\Delta G = \Delta H - T \times \Delta S \tag{8}$$

where K_a is the binding constant, R is the universal gas constant, and T is the temperature in Kelvins. The values of ΔH and ΔS were calculated from the slope and intercept of the linear plot of lnK_a vs. 1/T under the assumption that ΔH and ΔS are only very weakly dependent on temperature, which is usually valid. Human hemoglobin does not undergo structural degradation at the chosen measurement temperatures (298.15, 303.15, and 310.15 K).

4. Conclusions

The interaction models in the system of HHb–antipsychotics clozapine, ziprasidone, and sertindole have been reported for the first time. Different spectroscopic techniques have studied the drug's binding. The ground-state interactions have been studied by steady-state fluorescence and absorption spectroscopy. The extent of binding and corresponding thermodynamic parameters have also been determined from fluorescence experiments. The structural changes of proteins in the absence and presence of drugs have been studied by circular dichroism and synchronous fluorescence spectra. Theoretical docking analyses have predicted the probable binding site.

All drugs effectively quenched the intrinsic fluorescence of HHb by a static quenching mechanism under simulated pathophysiological conditions. The binding of clozapine, ziprasidone, and sertindole to HHb is unequivocally established, with an affinity of the order of magnitude of 10^3–10^4 M^{-1}, similar to the values obtained for the binding of most other drugs to this red blood cell protein. From the experimental data, we also conclude that HHb possesses one binding site for the examined antipsychotics. The binding was spontaneous and driven by negative enthalpies and favorable positive entropy contributions. In accordance with the previously described studies concerning the binding interactions of HHb with other small lipophylic heterocyclic drugs, the hydrophobic forces, with the help of electrostatic interactions or hydrogen bonds, play a significant role in stabilizing the HHb–antipsychotic complexes. Molecular docking studies further showed all three drugs bound HHb within its central cavity, consistent with the literature for hydrophobic ligands. Upon interaction, clozapine beneficially influenced the structural integrity properties of tetrameric HHb, based on the increase in α-helical content and the melting point, but reduced oxidant-induced protein deterioration. In contrast, sertindole and ziprasidone showed potentially harmful effects in the complex with HHb, primarily the acceleration of the spontaneous oxidation of an iron atom in the heme structure into the methemoglobin form that does not reversibly bind oxygen.

This work provides a comprehensive understanding of the interactions of well-prescribed second-generation antipsychotics with the physiologically important protein HHb as a biological target. Therefore, our study results may further help understand the pharmacology behavior of these lipophilic drugs and their clinical therapeutic potential, which depends on the drug's unbound concentration in the blood.

Author Contributions: Conceptualization, M.N., Č.M. and A.N.K.; methodology, investigation, and data collection and processing, L.P.A., N.G., I.C., A.M., M.K.R., S.M. and M.N.; writing—original draft preparation, L.P.A., I.C., N.G. and S.M.; supervision and preparation of the final manuscript, M.N., Č.M. and A.N.K. All authors have read and agreed to the published version of the manuscript.

Funding: This research was supported by the institutional funding project of the Faculty of Chemistry (contract no. 451-03-47/2023-01/200168) and the Institute for Chemistry, Technology, and Metallurgy (contract no. 451-03-47/2023-01/200026) financed by the Ministry of Science, Technological Development, and Innovation of the Republic of Serbia.

Institutional Review Board Statement: The study was conducted in accordance with the Declaration of Helsinki. Blood samples for hemoglobin isolation were obtained under the approval of the Ethical Committee for using human biological material in research at the University of Belgrade-Faculty of Chemistry (decision number/date of approval: 1-6/21).

Informed Consent Statement: Informed consent was obtained from the subject involved in the study. Written informed consent has been obtained from the patient to publish this paper.

Data Availability Statement: Not applicable.

Conflicts of Interest: The author, Lena Platanić Arizanović, is an employee of MDPI; however, she did not work for the International Journal of Molecular Sciences at the time of submission and publication. The other authors declare no conflicts of interest.

References

1. Hsia, C.C. Respiratory function of hemoglobin. *N. Engl. J. Med.* **1998**, *338*, 239–247. [CrossRef]
2. Lally, J.; MacCabe, J.H. Antipsychotic medication in schizophrenia: A review. *Br. Med. Bull.* **2015**, *114*, 169–179. [CrossRef] [PubMed]
3. Remington, G.; Lee, J.; Agid, O.; Takeuchi, H.; Foussias, G.; Hahn, M.; Fervaha, G.; Burton, L.; Powell, V. Clozapine's critical role in treatment resistant schizophrenia: Ensuring both safety and use. *Expert Opin. Drug Saf.* **2016**, *15*, 1193–1203. [CrossRef] [PubMed]
4. Komossa, K.; Rummel-Kluge, C.; Hunger, H.; Schwarz, S.; Bhoopathi, P.S.S.; Kissling, W.; Leucht, S. Ziprasidone versus other atypical antipsychotics for schizophrenia. *Cochrane Database Syst. Rev.* **2009**, *4*, CD006627. [CrossRef] [PubMed]
5. Siskind, D.; Gallagher, E.; Winckel, K.; Hollingworth, S.; Kisely, S.; Firth, J.; Correll, C.U.; Marteene, W. Does Switching Antipsychotics Ameliorate Weight Gain in Patients With Severe Mental Illness? A Systematic Review and Meta-analysis. *Schizophr. Bull.* **2021**, *47*, 948–958. [CrossRef]
6. Zoccali, R.A.; Bruno, A.; Muscatello, M.R.A. Efficacy and safety of sertindole in schizophrenia: A clinical review. *J. Clin. Psychopharmacol.* **2015**, *35*, 286–295. [CrossRef]
7. Wallace, S.M.; Reigelman, S. Uptake of acetazolamide by human erythrocytes in vitro. *J. Pharm. Sci.* **1977**, *66*, 729–731. [CrossRef]
8. Villa, C.H.; Anselmo, A.C.; Mitragotri, S.; Muzykantov, V. Red blood cells: Supercarriers for drugs, biologicals, and nanoparticles and inspiration for advanced delivery systems. *Adv. Drug Deliv. Rev.* **2016**, *106*, 88–103. [CrossRef]
9. Perutz, M.F.; Fermi, G.; Abraham, D.J.; Poyart, C.; Bursaux, E. Hemoglobin as a receptor of drugs and peptides: X-ray studies of the stereochemistry of binding. *J. Am. Chem. Soc.* **1986**, *108*, 1064–1078. [CrossRef]
10. Nedić, O.; Penezić, A.; Minić, S.; Radomirović, M.; Nikolić, M.; Ćirković Veličković, T.; Gligorijević, N. Food Antioxidants and Their Interaction with Human Proteins. *Antioxidants* **2023**, *12*, 815. [CrossRef]
11. Alpert, B.; Jameson, D.M.; Weber, G. Tryptophan emission from human haemoglobin and its isolated subunits. *J. Photochem. Photobiol.* **1980**, *31*, 1–4. [CrossRef] [PubMed]
12. Mueser, T.C.; Rogers, P.H.; Arnone, A. Interface sliding as illustrated by the multiple quaternary structures of liganded hemoglobin. *Biochemistry* **2000**, *39*, 15353–15364. [CrossRef]
13. Hirsch, R.E.; Nagel, R.L. Conformational studies of hemoglobins using intrinsic fluorescence measurements. *J. Biol. Chem.* **1981**, *256*, 1080–1083. [CrossRef] [PubMed]
14. Dohare, N.; Siddiquee, M.A.; Parray, M.D.; Kumar, A.; Patel, R. Esterase activity and interaction of human hemoglobin with diclofenac sodium: A spectroscopic and molecular docking study. *J. Mol. Recognit.* **2020**, *33*, e2841. [CrossRef]
15. Lakowicz, J.R. *Principles of Fluorescence Spectroscopy*; Springer: New York, NY, USA, 2006. [CrossRef]
16. Ware, W.R. Oxygen quenching of fluorescence in solution: An experimental study of the diffusion process. *J. Phys. Chem.* **1962**, *66*, 455–458. [CrossRef]
17. Cheng, H.; Liu, H.; Bao, W.; Zou, G. Studies on the interaction between docetaxel and human hemoglobin by spectroscopic analysis and molecular docking. *J. Photochem. Photobiol. B* **2011**, *105*, 126–132. [CrossRef] [PubMed]
18. Tunç, S.; Cetinkaya, A.; Duman, O. Spectroscopic investigations of the interactions of tramadol hydrochloride and 5-azacytidine drugs with human serum albumin and human hemoglobin proteins. *J. Photochem. Photobiol. B* **2013**, *120*, 59–65. [CrossRef]
19. Tunç, S.; Duman, O.; Bozoğlan, B.K. Studies on the interactions of chloroquine diphosphate and phenelzine sulfate drugs with human serum albumin and human hemoglobin proteins by spectroscopic techniques. *J. Lumines* **2013**, *140*, 87–94. [CrossRef]
20. Maurya, N.; Ud Din Parray, M.; Maurya, J.K.; Kumar, A.; Patel, R. Interaction of promethazine and adiphenine to human hemoglobin: A comparative spectroscopic and computational analysis. *Spectrochim. Acta A Mol. Biomol. Spectrosc.* **2018**, *199*, 32–42. [CrossRef]
21. Dufour, C.; Dangles, O. Flavonoid-serum albumin complexation: Determination of binding constants and binding sites by fluorescence spectroscopy. *Biochim. Biophys. Acta* **2005**, *1721*, 164–173. [CrossRef]
22. Undeland, I.; Kristinsson, H.G.; Hultin, H.O. Hemoglobin-mediated oxidation of washed minced cod muscle phospholipids: Effect of pH and hemoglobin source. *J. Agric. Food Chem.* **2004**, *52*, 4444–4451. [CrossRef] [PubMed]
23. Seal, P.; Sikdar, J.; Roy, A.; Haldar, R. Acetaminophen interacts with human hemoglobin: Optical, physical and molecular modeling studies. *J. Biomol. Struct. Dyn.* **2017**, *35*, 1307–1321. [CrossRef]
24. Miller, J.N. Recent advances in molecular luminescence analysis. *Proc. Anal. Div. Chem. Soc.* **1979**, *16*, 203–208.
25. Ross, P.D.; Subramanian, S. Thermodynamics of protein association reactions: Forces contributing to stability. *Biochemistry* **1981**, *20*, 3096–3102. [CrossRef]
26. Nishi, K.; Yamasaki, K.; Otagiri, M. Serum Albumin, Lipid and Drug Binding. *Subcell. Biochem.* **2020**, *94*, 383–397. [CrossRef] [PubMed]

27. Zhang, H.-M.; Wang, Y.-Q.; Jiang, M.-L. A fluorimetric study of the interaction of C.I. Solvent Red 24 with haemoglobin. *Dye. Pigm.* **2009**, *82*, 156–163. [CrossRef]
28. Greenfield, N.J. Using circular dichroism spectra to estimate protein secondary structure. *Nat. Protoc.* **2006**, *1*, 2876–2890. [CrossRef]
29. Kelly, S.M.; Jess, T.J.; Price, N.C. How to study proteins by circular dichroism. *Biochim. Biophys. Acta* **2005**, *1751*, 119–139. [CrossRef]
30. Stadler, A.M.; Garvey, C.J.; Bocahut, A.; Sacquin-Mora, S.; Digel, I.; Schneider, G.J.; Natali, F.; Artmann, G.M.; Zaccai, G. Thermal fluctuations of haemoglobin from different species: Adaptation to temperature via conformational dynamics. *J. R. Soc. Interface* **2012**, *9*, 2845–2855. [CrossRef]
31. Levy, A.; Kuppusamy, P.; Rifkind, J.M. Multiple heme pocket subconformations of methemoglobin associated with distal histidine interactions. *Biochemistry* **1990**, *29*, 9311–9316. [CrossRef]
32. Szebeni, J.; Winterbourn, C.C.; Carrell, R.W. Oxidative interactions between haemoglobin and membrane lipid. A liposome model. *Biochem. J.* **1984**, *220*, 685–692. [CrossRef]
33. Dalla Libera, A.; Scutari, G.; Boscolo, R.; Rigobello, M.P.; Bindoli, A. Antioxidant properties of clozapine and related neuroleptics. *Free Radic. Res.* **1998**, *29*, 151–157. [CrossRef]
34. Caruso, G.; Grasso, M.; Fidilio, A.; Tascedda, F.; Drago, F.; Caraci, F. Antioxidant Properties of Second-Generation Antipsychotics: Focus on Microglia. *Pharmaceuticals* **2020**, *13*, 457. [CrossRef] [PubMed]
35. Kosmachevskaya, O.V.; Topunov, A.F. Alternate and additional functions of erythrocyte hemoglobin. *Biochemistry* **2018**, *83*, 1575–1593. [CrossRef] [PubMed]
36. Vogel, F.; Gansmüller, R.; Leiblein, T.; Dietmaier, O.; Wassmuth, H.; Gründer, G.; Hiemke, C. The use of ziprasidone in clinical practice: Analysis of pharmacokinetic and pharmacodynamic aspects from data of a drug monitoring survey. *Eur. Psychiatry* **2009**, *24*, 143–148. [CrossRef] [PubMed]
37. Juruena, M.F.; de Sena, E.P.; de Oliveira, I.R. Sertindole in the Management of Schizophrenia. *J. Cent. Nerv. Syst. Dis.* **2011**, *3*, 75–85. [CrossRef]
38. Pizzino, G.; Irrera, N.; Cucinotta, M.; Pallio, G.; Mannino, F.; Arcoraci, V.; Squadrito, F.; Altavilla, D.; Bitto, A. Oxidative Stress: Harms and Benefits for Human Health. *Oxid. Med. Cell. Longev.* **2017**, *2017*, 8416763. [CrossRef]
39. Betigeri, S.; Thakur, A.; Raghavan, K. Use of 2,2′-azobis(2-amidinopropane) dihydrochloride as a reagent tool for evaluation of oxidative stability of drugs. *Pharm. Res.* **2005**, *22*, 310–317. [CrossRef]
40. Brinholi, F.F.; de Farias, C.C.; Bonifácio, K.L.; Higachi, L.; Casagrande, R.; Moreira, E.G.; Barbosa, D.S. Clozapine and olanzapine are better antioxidants than haloperidol, quetiapine, risperidone and ziprasidone in in vitro models. *Biomed. Pharmacother.* **2016**, *81*, 411–415. [CrossRef]
41. Gligorijević, N.; Vasović, T.; Lević, S.; Miljević, Č.; Nedić, O.; Nikolić, M. Atypical antipsychotic clozapine binds fibrinogen and affects fibrin formation. *Int. J. Biol. Macromol.* **2020**, *154*, 142–149. [CrossRef]
42. Šunderić, M.; Vasović, T.; Milčić, M.; Miljević, Č.; Nedić, O.; Nikolić, M.R.; Gligorijević, N. Antipsychotic clozapine binding to alpha-2-macroglobulin protects interacting partners against oxidation and preserves the anti-proteinase activity of the protein. *Int. J. Biol. Macromol.* **2021**, *183*, 502–512. [CrossRef]
43. Chang, W.H.; Lin, S.K.; Lane, H.Y.; Hu, W.H.; Jann, M.W.; Lin, H.N. Clozapine dosages and plasma drug concentrations. *J. Formos. Med. Assoc.* **1997**, *96*, 599–605.
44. Miller, D.D. Review and management of clozapine side effects. *J. Clin. Psychiatry* **2000**, *61* (Suppl. S8), 14–17; discussion 18–19.
45. Tentori, L.; Salvati, A.M. Hemoglobinometry in human blood. In *Methods in Enzymology*; Antonini, E., Rossi-Bernardi, L., Hiancone, E., Eds.; Academic Press: New York, NY, USA, 1981; pp. 707–715. [CrossRef]
46. Gligorijević, N.; Minić, S.; Radibratović, M.; Papadimitriou, V.; Nedić, O.; Sotiroudis, T.G.; Nikolić, M.R. Nutraceutical phycocyanobilin binding to catalase protects the pigment from oxidation without affecting catalytic activity. *Spectrochim. Acta A Mol. Biomol. Spectrosc.* **2021**, *251*, 119483. [CrossRef]
47. Minic, S.; Stanic-Vucinic, D.; Radomirovic, M.; Radibratovic, M.; Milcic, M.; Nikolic, M.; Cirkovic Velickovic, T. Characterization and effects of binding of food-derived bioactive phycocyanobilin to bovine serum albumin. *Food Chem.* **2018**, *239*, 1090–1099. [CrossRef]
48. Yokoyama, T.; Neya, S.; Tsuneshige, A.; Yonetani, T.; Park, S.-Y.; Tame, J.R.H. R-State Haemoglobin with Low Oxygen Affinity: Crystal Structures of Deoxy Human and Carbonmonoxy Horse Haemoglobin Bound to the Effector Molecule L35. *J. Mol. Biol.* **2006**, *356*, 790–801. [CrossRef]
49. Olsson, M.H.M.; Søndergaard, C.R.; Rostkowski, M.; Jensen, J.H. PROPKA3: Consistent Treatment of Internal and Surface Residues in Empirical pKa Predictions. *J. Chem. Theory Comput.* **2011**, *7*, 525–537. [CrossRef]
50. Phillips, J.C.; Hardy, D.J.; Maia, J.D.C.; Stone, J.E.; Ribeiro, J.V.; Bernardi, R.C.; Buch, R.; Fiorin, G.; Hénin, J.; Jiang, W.; et al. Scalable Molecular Dynamics on CPU and GPU Architectures with NAMD. *J. Chem. Phys.* **2020**, *153*, 044130. [CrossRef]
51. Pedretti, A.; Mazzolari, A.; Gervasoni, S.; Fumagalli, L.; Vistoli, G. The VEGA Suite of Programs: An Versatile Platform for Cheminformatics and Drug Design Projects. *Bioinformatics* **2021**, *37*, 1174–1175. [CrossRef]
52. Stewart, J.J.P. Optimization of Parameters for Semiempirical Methods VI: More Modifications to the NDDO Approximations and Re-Optimization of Parameters. *J. Mol. Model.* **2013**, *19*, 1–32. [CrossRef]

53. Stewart, J.J.P. MOPAC: A Semiempirical Molecular Orbital Program. *J. Comput. Aided Mol. Des.* **1990**, *4*, 1–105. [CrossRef] [PubMed]
54. Trott, O.; Olson, A.J. AutoDock Vina: Improving the Speed and Accuracy of Docking with a New Scoring Function, Efficient Optimization, and Multithreading. *J. Comput. Chem.* **2010**, *31*, 455–461. [CrossRef]
55. Lakowicz, J.R.; Weber, G. Quenching of fluorescence by oxygen. A probe for structural fluctuations in macromolecules. *Biochemistry* **1973**, *12*, 4161–4170. [CrossRef]
56. Bi, S.; Dinga, L.; Tiana, Y.; Songa, D.; Zhou, X.; Liua, X.; Zhanga, H. Investigation of the interaction between flavonoids and human serum albumin. *J. Mol. Struct.* **2004**, *703*, 37–45. [CrossRef]
57. Pace, N.C.; Scholtz, M.J.; Grimsley, G.R. Forces Stabilizing Proteins. *FEBS Lett.* **2014**, *588*, 2177–2184. [CrossRef]

International Journal of
Molecular Sciences

MDPI

Article

Proteomic Modulation in TGF-β-Treated Cholangiocytes Induced by Curcumin Nanoparticles

Elisa Ceccherini [1,*], Giovanni Signore [1,2], Lorena Tedeschi [1], Federico Vozzi [1], Nicoletta Di Giorgi [1], Elena Michelucci [1,3], Antonella Cecchettini [1,4,†] and Silvia Rocchiccioli [1,†]

1 Institute of Clinical Physiology, National Research Council, 56124 Pisa, Italy; giovanni.signore@unipi.it (G.S.); tedeschi@ifc.cnr.it (L.T.); vozzi@ifc.cnr.it (F.V.); digiorgi@ifc.cnr.it (N.D.G.); elena.michelucci@pi.iccom.cnr.it (E.M.); antonella.cecchettini@unipi.it (A.C.); silvia.rocchiccioli@ifc.cnr.it (S.R.)
2 Biochemistry Unit, Department of Biology, University of Pisa, 56123 Pisa, Italy
3 Institute of Chemistry of Organometallic Compounds, National Research Council, 56124 Pisa, Italy
4 Department of Clinical and Experimental Medicine, University of Pisa, 56126 Pisa, Italy
* Correspondence: ceccherini@ifc.cnr.it; Tel.: +39-050-315-3187
† These authors contributed equally to this work.

Abstract: Curcumin is a natural polyphenol that exhibits a variety of beneficial effects on health, including anti-inflammatory, antioxidant, and hepato-protective properties. Due to its poor water solubility and membrane permeability, in the present study, we prepared and characterized a water-stable, freely dispersible nanoformulation of curcumin. Although the potential of curcumin nanoformulations in the hepatic field has been studied, there are no investigations on their effect in fibrotic pathological conditions involving cholangiocytes. Exploiting an in vitro model of transforming growth factor-β (TGF-β)-stimulated cholangiocytes, we applied the Sequential Window Acquisition of All Theoretical Mass Spectra (SWATH-MS)-based quantitative proteomic approaches to study the proteome modulation induced by curcumin nanoformulation. Our results confirmed the well-documented anti-inflammatory properties of this nutraceutic, highlighting the induction of programmed cell death as a mechanism to counteract the cellular damages induced by TGF-β. Moreover, curcumin nanoformulation positively influenced the expression of several proteins involved in TGF-β-mediated fibrosis. Given the crucial importance of deregulated cholangiocyte functions during cholangiopathies, our results provide the basis for a better understanding of the mechanisms associated with this pathology and could represent a rationale for the development of more targeted therapies.

Keywords: cholangiocytes; curcumin; nanoparticles; TGF-β; proteomics; LC-MS/MS

1. Introduction

Curcumin (Cur) is a natural polyphenolic pigment that is extracted from turmeric with a broad safety profile [1]. This phytochemical has been used as an anti-inflammatory, anti-oxidant, anti-bacterial, anticarcinogenic, antifungal, and hepatoprotective agent [2]. Despite several benefits, Cur has limited water solubility and membrane permeability, and it is unstable at high temperatures when exposed to light and oxidative conditions [3]. After oral administration, it undergoes a fast first-pass metabolism and is quickly eliminated, thus reducing its bioavailability [4].

Nanoparticles (NPs) have a great impact on future medical treatments for many diseases. Compared to conventional therapies, drug-loaded NPs possess several advantages, including a prolonged circulation half-life and the elimination of side effects [5,6]. Over the last decade, scientific research has investigated the therapeutic potential of CurNPs in various pathological conditions, with the aim of overcoming their limitations and improving their pharmacologic profile. The hepato-protective effects of CurNPs have been partially elucidated, but relevant data from the literature also involve the anti-carcinogenic activity of CurNPs through their antioxidant, anti-inflammatory, and pro-apoptotic activities, as demonstrated in a mouse model of diethylnitrosamine-induced liver carcinoma [7]. Similarly, Mona and colleagues reported the hepato-protective effects of CurNPs treatment in HepG2 and Huh-7 cancer cells due to their viability reduction and apoptosis increase [8]. Recently, the anti-liver fibrosis activity of CurNPs has also been investigated. The phytochemical was able to maintain its hepatic function and architecture during fibrosis development in a CCl_4-induced liver fibrosis mouse model by the direct inhibition of fibrosis-mediating proteins (e.g., the platelet-derived growth factor receptor beta, TIMP metallopeptidase inhibitor 1, Toll-like receptor 9 and TGF-β [9]. In another study, Ji and colleagues correlated the anti-fibrotic effect of CurNPs to their anti-inflammatory activity and to the up-regulation of hepatocyte growth factors and matrix metalloprotease-2 [10]. Although the potentiality of CurNPs in the hepatic field has been studied, there are no investigations on their effect in the fibrotic pathological conditions involving cholangiocytes.

TGF-β is a cytokine that modulates several cellular processes, including proliferation, migration, invasion, angiogenesis, and immune responses [11]. During hepatic fibrosis, TGF-β promotes inflammation and regulates the hepatocyte's behavior and activation of human stellate cells (HSCs), which could represent the main source of the extracellular matrix [12–17].

The main purpose of this work was to prepare and characterize a water stable, including the freely dispersible nanoformulation of curcumin, and to study the proteome modulation that it induces in a well-established model of in vitro TGF-β-stressed cholangiocytes [18].

2. Results

2.1. CurNPs Synthesis and Characterization

First, we sought to set up an automated, robust method for the reproducible preparation of CurNPs. To this end, we adopted a nanoprecipitation approach using DMSO-dissolved curcumin and water as a non-solvent. A microfluidic-based protocol was implemented to allow for the preparation of CurNPs with a reproducible, uniform size and with the possibility to scale up the process. Different volume ratios between DMSO and water were tested, leading to CurNPs preparations 1:79 v/v ($CurNP_a$), 1:158 v/v ($CurNP_b$), 1:316 v/v ($CurNP_c$), 1:625 v/v ($CurNP_d$), and 1:1280 v/v ($CurNP_e$). In all cases, these distributions were monodispersed with a relatively narrow distribution. The size and dispersion (evaluated by intensity distribution on DLS) obtained for the three preparations were 158 nm $\pm$ 3 nm ($CurNP_a$), 120 nm $\pm$ 1 nm ($CurNP_b$), 97 nm $\pm$ 2 nm ($CurNP_c$), 95 nm $\pm$ 1 nm ($CurNP_d$), and 97 nm $\pm$ 9 nm ($CurNP_e$), indicating a slow, yet noticeably decreasing trend in CurNPs diameter with an increasing water/DMSO ratio. CurNPs were stabilized at approximately 97 nm for any water/DMSO ratio greater than 316:1, while larger sizes were obtained at lower ratios. The size distributions for the synthesized nanoparticles are reported in Figure S1.

With the aim of using smaller, homogeneous, and highly concentrated CurNPs in subsequent experiments, attention was focused on $CurNP_b$, $CurNP_c$, and $CurNP_d$. In fact, $CurNP_e$ was too diluted to be of practical use, while $CurNP_a$ was too polydisperse.

2.2. CurNPs Dispersion Characterization

The solutions of curcumin in DMSO maintained the same absorbance spectra over at least 48 h, while fluorescence decreased over time, reaching about 70% of the initial value in 48 h (Figure S2a–d). Curcumin in the DMSO was added to the culture medium or PBS to not produce real/stable solutions even if an amount of DMSO and serum was added. The absorbance and fluorescence decreased, and the discoloring of these solutions was evident (Figure S2e–h).

CurNPs were produced instead of a more stable dispersion without any sign of precipitation over the hours. In order to exactly quantify the number of CurNPs per volume unit of the stock solution for each preparation, the absorbance of serial dilutions of $CurNP_b$, $CurNP_c$, and $CurNP_d$ in PBS was evaluated, and an "equivalent concentration" [μM] of curcumin was calculated from the standard curve of Curcumin in DMSO. $CurNP_d$ (1:615) preparation was 13.1 μM, $CurNP_c$ (1:316) was 55.6 μM and $CurNP_b$ (1:158) 187.4 μM. $CurNP_b$ was used as a stock solution to obtain the desired amount of CurNPs for the experiment with cells.

With the aim of assessing the behavior of CurNPs in the culture medium, the optical properties of their dispersions in the culture medium, with different amounts of the fetal bovine serum (FBS), were evaluated over time. In all the conditions tested, a reduction in the fluorescence emission was to be less than 40%; this is slightly higher than the reduction in fluorescence of curcumin in DMSO (30%) in the same interval (Figure S2i). The administration of CurNPs instead of "free" curcumin ensured that a "nominal" amount of curcumin in the medium could be reached (till 50 μM), and cells were not affected by the massive precipitation of aggregates.

2.3. *Sample Collection*

Primary human cholangiocytes were treated for 48h with TGF-β at a concentration of 10 ng/mL, in accordance with our previous study [18]. Another sample group was obtained by treating cholangiocytes with TGF-β 10 ng/mL for 24h before adding CurNPs 10 μM for 24 h. The CurNPs concentration and treatment time were determined by applying a viability cut-off greater than 70%, as shown in Figure 1. The control group was obtained by growing cholangiocytes in their specific culture medium with 1% FBS. At the end of the experiments, the cell pellets were collected for subsequent liquid chromatography-tandem mass spectrometry (LC-MS/MS) analysis.

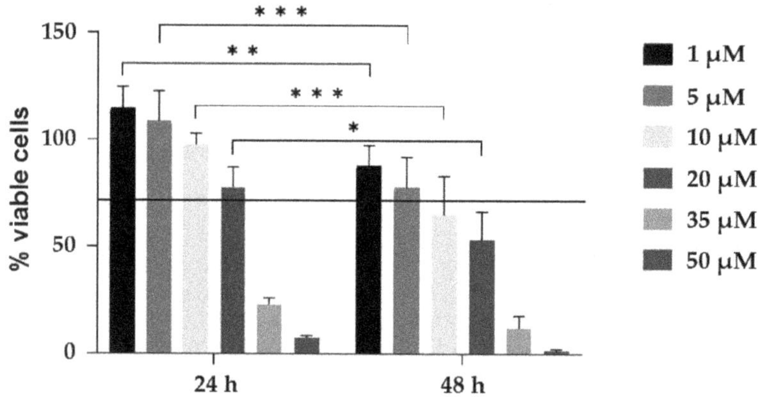

Figure 1. Cytotoxic effect of CurNPs-treated with cholangiocytes for 24 h and 48 h. Data were obtained from 3 independent experiments and are presented as the percentage of viable cells compared to control cells. Statistical analysis was performed with two-way ANOVA and Tukey's multiple comparison test; * $p \leq 0.05$, ** $p \leq 0.01$, *** $p \leq 0.001$.

2.4. *IL-6 Dosage*

The anti-inflammatory effect exerted by curcumin in multiple pathologic conditions, including liver diseases, is well known [19,20]. Hence, we evaluated interleukin-6 (IL-6) in the culture medium for each condition tested (Figure 2). In TGF-β-treated cholangiocytes, IL-6 levels significantly increased compared to the control cells; conversely, CurNPs seemed to reduce the levels of TGF- β-induced inflammation.

2.5. *Hypothesis Free Proteomics and Bioinformatic Analysis*

Human cholangiocytes were subjected to protein extraction, reduction, alkylation, and digestion with trypsin, and the resulting peptide mixture was prepared for LC-MS/MS analysis, thus quantifying 2479 proteins. To obtain an overview of biological significance, these proteins were categorized according to their biological function using a David bioinformatic database. As reported in Figure 3, the enriched biological processes were primarily related to RNA metabolism, infections, proteins and amino acid metabolism, and cell cycle. Interestingly, the immune response and interleukin signaling were also identified.

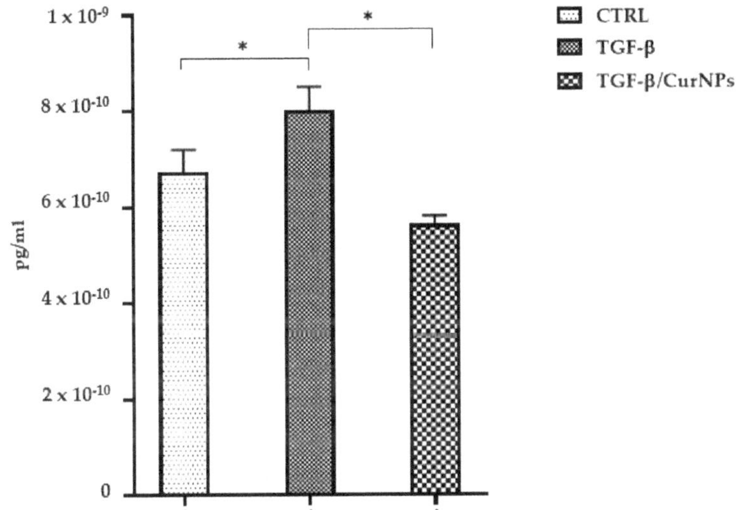

Figure 2. Levels of IL-6 in culture media. Data represent the mean of 3 independent experiments. Statistical analysis was performed with one-way ANOVA and Holm–Sidak's multiple comparison test; * $p \leq 0.05$.

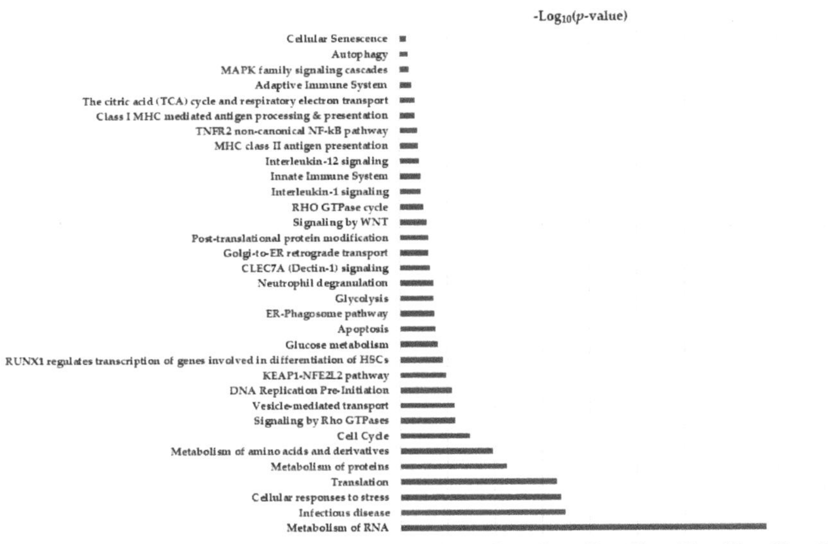

Figure 3. Biological functions of deregulated proteins analyzed using the David database; the significant ones ($p \leq 0.05$) were reported.

To identify the dysregulated proteins during TGF-β and TGF-β/CurNPs treatments, a comparative analysis with the control cells and TGF-β-treated cholangiocytes was performed, respectively.

In TGF-β-treated cholangiocytes, a total of 28 proteins were significantly dysregulated with respect to the control cells greater than ±2.5-fold. In particular, data analysis showed 24 upregulated and 4 downregulated proteins following 48 h of treatment. Among them, fibronectin and dynamin-3 were the most deregulated proteins, with a fold-change (FC) of 295.85 and 0.01, respectively (Table 1).

Table 1. List of proteins that displayed more than a ±2.5-FC in TGF-β-treated cholangiocytes following 48 h treatment. The table reports the degrees of variation (evaluated as fold change) compared to the control cells and *p*-value. * Although this protein did not respect the established cut-off for expression levels, it was nonetheless indicated for its biological relevance and its significant down-regulation compared to the control cells.

Protein Name	Protein ID	FC	*p*-Value
Fibronectin	P02751	295.85	0.0045
Cullin-3	Q13618	13.69	0.0233
Tensin-1	Q9HBL0	7.15	0.008
Haloacid dehalogenase-like hydrolase domain-containing protein 2	Q9H0R4	6.05	0.0055
Palladin	Q8WX93	5.03	0.001
Bcl-2-like protein 12	Q9HB09	4.77	0.0042
Protein MTSS 1	O43312	4.51	0.0026
TSC22 domain family protein 1	Q15714	4.16	0.0016
Cytospin-A	Q69YQ0	3.19	0.0167
Phosphatidylcholine translocator ABCB4	P21439	3.10	0.049
Centromere protein F	P49454	3.04	0.0069
Four and a half LIM domains protein 2	Q14192	3.01	0.0036
High mobility group protein HMGI-C	P52926	3.00	0.0074
Protein PRRC1	Q96M27	3.00	0.0102
Zinc finger protein 746	Q6NUN9	2.95	0.0376
Protein FAM114A2	Q9NRY5	2.94	0.0367
Keratin, type II cytoskeletal 1	P04264	2.90	0.0210
Fos-related antigen 1	P15407	2.85	0.0212
Calpain small subunit 1	P04632	2.66	0.0266
SH3 and PX domain-containing protein 2A	Q5TCZ1	2.65	0.0075
Transducin-like enhancer protein 3	Q04726	2.62	0.0089
Zinc finger protein Helios	Q9UKS7	2.60	0.0036
Tetratricopeptide repeat protein 24	A2A3L6	2.57	0.0036
Moesin	P26038	2.51	0.0003
Kinesin-like protein KIF11 *	P52732	0.41	0.0132
Protein HEXIM1	O94992	0.40	0.0228
E3 ubiquitin-protein ligase BRE1A	Q5VTR2	0.30	0.0001
Cornifin-A	P35321	0.25	0.0009
Dynamin-3	Q9UQ16	0.01	<0.0001

The data analysis of TGF-β/CurNPs-treated cholangiocytes highlighted 10 significantly deregulated proteins compared to TGF-β-treated (48 h) cells. In particular, we showed eight upregulated and two downregulated proteins (Table 2). Among them, dynamin-3 and Ras-related protein Rab-15 were the most deregulated proteins, with a FC of 248.1 and 0.0014, respectively.

2.6. Comparative Analysis of TGF-β and TGF-β/CurNPs Dysregulated Proteins

In order to understand if some deregulated proteins in TGF-β-treated cells were also modulated by curNPs treatment, a comparative analysis was performed. Surprisingly, only three proteins (E3 ubiquitin-protein ligase BRE1A, kinesin-like protein KIF11, and dynamin-3) were shown to be common between the treatments (Table 3). As shown in

Figure 4, the protein expression showed opposite trends. Indeed, these proteins were up-regulated during TGF-β/CurNPs treatment while they were down-regulated with TGF-β alone, showing a significant variation between the two groups.

Table 2. List of proteins that displayed more than ±2.5-FC in TGF-β/CurNPs-treated cholangiocytes. The table reports the degrees of variation (evaluated as FC) compared to TGF-β-treated cells (48 h).

Protein Name	Protein ID	FC	*p*-Value
Dynamin-3	Q9UQ16	248.1	0.0371
Kinesin-like protein KIF11	P52732	4.60	0.0081
E3 ubiquitin-protein ligase BRE1A	Q5VTR2	3.93	0.0462
Histone H1.2	P16403	2.94	0.0062
Nucleobindin-1	Q02818	2.79	0.03
Mannosyl-oligosaccharide glucosidase	Q13724	2.75	0.0018
Thioredoxin domain-containing protein 15	Q96J42	2.69	0.0086
Bromodomain adjacent to zinc finger domain protein 1A	Q9NRL2	2.67	0.0485
Fatty acid-binding protein, adipocyte	P15090	0.012	0.0001
Ras-related protein Rab-15	P59190	0.0014	0.0002

Table 3. List of proteins in common between TGF-β and TGF-β/CurNPs-treated cholangiocytes. The table reports the degrees of variation (evaluated as FC) compared to the control and TGF-β for TGF-β-treated cells and TGF-β/CurNPs-treated cells, respectively.

Protein Name	Protein ID	FC	
		TGF-β	**TGF-β/CurNPs**
E3 ubiquitin-protein ligase BRE1A	Q5VTR2	0.30	3.93
Kinesin-like protein KIF11	P52732	0.41	4.60
Dynamin-3	Q9UQ16	0.01	248.1

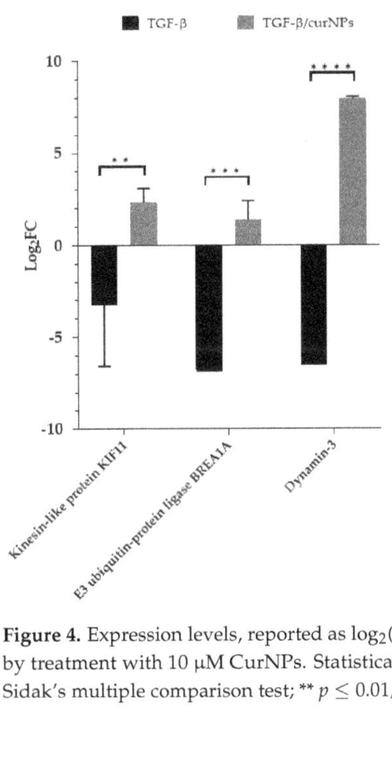

Figure 4. Expression levels, reported as $\log_2$(FC), of proteins deregulated by TGF-β and modulated by treatment with 10 μM CurNPs. Statistical analysis was performed with two-way ANOVA and Sidak's multiple comparison test; ** $p \leq 0.01$, *** $p \leq 0.001$, **** $p \leq 0.0001$.

3. Discussion

Nutraceuticals play an important role in the management of multiple pathological conditions [21]. Among them, curcumin has been demonstrated with several properties, including antioxidant, anti-inflammatory, and anti-proliferative activity inhibiting multiple targets and molecular pathways [20]. However, the therapeutic use of curcumin is strongly limited by its low water solubility, which impairs its bioavailability. Currently, nanomedicine has overcome these obstacles improving curcumin pharmacokinetics, efficacy, and cellular uptake. In recent years, the anti-carcinogenic and anti-fibrotic activities of curNPs have been investigated in the hepatic field [6–9]; however, no study has evaluated its potentiality in the fibrotic pathological conditions that affect cholangiocytes.

In the present study, we synthesized, characterized, and identified a phototherapy formulation that was CurNPs-based (Supplementary Materials, Figures S1 and S2). Following the cytotoxicity assay (Figure 1), 10 μM CurNPs was selected as the best concentration to investigate the potential effects of curcumin in a well-established in vitro model of TGFβ-damaged cholangiocytes, applying a SWATH-MS-based quantitative proteomic approach [18]. In TGFβ-treated cholangiocytes, we highlighted the up-regulation of several proteins directly involved in the cell cycle, such as the Bcl-2-like protein 12, centromere protein F, and Proline Rich Coiled-Coil 1 protein (PRRC1) (Table 1). Moreover, palladin, cullin-3, tensin-1, and moesin, which interacted with actin filaments, were also up-regulated following TGF-β treatment (Table 1). The role of the TGF-β family in controlling many aspects of cellular behavior, including proliferation, apoptosis, dormancy, autophagy, and senescence, is well known [11,22]. According to our data, the enriched biological processes of deregulated proteins in TGF-β-treated cholangiocytes were primarily related to the cell cycle, especially to cell proliferation and mitosis (Figure 5). Thus, our proteomic analysis, together with the analysis of biological processes, seemed to support the idea that TGF-β perturbed the cholangiocytes' choice between proliferation, cell-cycle arrest, and apoptosis, promoting cell survival and proliferation.

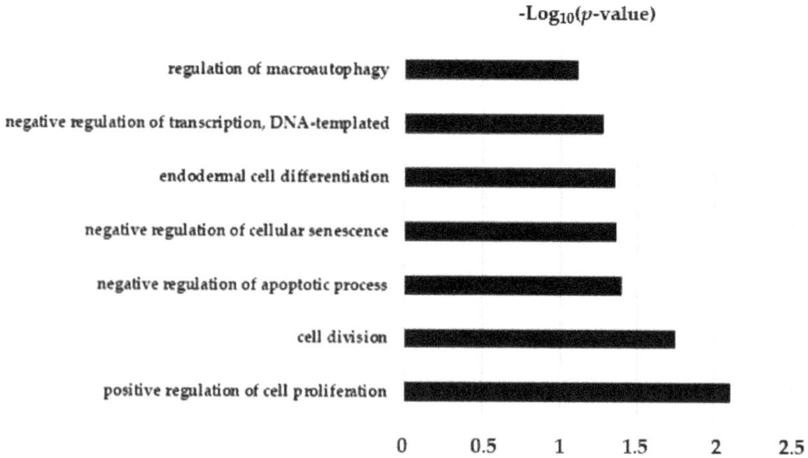

Figure 5. Biological function analysis of differentially expressed proteins in TGF-β-treated cholangiocytes compared to the control cells.

In addition, TGF-β induced a more than 290-fold increase in the fibronectin level compared to the control cells, which could be related to the well-documented pro-fibrotic properties of this cytokine [23]. Globally, our data indicated the pro-inflammatory and pro-fibrotic role of TGF-β in human cholangiocytes, together with its ability to perturb the cell cycle and related cytoskeleton dynamic.

In our proteomic analysis, we also identified cornifin-A as a marked down-regulated protein (FC = 0.25) compared to the control cells. Cornifin-A is part of a small proline-rich protein (SPRR) and was recently identified in the biliary epithelium. Demetris and colleagues reported that SPRR2A overexpression within in vitro biliary epithelial cells increased resistance to oxidative injury and promoted wound restitution by enhancing the migration and acquisition of mesenchymal characteristics [24]. Similarly, Nozaki and colleagues showed increased SPRR2 expression in the biliary epithelium of mice after bile duct ligation: a surgical method used to induce liver fibrosis [25]. Although these sporadic data could suggest the decisive role of SPPR2A in the function of the biliary barrier, no data elucidated the role of cornifin-A (SPPR1A). TGF-β could directly interfere with the function of the biliary barrier through the mediation of cornifin-A, but further investigations were necessary to elucidate this aspect.

The global proteome analysis of TGF-β/CurNPs-treated cholangiocytes highlighted interesting information regarding curNPs activity. As reported in Figure 6, these nutraceutic-modulated different pathways included antigen presentation, cell cycle, triglyceride metabolism, and post-translational protein modifications.

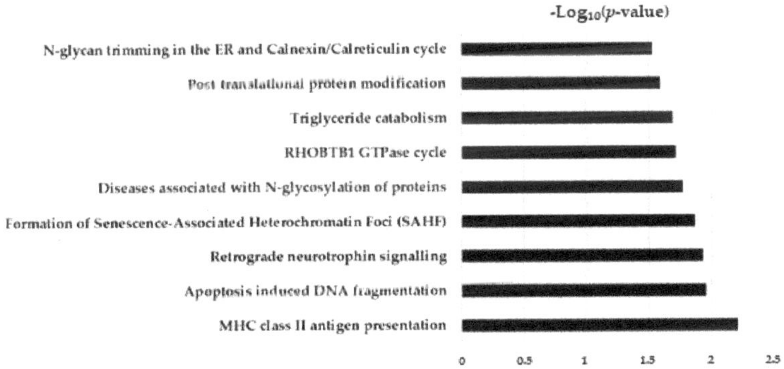

Figure 6. Pathway analysis of differentially expressed proteins in TGF-β/CurNPs-treated cholangiocytes.

CurNPs induced the up-regulation of histone H1.2 (FC = 2.94), which plays several important functions in multiple cellular processes, including apoptosis, autophagy, senescence, cell cycle control, and gene transcription, regulating chromatin dynamics [26,27]. According to the literature, our results could indicate the protective role of CurNPs through the induction of programmed cholangiocyte death, which acts as a barrier to further cellular damage by suppressing the growth of TGF-β-stressed cells.

In our previous work, we reported the impairment of Ca^{2+} homeostasis in human cholangiocytes under TGF-β treatment through the downregulation of multiple associated proteins such as calumin, calcium/calmodulin-dependent protein kinase type II subunit delta, caveolin-1, and NipSnap1 [18]. Human cholangiocytes have primary cilia with sensory functions that are linked to numerous receptors and ion-conducting channels and are involved in several cellular mechanisms, including calcium regulation and bile secretion [28–31]. Interestingly, in the present study, CurNPs determined the marked up-regulation of three proteins (kinesin-like protein KIF11, thioredoxin domain-containing protein 15, and dynamin-3) which were part of a primary cilium structure and the functions of most eukaryotic cells and probably also cholangiocytes. In human retinal pigment epithelial cells, the siRNA knockdown of the thioredoxin domain-containing protein 15 exerted alterations in ciliogenesis, with a significant reduction in ciliated cells compared to the control cells [32]. The kinesin-like protein KIF11 is primarily known as a mitotic regulator [33]; however, a non-mitotic role has recently emerged, suggesting its association with the primary cilia [34]. In that study, Zalenski and colleagues demonstrated how the kinesin-like protein KIF11 was localized in the basal bodies of primary cilia in multiple cell types, regulating cilia dynamics. Moreover, decreased protein levels were correlated to an increase in the number of ciliated cells, while the depletion of kinesin-like protein KIF11 determined a modification in the kinetics of cilia disassembly and increased the cilium length. In primary sclerosing cholangitis (PSC), which is a chronic and progressive cholestatic disease, cholangiocytes showed longer primary cilia compared to normal cholangiocytes, and this phenomenon impaired their physiological functions [35]. Although the thioredoxin domain-containing protein 15 expression was not perturbed by TGF-β, kinesin-like protein KIF11 levels were greatly reduced compared to the control cells (Figure 4), suggesting the possible role of TGF-β in the deregulation of cholangiocytes' cilia function. Interestingly, CurNPs treatment reverted the effect of TGF-β on kinesin-like protein KIF11 expression (Figure 4) and also promoted increased levels of thioredoxin domain-containing protein 15 (Table 2). These data support the idea that CurNPs could restore the physiological function of cilia epithelium that was impaired in TGF-β-stressed cholangiocytes. In many cell types, the proximal part of the primary cilium presented an invagination of the plasma membrane, also known as a ciliary pocket, which is a specialized site for endocytosis [36]. In TGF-β-treated cholangiocytes, we reported a marked reduction in dynamin-3 levels (Table 1), which is one of the major proteins involved in endocytosis [37]. Recently, Inokawa and colleagues reported the decreased expression of dynamin-3 in hepatocellular carcinoma tissue [38]. Moreover, patients with reduced levels of this protein in these tumor tissues exhibit worse prognoses compared to patients without a decreased expression. Our results demonstrated that CurNPs reverted the effect of increasing TGF-β in the dynamin-3 levels of human cholangiocytes (Figure 4). These data suggest that nutraceutic positively modulates the endocytic processes that would be compromised in TGF-β-treated cholangiocytes, restoring the dynamin-3 levels whose reduction is also correlated with malignancy.

Considering the well-documented anti-inflammatory properties of curcumin in liver diseases [19], in the present study, we investigated this effect in TGF-β-treated cholangiocytes. As reported in Figure 2, CurNPs significantly reduced the level of IL-6 that was increased following TGF-β treatment. Interestingly, our proteomic analysis highlighted the downregulation of E3 ubiquitin-protein ligase BRE1A in TGF-β-treated cholangiocytes and an opposite trend following CurNPs treatment (Figure 4). Recent findings correlated the overexpression of E3 ubiquitin-protein ligase BRE1A with the reversion of the TGF-β-induced activation of fibrotic proteins (VEGFA, TNF-α, and IL-6) in a human hepatic stellate cell line LX-2 [39]. According to these findings, our data could indicate the potential anti-fibrotic activity of CurNPs by increasing E3 ubiquitin-protein ligase BRE1A levels and reducing fibrotic mediators (i.e., IL-6).

In conclusion, we highlighted several aspects of curNPs activity on human cholangiocytes whose functions were perturbed during TGF-β stress, as reported in Figure 7.

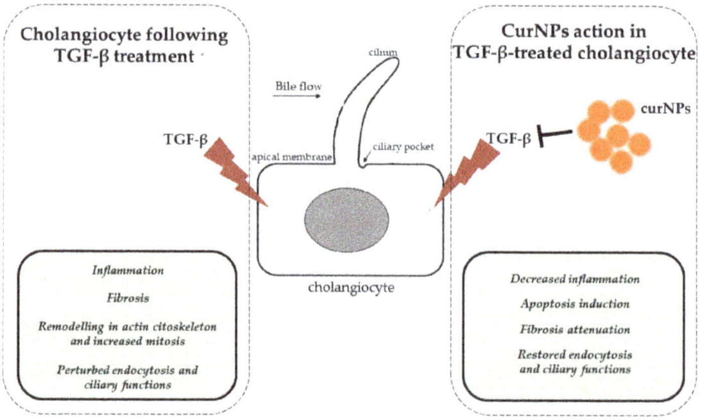

Figure 7. Representative image of the possible protective role of CurNPs in TGF-β-treated cholangiocytes.

In particular, we confirmed the anti-inflammatory properties of this nutraceutic and the induction of programmed cell death to counteract the cellular damages induced by TGF-β. In human cholangiocytes, this cytokine is also responsible for the fibrotic processes, and CurNPs could attenuate this condition by reverting the expression of key mediators. In our study, TGF-β seemed to perturbate the cholangiocyte cilia functions and endocytic processes that could be restored by CurNPs. Given the crucial importance of deregulated cholangiocyte functions during cholangiopathies [40], our results lay the basis for a better understanding of the pathophysiology of cholangiocytes and could represent a rationale for the development of more targeted therapies.

4. Materials and Methods

4.1. Materials

Primary human cholangiocytes (catalog number: 36755-12) and their specific growth medium without serum were purchased from Celprogen (Torrance, CA, USA). Curcumin (Merck, Rahway, NJ, USA), dimethyl sulfoxide (Merck, Rahway, NJ, USA), Ethanol absolute (Merck, Rahway, NJ, USA), fetal bovine serum (FBS, Gibco®, Carlsbad, CA, USA), Dulbecco's Modified Eagle Medium (DMEM, Gibco®, Grand Island, NY, USA), Penicillin-Streptomycin (Gibco®, Grand Island, NY, USA), Trypsin-EDTA (Gibco®, Grand Island, NY, USA), Sirius Red/Fast Green Staining Collagen (Chondrex, Redmond, WA, USA), fertilized chicken eggs (PT. Lestari Farm, Sragen, Indonesia), distilled water (IPHA Laboratories, Bandung, Indonesia), ethyl acetate (Brataco, Jakarta, Indonesia), ascorbic acid, PBS pH 7.4, and 1,1-diphenyl-2-picrylhydrazyl (DPPH) were purchased from Sigma-Aldrich (St. Louis, MO, USA). TGF-β (T7039) was purchased from Merck. Formic acid (5.33002), trifluoroacetic acid (80457), and ammonium bicarbonate (Ambic, 40867), all eluent additives for LC-MS, were purchased, respectively, from Merck (Darmstadt, Germany), Sigma-Aldrich (St. Louis, MO, USA) and Fluka Analytical (Sigma-Aldrich, St. Louis, MO, USA). Acetonitrile (1.00029) hypergrade solvent LiChrosolv for LC-MS was bought from Merck (Darmstadt, Germany), while sodium deoxycholate (D6750) and albumin from bovine serum (A7906) from Sigma-Aldrich (St. Louis, MO, USA). Reagent A (23228) and reagent B (1859078) for Pierce BCA (bicinchoninic acid) Protein Assay were purchased from Thermo Scientific (Rockford, IL, USA). Iodoacetamide (RPN6302V), dithiothreitol (D1 309.0010), and trypsin modified sequencing grade (11418033001) were obtained, respectively, from GE Healthcare (Chicago, IL, USA), Duchefa Biochemie (Haarlem, The Netherlands) and Roche (Indianapolis, IN, USA). Mobicol spin columns (M1003) and their filters (M2110) 10 μm pore size were purchased from Mo Bi Tec (Goettingen, Germany), while VersaFlash spherical C18 bonded flash silica from Supelco Analytical (Bellefonte, PA, USA). Milli-Q deionized water was filtered on a Millipak filter (0.22 μm, MPGL040001) and purified on an LC-Pak cartridge (C18, LCPAK0001); all Millipore, Bedford, MA, USA). Beta-galactosidase digested (4333606) was obtained from AB Sciex (Framingham, MA, USA).

4.2. CurNPs Synthesis

CurNPs were prepared by nanoprecipitation in an Asia Syrris microfluidic setup (Syrris, Royston, UK). Briefly, a Cur solution in DMSO (10 mM) was mixed with deionized water at 1:158, 1:316, and 1:625 volume ratios using a 250 μL microfluidic mixer chip. The total flow rate was maintained at 50 μL/min, and the temperature was controlled at 25 °C. An eluted suspension was collected and stored at 4 °C until analysis. DLS analyses were performed on a Malvern Zetasizer ZS90 dynamic light scattering, using default settings. Nanoparticles were stable for more than 80 days without appreciable signs of degradation, as indicated by the DLS analyses performed on aged samples.

4.3. CurNPs Stability in Aqueous Media

The concentration and the stability of CurNPs, once added to the PBS or culture medium (with different percentages of FBS), were evaluated by UV-VIS spectrophotometric and spectrofluorimetric analysis performed on a Tecan Infinite M200 PRO plate reader, using Corning®96 Well Clear Flat Bottom UV-Transparent Microplates and 200 μL of each solution.

4.4. CurNPs Cytotoxicity Assay

In total, 2000 human cholangiocytes per well were seeded in 96-well cell culture plates and incubated in a humidified atmosphere with 5% CO_2 for 24 h. The culture medium was then replaced with CurNPs stock dispersion diluted in a medium supplemented with 1% FBS to achieve a final concentration of 50 μM, 35 μM, 20 μM, 10 μM, 5 μM and 1 μM. The cells were treated for 24 h and 48 h. Finally, 10 μL of the CCK-8 reagent (Sigma-Aldrich, St. Louis, MO, USA) was added into each well, and OD at 450 nm was measured using a microplate reader (FLUOstar Omega, BMG LABTECH, Ortenberg, Germany) after 2 h incubation at 37 °C.

4.5. Sample Preparation for Mass Spectrometry-Based Proteomics

The cell pellets were treated with 1% sodium deoxycholate in Ambic 50 mM, were sonicated for 5 min (five cycles of 20 s with an interval between cycles of 40 s on ice), and then clarified by centrifugation at 16,000× *g* for 10 min at 4 °C. The bicinchoninic acid assay was used to determine the protein concentration by using serum albumin as a standard. For each sample, 100 μg of protein was reduced with dithiothreitol (10 mM, for 30 min at 65 °C) and alkylated using iodoacetamide (20 mM, 30 min at 37 °C in dark conditions). Protein digestion was performed using trypsin (*w*/*w* ratio 1:50) at 37 °C for 18 h. The samples were incubated with 10% trifluoroacetic acid for 10 min at 37 °C to quench the trypsin reaction and remove sodium deoxycholate by acid precipitation. The samples were centrifuged at 16,000× g for 10 min and subsequently desalted with Mobicol spin columns equipped with 10 μm pore size filters and filled with a C18 stationary phase. The peptide mixtures were evaporated to dryness under a vacuum at 36 °C (Speed Vac Concentrator, Savant Instruments Inc., Farmingdale, NY, USA) before being resuspended in 50 μL of 5/95 CH_3CN/0.1% HCOOH to achieve a final peptide concentration of 2 μg/μL and were finally transferred in plastic vials for LC-MS/MS analysis.

4.6. LC-MS/MS Analysis

Sample analyses were performed using a micro-HPLC (Eksigent Ekspert microLC 200) combined with a Triple TOF 5600 mass spectrometer equipped with a Turbo Ion Spray probe as the ion source (both AB Sciex, Concord, ON, Canada). Five μL of each sample/pool, set in an autosampler at 8 °C, was injected onto a C18 Jupiter column (150 mm × 0.3 mm i.d., 4 μm particle size, 90 Å), thermostatted at 30 °C and equipped with a micro trap C18 (10 mm × 0.3 mm) (both Phenomenex, Torrance, CA, USA). The flow rate was set at 5 μL/min, and the mobile phases A and B were, respectively, H_2O and $CHCN_3$, both with 0.1% HCOOH. The elution program was: 0 min, 5% B; 1 min, 5% B; 51 min, 22% B; 51.5 min, 90% B; 53.5 min, 90% B; 54 min, 5% B; 60 min, 5%B. Chromatographic performances and TOF accuracy were evaluated using an intra-run injection (5 μL) of beta-galactosidase 100 fmol/μL. The mass spectrometer was set in the positive ion mode, and the operation conditions of the ion source were the following: ion spray voltage floating 5.5 kV, probe temperature 150 °C, curtain gas 25 psi, ion source gas 1 and gas 2, respectively 30 and 20 psi, declustering potential 100 V. N_2, as an inert gas, was used for CID (Collision Induced Dissociation) experiments.

A sample pool was analyzed (double injection) with an information-dependent acquisition (IDA) tandem mass spectrometry method based on an MS1 survey scan from which the 20 most abundant precursor ions were selected for subsequent CID fragmentation. The MS1 survey and MS2 scans were acquired over a mass range of 250–1250 m/z and 100–1500 m/z, with an accumulation time of 250 and 100 milliseconds, respectively. CID experiments were carried out using rolling collision energy that was automatically calculated according to the m/z and the charge in of the candidate ion, with a collision energy spread (CES) of 5 V. Precursor ions with a charge state from 2^+ to 5^+ were chosen to trigger the MS/MS experiments. The isolation width for precursor ion selection was set at 0.7 m/z on the Q1.

IDA results were used to generate the SWATH data-independent acquisition (DIA) method for protein quantitation. The SWATH acquisitions for each sample were performed over 50 overlapping isolation mass windows of variable length (minimum window width 3 Da, window overlap 1 Da) depending on the peptide density distribution along the entire mass range of 250–1250 m/z. Precursor ion activation was performed by CID as described before; however, an acquisition mass range of 230–1500 m/z and a charge state of 2^+ were chosen. An accumulation time of 100 milliseconds for MS1 and 50 milliseconds for MS2 scans resulted in an overall duty cycle of 2.6 s (~15 points per elution peak).

4.7. IL-6 Dosage

According to the manufacturer's protocol, IL-6 levels were determined using the Roche Elecsys Kit (09015604190). In total, 200 µL of the culture media underwent the first incubation with IL-6 specific antibodies, followed by a second incubation with IL-6 specific antibodies labeled with ruthenium complexes. After that, complexes were magnetically captured, and an electric voltage application determined the chemiluminescent emission that was directly proportional to the IL-6 concentration.

4.8. Data Processing and Analysis

SWATH data analysis was carried out with DIA-NN (https://github.com/vdemichev/DiaNN) version 1.8, the free universal software for DIA proteomics data processing by V. Demichev, M. Ralser, and K. S. Lilley (The Francis Crick Institute, Molecular Biology of Metabolism laboratory, London, UK and Department of Biochemistry and The Milner Therapeutics Institute, University of Cambridge, Cambridge, UK).

First, SWATH raw file.wiff were converted into .wiff.dia and then used to generate the spectral library utilizing the following parameters: FASTA digest for library-free search/library generation enabled; Human Swiss-Prot FASTA database of 20404 reviewed, non-redundant protein species (downloaded on 4 January 2023); deep learning-based spectra, RTs and IMs prediction enabled; protease = trypsin; missed cleavages = 2; maximum number of variable modifications = 3; N-term M excision, C carbamidomethylation as fixed modification and M oxidation as variable modification all enabled; peptide length range = 7–30; precursor charge range = 1–4; precursor m/z range = 250–1250; fragment ion m/z range = 230–1500; generate spectral library enabled; precursor FDR = 1%; mass accuracy was determined automatically (mass accuracy = 0, MS1 accuracy = 0); use isotopologues, MBR (match between run) and remove likely interferences all enabled; neural network classifier = single-pass mode; protein inference = genes; quantification strategy = any LC (high accuracy); cross-run normalization = RT-dependent; library generation = smart profiling; where not specified, the standard settings were used.

For the protein quantitation step on files.wiff.dia, we kept the same parameters used above for spectral library generation, with only these exceptions: in the spectral library, the report-lib.tsv file generated was selected in the previous step; generate the spectral library, FASTA digest for library-free search/library generation, and deep learning-based spectra, RTs, and IMs prediction were all disabled; quantities matrices were enabled.

The report-quantification.pg_matrix output file of DIA-NN, reporting the $\log_2$ transformed cross-run normalized protein areas (obtained as the sum of the intensity for the top 3 precursor ions identified at 1% FDR and cross-run normalized), was used for further data analysis.

For each protein, the FC value was calculated as the ratio between the mean expression in TGF-β-treated cells and the mean expression in the control cells; and the mean expression in TGF-β/CurNPs-treated cells and the mean expression in TGF-β-treated cells. Proteins were considered differentially expressed when FCs were higher than 2.5 (FC $\leq 1/2.5$ or FC ≥ 2.5). The values between different conditions were analyzed by two-way ANOVA followed by post hoc multiple comparison tests using GraphPad Prism version 8.0.0

for Windows (San Diego, CA, USA), and a p-value ≤ 0.05 was considered statistically significant. Biological significance and pathway analysis were conducted using DAVID Bioinformatics Resources 6.8 and the Reactome database, respectively.

Supplementary Materials: The following supporting information can be downloaded at: https://www.mdpi.com/article/10.3390/ijms241310481/s1.

Author Contributions: Conceptualization, E.C. and A.C.; methodology, E.C., N.D.G., E.M., G.S. and L.T.; resources, S.R.; data curation, E.C., G.S., E.M. and L.T.; writing—original draft preparation, E.C. and A.C.; writing—review and editing, E.C., S.R., G.S., F.V., L.T., E.M., A.C. and N.D.G.; supervision, A.C.; project administration, S.R.; funding acquisition, S.R. All authors have read and agreed to the published version of the manuscript.

Funding: This research was funded by the AIRCS-Italian Association for Sclerosing Cholangitis Research.

Institutional Review Board Statement: Not applicable.

Informed Consent Statement: Not applicable.

Data Availability Statement: The raw data supporting the conclusions of this article will be made available by the authors, without undue reservation.

Conflicts of Interest: The authors declare no conflict of interest.

References

1. Lv, G.-P.; Hu, D.-J.; Zhou, Y.-Q.; Zhang, Q.-W.; Zhao, J.; Li, S.-P. Preparation and Application of Standardized Typical Volatile Components Fraction from Turmeric (Curcuma Longa L.) by Supercritical Fluid Extraction and Step Molecular Distillation. *Molecules* **2018**, *23*, 1831. [CrossRef] [PubMed]
2. Fu, Y.-S.; Chen, T.-H.; Weng, L.; Huang, L.; Lai, D.; Weng, C.-F. Pharmacological Properties and Underlying Mechanisms of Curcumin and Prospects in Medicinal Potential. *Biomed. Pharmacother.* **2021**, *141*, 111888. [CrossRef] [PubMed]
3. Priyadarsini, K. The Chemistry of Curcumin: From Extraction to Therapeutic Agent. *Molecules* **2014**, *19*, 20091–20112. [CrossRef] [PubMed]
4. Anand, P.; Kunnumakkara, A.B.; Newman, R.A.; Aggarwal, B.B. Bioavailability of Curcumin: Problems and Promises. *Mol. Pharm.* **2007**, *4*, 807–818. [CrossRef] [PubMed]
5. Nikezić, A.V.V.; Bondžić, A.M.; Vasić, V.M. Drug Delivery Systems Based on Nanoparticles and Related Nanostructures. *Eur. J. Pharm. Sci.* **2020**, *151*, 105412. [CrossRef]
6. Felice, F.; Piras, A.M.; Rocchiccioli, S.; Barsotti, M.C.; Santoni, T.; Pucci, A.; Burchielli, S.; Chiellini, F.; Ucciferri, N.; Solaro, R.; et al. Endothelial Progenitor Cell Secretome Delivered by Novel Polymeric Nanoparticles in Ischemic Hindlimb. *Int. J. Pharm.* **2018**, *542*, 82–89. [CrossRef]
7. Mohammed, E.; El-Beih, N.; El-Hussieny, E.; El-Ahwany, E.; Hassan, M.; Zoheiry, M. Effects of Free and Nanoparticulate Curcumin on Chemically Induced Liver Carcinoma in an Animal Model. *Arch. Med. Sci.* **2021**, *17*, 218–227. [CrossRef]
8. Atia, M.M.; Abdel-Tawab, H.S.; Mostafa, A.M.; Mobarak, S.A. Nanocurcumin and Curcumin Prevent N, N′-Methylenebisacrylamide-Induced Liver Damage and Promotion of Hepatic Cancer Cell Growth. *Sci. Rep.* **2022**, *12*, 8319. [CrossRef]
9. Elzoheiry, A.; Ayad, E.; Omar, N.; Elbakry, K.; Hyder, A. Anti-Liver Fibrosis Activity of Curcumin/Chitosan-Coated Green Silver Nanoparticles. *Sci. Rep.* **2022**, *12*, 18403. [CrossRef]
10. Wang, J.; Pan, W.; Wang, Y.; Lei, W.; Feng, B.; Du, C.; Wang, X. Enhanced Efficacy of Curcumin with Phosphatidylserine-Decorated Nanoparticles in the Treatment of Hepatic Fibrosis. *Drug Deliv.* **2018**, *25*, 1–11. [CrossRef]
11. Tie, Y.; Tang, F.; Peng, D.; Zhang, Y.; Shi, H. TGF-Beta Signal Transduction: Biology, Function and Therapy for Diseases. *Mol. Biomed.* **2022**, *3*, 45. [CrossRef]
12. Meng, X.; Nikolic-Paterson, D.J.; Lan, H.Y. TGF-β: The Master Regulator of Fibrosis. *Nat. Rev. Nephrol.* **2016**, *12*, 325–338. [CrossRef]
13. Santos-Laso, A.; Munoz-Garrido, P.; Felipe-Agirre, M.; Bujanda, L.; Banales, J.M.; Perugorria, M.J. New Advances in the Molecular Mechanisms Driving Biliary Fibrosis and Emerging Molecular Targets. *CDT* **2017**, *18*, 908–920. [CrossRef]
14. Fabregat, I.; Caballero-Díaz, D. Transforming Growth Factor-β-Induced Cell Plasticity in Liver Fibrosis and Hepatocarcinogenesis. *Front. Oncol.* **2018**, *8*, 357. [CrossRef]
15. Zhang, K.; Zhang, M.; Luo, Z.; Wen, Z.; Yan, X. The Dichotomous Role of TGF-β in Controlling Liver Cancer Cell Survival and Proliferation. *J. Genet. Genom.* **2020**, *47*, 497–512. [CrossRef]
16. Dewidar, B.; Soukupova, J.; Fabregat, I.; Dooley, S. Meindl-Beinker TGF-β in Hepatic Stellate Cell Activation and Liver Fibrogenesis—Updated 2019. *Cells* **2019**, *8*, 1419. [CrossRef]

17. Tang, L.-Y.; Heller, M.; Meng, Z.; Yu, L.-R.; Tang, Y.; Zhou, M.; Zhang, Y.E. Transforming Growth Factor-β (TGF-β) Directly Activates the JAK1-STAT3 Axis to Induce Hepatic Fibrosis in Coordination with the SMAD Pathway. *J. Biol. Chem.* **2017**, *292*, 4302–4312. [CrossRef]
18. Ceccherini, E.; Di Giorgi, N.; Michelucci, E.; Signore, G.; Tedeschi, L.; Vozzi, F.; Rocchiccioli, S.; Cecchettini, A. Biological Effects of Transforming Growth Factor Beta in Human Cholangiocytes. *Biology* **2022**, *11*, 566. [CrossRef]
19. Farzaei, M.; Zobeiri, M.; Parvizi, F.; El-Senduny, F.; Marmouzi, I.; Coy-Barrera, E.; Naseri, R.; Nabavi, S.; Rahimi, R.; Abdollahi, M. Curcumin in Liver Diseases: A Systematic Review of the Cellular Mechanisms of Oxidative Stress and Clinical Perspective. *Nutrients* **2018**, *10*, 855. [CrossRef]
20. Hewlings, S.; Kalman, D. Curcumin: A Review of Its Effects on Human Health. *Foods* **2017**, *6*, 92. [CrossRef]
21. Devi, A.; Chennakesavulu, S.; Suresh, C.; Reddy, A.B.M. Nutraceuticals and Their Role in Human Health and Disease. In *Functional Food and Human Health*; Rani, V., Yadav, U.C.S., Eds.; Springer Singapore: Singapore, 2018; pp. 375–403, ISBN 9789811311222.
22. Zhang, Y.; Alexander, P.B.; Wang, X.-F. TGF-β Family Signaling in the Control of Cell Proliferation and Survival. *Cold Spring Harb. Perspect. Biol.* **2017**, *9*, a022145. [CrossRef] [PubMed]
23. Fabregat, I.; Moreno-Càceres, J.; Sánchez, A.; Dooley, S.; Dewidar, B.; Giannelli, G.; ten Dijke, P. The IT-LIVER Consortium TGF-β Signalling and Liver Disease. *FEBS J.* **2016**, *283*, 2219–2232. [CrossRef] [PubMed]
24. Demetris, A.J.; Specht, S.; Nozaki, I.; Lunz, J.G.; Stolz, D.B.; Murase, N.; Wu, T. Small Proline-Rich Proteins (SPRR) Function as SH3 Domain Ligands, Increase Resistance to Injury and Are Associated with Epithelial–Mesenchymal Transition (EMT) in Cholangiocytes. *J. Hepatol.* **2008**, *48*, 276–288. [CrossRef]
25. Nozaki, I.; Lunz, J.G.; Specht, S.; Stolz, D.B.; Taguchi, K.; Subbotin, V.M.; Murase, N.; Demetris, A.J. Small Proline-Rich Proteins 2 Are Noncoordinately Upregulated by IL-6/STAT3 Signaling after Bile Duct Ligation. *Lab. Investig.* **2005**, *85*, 109–123. [CrossRef] [PubMed]
26. Konishi, A.; Shimizu, S.; Hirota, J.; Takao, T.; Fan, Y.; Matsuoka, Y.; Zhang, L.; Yoneda, Y.; Fujii, Y.; Skoultchi, A.I.; et al. Involvement of Histone H1.2 in Apoptosis Induced by DNA Double-Strand Breaks. *Cell* **2003**, *114*, 673–688. [CrossRef]
27. Funayama, R.; Saito, M.; Tanobe, H.; Ishikawa, F. Loss of Linker Histone H1 in Cellular Senescence. *J. Cell Biol.* **2006**, *175*, 869–880. [CrossRef]
28. Masyuk, A.I.; Masyuk, T.V.; Splinter, P.L.; Huang, B.Q.; Stroope, A.J.; LaRusso, N.F. Cholangiocyte Cilia Detect Changes in Luminal Fluid Flow and Transmit Them Into Intracellular Ca2+ and CAMP Signaling. *Gastroenterology* **2006**, *131*, 911–920. [CrossRef]
29. Maroni, L.; Haibo, B.; Ray, D.; Zhou, T.; Wan, Y.; Meng, F.; Marzioni, M.; Alpini, G. Functional and Structural Features of Cholangiocytes in Health and Disease. *Cell. Mol. Gastroenterol. Hepatol.* **2015**, *1*, 368–380. [CrossRef]
30. Gradilone, S.A.; Masyuk, A.I.; Splinter, P.L.; Banales, J.M.; Huang, B.Q.; Tietz, P.S.; Masyuk, T.V.; LaRusso, N.F. Cholangiocyte Cilia Express TRPV4 and Detect Changes in Luminal Tonicity Inducing Bicarbonate Secretion. *Proc. Natl. Acad. Sci. USA* **2007**, *104*, 19138–19143. [CrossRef]
31. LaRusso, N.F.; Masyuk, T.V. The Role of Cilia in the Regulation of Bile Flow. *Dig. Dis.* **2011**, *29*, 6–12. [CrossRef]
32. Ciliopathy Working Group; Shaheen, R.; Szymanska, K.; Basu, B.; Patel, N.; Ewida, N.; Faqeih, E.; Al Hashem, A.; Derar, N.; Alsharif, H.; et al. Characterizing the Morbid Genome of Ciliopathies. *Genome Biol.* **2016**, *17*, 242. [CrossRef]
33. Cochran, J.C.; Sontag, C.A.; Maliga, Z.; Kapoor, T.M.; Correia, J.J.; Gilbert, S.P. Mechanistic Analysis of the Mitotic Kinesin Eg5. *J. Biol. Chem.* **2004**, *279*, 38861–38870. [CrossRef]
34. Zalenski, A.A.; Majumder, S.; De, K.; Venere, M. An Interphase Pool of KIF11 Localizes at the Basal Bodies of Primary Cilia and a Reduction in KIF11 Expression Alters Cilia Dynamics. *Sci. Rep.* **2020**, *10*, 13946. [CrossRef]
35. Masyuk, T.V.; Masyuk, A.I.; LaRusso, N.F. TGR5 in the Cholangiociliopathies. *Dig. Dis.* **2015**, *33*, 420–425. [CrossRef]
36. Pedersen, L.B.; Mogensen, J.B.; Christensen, S.T. Endocytic Control of Cellular Signaling at the Primary Cilium. *Trends Biochem. Sci.* **2016**, *41*, 784–797. [CrossRef]
37. Perrais, D. Cellular and Structural Insight into Dynamin Function during Endocytic Vesicle Formation: A Tale of 50 Years of Investigation. *Biosci. Rep.* **2022**, *42*, BSR20211227. [CrossRef]
38. Inokawa, Y.; Nomoto, S.; Hishida, M.; Hayashi, M.; Kanda, M.; Nishikawa, Y.; Takeda, S.; Fujiwara, M.; Koike, M.; Sugimoto, H.; et al. Dynamin 3: A New Candidate Tumor Suppressor Gene in Hepatocellular Carcinoma Detected by Triple Combination Array Analysis. *OncoTargets Ther.* **2013**, *6*, 1417–1424. [CrossRef]
39. Chen, S.; Dai, X.; Li, H.; Gong, Y.; Zhao, Y.; Huang, H. Overexpression of Ring Finger Protein 20 Inhibits the Progression of Liver Fibrosis via Mediation of Histone H2B Lysine 120 Ubiquitination. *Hum. Cell* **2021**, *34*, 759–770. [CrossRef]
40. Banales, J.M.; Huebert, R.C.; Karlsen, T.; Strazzabosco, M.; LaRusso, N.F.; Gores, G.J. Cholangiocyte Pathobiology. *Nat. Rev. Gastroenterol. Hepatol.* **2019**, *16*, 269–281. [CrossRef]

International Journal of
Molecular Sciences

MDPI

Article

Effects of L-Cysteine and γ-Aminobutyric Acid Treatment on Postharvest Quality and Antioxidant Activity of Loquat Fruit during Storage

Huifen Zhang †, Jing Pu †, Han Liu, Miao Wang, Ying Du, Xiaofu Tang, Xian Luo, Yongqing Wang and Qunxian Deng *

College of Horticulture, Sichuan Agricultural University, Chengdu 611130, China; hfcheung@sicau.edu.cn (H.Z.); pg7191127@163.com (J.P.); 13088016648@163.com (H.L.); wangmiao202102@163.com (M.W.); yj18987246849@163.com (Y.D.); tangxf152@163.com (X.T.); lawxian@aliyun.com (X.L.); yqw14@sicau.edu.cn (Y.W.)
* Correspondence: dqxlwj@sina.com.cn; Tel.: +86-13551551617
† These authors contributed equally to this work.

Abstract: Sichuan is the China's leading producer of loquat, with the largest cultivation area and yield ranked first in China. Loquat is a seasonal fruit highly appreciated by consumers; however, the fruit is prone to browning and lignification after harvest, affecting its storage quality. The effects of L-Cysteine (L-Cys, 0.01, 0.05, 0.1, 0.2%) and γ-aminobutyric acid (GABA, 0.025, 0.05, 0.075, 0.1%) on the sensory quality and antioxidant activity of loquat fruit during cold storage at 4 °C for 35 days and simulated shelf life for 5 days were investigated. The results showed that after 40 days of storage, compared with the control, 0.05% L-Cys and 0.05% GABA treatment of 'Zaozhong No. 6' loquat fruit effectively reduced the weight loss rate, browning index, decay index, respiratory rate, firmness, and lignin content and slowed the decreases in total soluble solids, soluble sugar, titratable acidityand vitamin C contents. The application of 0.05% L-Cys and 0.05% GABA significantly increased the contents of total phenols, total flavonoids, flavanols, and carotenoids; delayed the increase of relative electric conductivity, MDA, POD, and PPO activities; and significantly enhanced the activities of SOD and CAT, DPPH free radical scavenging ability, and FRAP, thereby improving antioxidant capacity. In summary, 0.05% L-Cys and 0.05% GABA treatment promotes the quality of loquat fruit after 40 days of storage, and significantly enhances antioxidant capacity, thus delaying senescence after harvest.

Keywords: loquat; L-Cysteine; γ-aminobutyric acid; fruit quality; antioxidant system; cold storage; shelf life

1. Introduction

Loquat (*Eriobotrya japonica* Lindl.) is a subtropical tree native to China. The non-climacteric fruit is soft and juicy, with a unique flavor and rich in a variety of nutrients and functional components, such as carotenoids, phenols, and other bioactive components. According to traditional Chinese medicine, the fruit has the effects of moistening the lung, relieving thirst, and clearing heat [1], and it is becoming increasingly popular among consumers; however, it is difficult to transport the fruit to market. The fruit is harvested at a time of year when the temperature is high and the weather is often rainy. With thin fruit skin, mechanical damage to loquats can result in rot and deterioration [2]. Previous studies tested treatments with chitosan coating [3], different biomaterials [4], heat treatment [5], and 1-MCP [6] to promote the synthesis of stress-resistant substances by inducing changes in the secondary metabolic pathways of the fruit, thereby delaying fruit senescence and maintaining fruit quality. Recently, natural amino acids have been widely applied in agricultural production as safe exogenous treatments. Exogenous amino acid treatment can regulate growth and development and can improve the quality of fruits and vegetables [7] and enhance the stress resistance of fruits [8].

Int. J. Mol. Sci. **2023**, *24*, 10541. https://doi.org/10.3390/ijms241310541 https://www.mdpi.com/journal/ijms

Treatment with L-Cysteine (L-Cys), a thiol compound, can have anti-browning and anti-oxidation effects [9], which can alter plant growth and development [10]. L-Cys can react with quinones produced by enzymatic reactions to inhibit the formation of browning [11]. L-Cys can directly act on polyphenol oxidase (PPO) to inhibit enzymatic browning by reducing its activity [12]. Zhou et al. [13] found that 0.8 mg·mL^{-1} L-Cys could prevent enzymatic browning during pear-juice processing, and Wu et al. [14] found that L-Cys can reduce the browning of fresh wet noodles by reducing pH, inhibiting PPO activity and polyphenol oxidation. L-Cys can significantly slow the browning of fresh-cut lettuce and greatly improve antioxidant capacity [15]. Numerous studies have shown that exogenous L-Cys can effectively regulate the enzymatic browning of fresh-cut fruits and vegetables. Nevertheless, little is known about the role of L-Cys on loquat quality during storage.

γ-aminobutyric acid (GABA) is a four-carbon, non-protein amino acid that is widely found in animals and plants, and has important effects on plant growth and resistance [16]. GABA is a natural compound that is considered safe and may be beneficial to human health [17]. Studies have shown that GABA treatment increased the content of proline and endogenous GABA and enhanced the activity of Δ^1-pyrroline-5-carboxylate synthetase (P5CS), ornithine δ-aminotransferase (OAT) and glutamate decarboxylase (GAD)) to effectively limit chilling injury of peach fruit [18]. Wang et al. [19] found that 20 mM GABA treatment effectively delayed the increase of CI index and MDA content in banana peel, promoted the accumulation of total phenols and proline, and enhanced the antioxidant capacity of banana fruit. Another study on kiwifruit found that GABA-treated kiwifruit fruit had higher levels of AsA, which effectively protected against cell membrane damage [20].

Loquat cultivation has great commercial and economic value. However, the delicate fruit is at risk for quality deterioration due to its susceptibility to mechanical damage, decay, and lignification. Based on previous reports, exogenous L-Cys and GABA treatments can reduce CI, retard browning, and enhance antioxidant capacity of fruits and vegetables, thus maintaining fruit quality traits during storage, but the effects of exogenous L-Cys and GABA treatments on loquat storage quality have not been reported. Therefore, this study aimed to evaluate the effects of these treatments on postharvest quality and antioxidant capacity of loquat fruits during cold storage, to assess its application to loquat storage to improve post-harvest quality, and to provide theoretical support for postharvest preservation of loquat.

2. Results

2.1. *Effects of L-Cys and GABA Treatments on Weight Loss Rate, Browning Index, Decay Index, and Respiration Intensity*

Fruit weight loss rate is an important index to evaluate fruit freshness. As shown in Table 1, the weight loss rate of loquat fruit increased during storage. After 21 days of storage, the weight loss rates of L-Cys and GABA groups were significantly lower than that of the control group, with the most significant results for the 0.05% L-Cys and 0.05% GABA treatment. At the end of storage, the weight loss rates of loquat fruit treated with 0.05% L-Cys and 0.05% GABA were significantly lower, by 17.5% and 13.5%, respectively, than that of the water control ($p < 0.05$). This showed that these two exogenous treatments could inhibit the increase in weight loss rate of loquat fruit.

Throughout storage, the browning index of loquat fruit gradually increased (Table 1). After 7 days of storage, the browning index of the control group was significantly higher than those of the L-Cys and GABA treatments ($p < 0.05$), and the browning index was significantly lower for the 0.05% L-Cys and 0.05% GABA treatments ($p < 0.05$). After 40 days of storage, the browning index of the control group was 71%, while those for the 0.05% L-Cys and 0.05% GABA treatments were 49.83% and 48.17%, respectively. Thus, these two treatments acted to inhibit the browning of loquat fruit.

Table 1. Effects of different concentrations of L-Cys and GABA treatments on weight loss rate, browning index, decay index, and respiratory rate of loquat during storage. Values of three replicates are expressed as the means ± SD. Different lowercase letters (a–c) in the table denote significant differences between sampling dates for each treatment by Duncan's multiple range test ($p < 0.05$).

			Storage Time						
	Treatments	**Concentration**	**0 Day**	**7 Days**	**14 Days**	**21 Days**	**28 Days**	**35 Days**	**40 Days**
Weight loss rate (%)	Control	Water	0	0.77 ± 0.23a	1.21 ± 0.18a	1.62 ± 0.10a	2.54 ± 0.12a	3.51 ± 0.05a	4.74 ± 0.12a
	L-Cys	0.01%	0	0.58 ± 0.10a	0.94 ± 0.04a	1.40 ± 0.08ab	1.75 ± 0.08cd	3.15 ± 0.11abc	4.29 ± 0.20bc
		0.05%	0	0.45 ± 0.09a	0.87 ± 0.09a	1.10 ± 0.02c	1.56 ± 0.10d	2.94 ± 0.06bc	3.91 ± 0.05c
		0.1%	0	0.54 ± 0.05a	1.11 ± 0.06a	1.37 ± 0.04b	2.15 ± 0.13b	3.10 ± 0.14abc	4.32 ± 0.10abc
		0.2%	0	0.65 ± 0.09a	1.18 ± 0.09a	1.43 ± 0.05ab	2.22 ± 0.08b	3.19 ± 0.15abc	4.27 ± 0.12bc
	GABA	0.025%	0	0.62 ± 0.04a	1.03 ± 0.09a	1.21 ± 0.11bc	2.19 ± 0.13b	3.28 ± 0.11ab	4.39 ± 0.12ab
		0.05%	0	0.41 ± 0.06a	0.89 ± 0.09a	1.19 ± 0.03bc	1.82 ± 0.01cd	2.86 ± 0.11c	4.10 ± 0.18bc
		0.075%	0	0.66 ± 0.14a	1.00 ± 0.14a	1.38 ± 0.04b	2.18 ± 0.06b	3.05 ± 0.19bc	4.44 ± 0.09ab
		0.1%	0	0.64 ± 0.02a	1.10 ± 0.11a	1.36 ± 0.10b	1.20 ± 0.03bc	3.13 ± 0.13abc	4.51 ± 0.09ab
Browning index (%)	Control	Water	0	7.61 ± 1.18a	11.11 ± 1.39a	27.08 ± 2.08a	44.17 ± 0.17a	60.42 ± 5.51a	71.00 ± 2.02a
	L-Cys	0.01%	0	5.55 ± 2.78ab	4.83 ± 0.35c	18.89 ± 1.94bcd	38.33 ± 0.83ab	55.83 ± 1.27ab	61.83 ± 2.04b
		0.05%	0	2.78 ± 1.47ab	4.10 ± 0.54c	15.56 ± 0.56d	28.33 ± 1.67c	38.89 ± 2.00c	49.83 ± 0.63c
		0.1%	0	6.93 ± 1.40ab	7.50 ± 1.73abc	19.16 ± 1.27bcd	35.28 ± 1.21b	53.06 ± 1.55ab	59.28 ± 1.62b
		0.2%	0	3.15 ± 1.61ab	8.89 ± 0.56ab	21.94 ± 1.55bc	42.22 ± 1.47a	52.78 ± 2.78ab	59.17 ± 2.20b
	GABA	0.025%	0	5.26 ± 1.57ab	7.07 ± 1.26bc	17.78 ± 1.11cd	39.72 ± 1.21ab	54.72 ± 2.90ab	60.53 ± 2.22b
		0.05%	0	1.67 ± 1.67b	5.56 ± 1.39bc	15.56 ± 0.56d	28.06 ± 1.55c	38.33 ± 0.83c	48.17 ± 1.41c
		0.075%	0	3.04 ± 1.54ab	6.67 ± 1.67bc	20.00 ± 0.00bcd	35.00 ± 2.89b	46.67 ± 1.67bc	58.33 ± 1.67b
		0.1%	0	3.04 ± 1.54ab	6.06 ± 0.97bc	23.33 ± 1.67ab	41.67 ± 4.41ab	50.56 ± 4.44b	57.72 ± 0.96b
Decay index (%)	Control	Water	0	5.14 ± 0.60a	12.44 ± 1.44a	20.00 ± 2.89a	30.28 ± 4.72a	35.33 ± 2.17a	49.36 ± 1.78a
	L-Cys	0.01%	0	2.10 ± 1.20ab	9.72 ± 1.39abc	13.33 ± 0.83bc	23.33 ± 2.93abc	28.05 ± 3.74bc	42.00 ± 1.75b
		0.05%	0	0.72 ± 0.72b	6.93 ± 1.40c	11.11 ± 2.00c	20.56 ± 2.42 bc	26.67 ± 1.67bc	34.17 ± 2.20c
		0.1%	0	0.72 ± 0.72b	11.06 ± 0.82 ab	14.72 ± 1.21abc	23.33 ± 1.67abc	30.83 ± 1.27abc	42.67 ± 0.51ab
		0.2%	0	2.39 ± 1.45ab	9.61 ± 0.66abc	14.17 ± 0.83abc	23.33 ± 1.67abc	32.78 ± 1.47ab	44.50 ± 2.30ab
	GABA	0.025%	0	4.45 ± 0.28a	10.28 ± 1.21abc	11.67 ± 0.83c	25.28 ± 2.65abc	31.11 ± 1.94abc	45.61 ± 1.22ab
		0.05%	0	0.72 ± 0.72b	7.55 ± 1.68bc	10.83 ± 0.83c	18.89 ± 1.11c	25.00 ± 0.00c	34.44 ± 0.56c
		0.075%	0	3.33 ± 1.67ab	10.00 ± 0.00abc	16.67 ± 1.67abc	28.44 ± 2.51ab	30.00 ± 2.89abc	40.83 ± 1.59bc
		0.1%	0	3.05 ± 1.55ab	11.58 ± 0.80a	18.72 ± 3.20ab	23.61 ± 1.39abc	30.83 ± 1.27abc	42.22 ± 1.47b
Respiratory rate (mg·kg^{-1}·h)	Control	Water	83.24 ± 7.96	81.57 ± 1.58a	79.81 ± 0.68a	76.78 ± 1.07a	73.45 ± 1.11a	70.54 ± 0.72a	67.05 ± 0.85a
	L-Cys	0.01%	83.24 ± 7.96	79.00 ± 3.19abc	78.04 ± 3.78ab	75.54 ± 1.18a	71.29 ± 1.22ab	68.94 ± 0.60a	66.03 ± 0.78ab
		0.05%	83.24 ± 7.96	73.83 ± 1.36bc	72.67 ± 0.52ab	68.26 ± 1.40b	67.54 ± 0.69bc	64.84 ± 0.34b	63.04 ± 0.25c
		0.1%	83.24 ± 7.96	78.36 ± 2.02abc	72.64 ± 4.04ab	76.16 ± 0.70a	72.23 ± 1.24a	69.50 ± 0.62a	66.19 ± 1.62ab
		0.2%	83.24 ± 7.96	79.30 ± 0.61abc	79.62 ± 3.69aab	74.60 ± 1.45a	72.98 ± 1.20a	70.31 ± 1.50a	66.90 ± 0.85a
	GABA	0.025%	83.24 ± 7.96	80.32 ± 2.86ab	78.45 ± 1.19a	74.28 ± 2.28a	72.00 ± 1.20a	70.55 ± 1.45a	65.20 ± 0.49abc
		0.05%	83.24 ± 7.96	72.20 ± 2.01c	70.44 ± 1.53b	67.81 ± 1.47b	66.10 ± 1.43b	63.70 ± 1.10b	62.33 ± 0.37bc
		0.075%	83.24 ± 7.96	77.75 ± 1.94abc	75.82 ± 0.62ab	78.47 ± 2.46a	72.60 ± 1.65a	71.85 ± 0.65a	65.62 ± 1.35ab
		0.1%	83.24 ± 7.96	78.59 ± 2.67abc	78.42 ± 0.44a	75.68 ± 0.50a	72.14 ± 1.71a	70.70 ± 0.90a	65.35 ± 1.19abc

Decay index is an important index of the effect of fruit storage and preservation. Loquat fruit can easily rot during storage, reducing its quality. With the extension of storage time, the decay index of loquat fruit increased (Table 1). After 40 days of storage, the decay index of loquat fruit in the control group was 49.36%, while the decay index values of loquat fruit in the 0.05% L-Cys and 0.05% GABA treatment groups were 34.17% and 34.44%, respectively. These decay index values were significantly lower than that of the control group ($p < 0.05$), indicating that these two treatments effectively inhibited the decay of loquat fruit during storage.

The respiration rate of loquat fruit decreased during storage, as shown in Table 1. After 7 days of storage, the respiration rate of the control group was significantly higher than those of the L-Cys and GABA treatments ($p < 0.05$). Both treatments inhibited the respiration of the fruit, thereby delaying fruit senescence, with 0.05% L-Cys and 0.05% GABA treatments exhibiting the most significant inhibitory effects on respiration rate. Again, these results show that L-Cys and GABA treatments can effectively inhibit fruit respiration and delay the decomposition and consumption of organic matter.

2.2. *Effects of L-Cys and GABA Treatments on TSS, SS, TA, and Vitamin C Content*

As shown in Figure 1, loquat fruits treated with 0.05% L-Cys and 0.05% GABA had significantly higher TSS, SS, TA, and vitamin C levels compared to the water control group ($p < 0.05$). Specifically, the TSS levels (Figure 1A) were 32.02% and 20.95% higher for the 0.05% L-Cys and 0.05% GABA groups, respectively, while soluble sugar levels (Figure 1B) were 11.80% and 9.05% higher, and the titratable acid content (Figure 1C) was 43.49% and 43.56% higher, respectively. The vitamin C content (Figure 1D) was 7.34% and 7.91% higher, respectively, while the other L-Cys and GABA treatment groups did not show significant differences but were significantly higher than the control group. Additionally, after 40 days of storage, the application of L-Cys and GABA significantly inhibited the decrease in TSS, SS, TA, and vitamin C content.

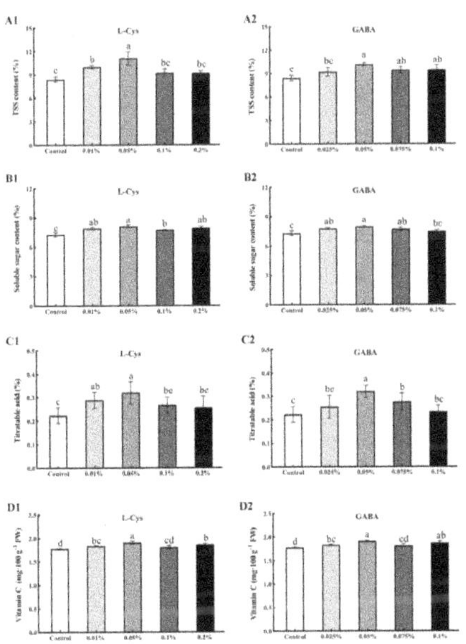

Figure 1. Effects of different concentrations of L-Cys and GABA treatments on TSS (**A**), SS (**B**), TA (**C**), and vitamin C content (**D**) of loquat fruit after 40 days of storage. A1, B1, C1, D1 and A2, B2, C2, D2 indicate the effect of L-Cys and GABA, respectively. Each column represents mean ± standard deviation (n = 3), and a *t* test was performed at the 0.05 significance level ($p < 0.05$). Different letters (a, b, c, d) indicate significant differences, and the same letter indicates no significant difference.

2.3. *Effects of L-Cys and GABA Treatments on Firmness and Lignin Content of Loquat Fruit after Harvest*

The trends of firmness and lignin content were consistent during loquat storage, and L-Cys and GABA treatments delayed the increase of firmness and lignin in postharvest loquat fruits. The firmness (Figure 2A) and lignin (Figure 2B) content of loquat fruits treated with 0.05% L-Cys and 0.05% GABA were the lowest, and significantly lower than those of the control ($p < 0.05$). Firmness decreased by 24.49% and 22.80%, and lignin content decreased by 22.10% and 25.54% for the 0.05% L-Cys and 0.05% GABA treatments, respectively, indicating that L-Cys and GABA treatments effectively slowed the lignification and senescence of loquat fruits during storage.

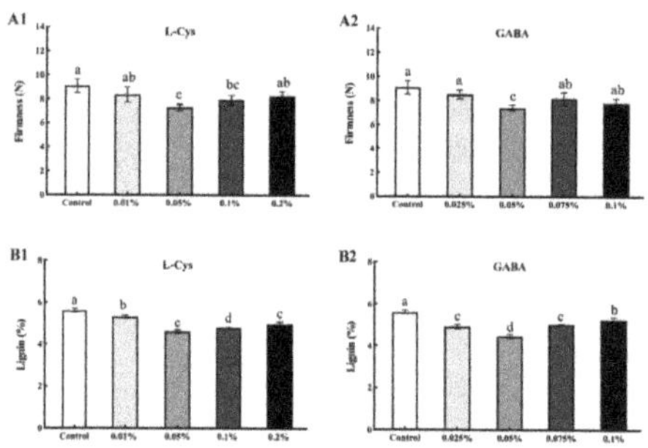

Figure 2. Effects of different concentrations of L-Cys and GABA treatments on fruit firmness (**A**) and lignin (**B**) of loquat fruit after 40 days of storage. A1, B1 and A2, B2 indicate the effect of L-Cys and GABA, respectively. Each column represents mean ± standard deviation (n = 3), and a *t* test was performed at the 0.05 significance level ($p < 0.05$). Different letters (a, b, c, d) indicate significant differences, and the same letter indicates no significant difference.

2.4. *Effects of L-Cys and GABA on Relative Electric Conductivity and MDA Content of Loquat Fruit after Harvest*

Relative electric conductivity is an important indicator of cell membrane permeability. As shown in Figure 3A, the relative electric conductivity was significantly lower by 11.30% and 21.45%, respectively, for loquat fruit treated with 0.05% L-Cys and 0.05% GABA compared to the control ($p < 0.05$). These results indicated that L-Cys and GABA treatments inhibited the increase in relative electric conductivity for better maintenance of the integrity of the cell membrane.

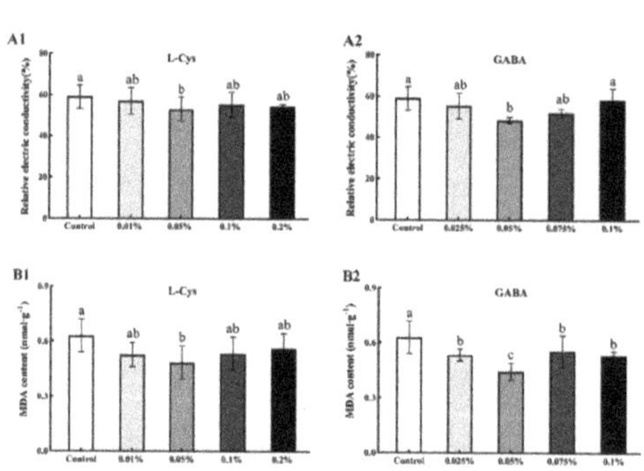

Figure 3. Effects of different concentrations of L-Cys and GABA treatments on relative electric conductivity (**A**) and MDA content (**B**) of loquat fruit after 40 days of storage. A1, B1 and A2, B2 indicate the effect of L-Cys and GABA, respectively. Each column represents mean ± standard deviation (n = 3), and a *t* test was performed at the 0.05 significance level ($p < 0.05$). Different letters (a, b, c) indicate significant differences, and the same letter indicates no significant difference.

Malondialdehyde (MDA) measurement indicates the level of lipid peroxidation in the cell membrane. As shown in Figure 3B, treatments of 0.05% L-Cys and 0.05% GABA resulted in MDA values 28.99% and 40.95% lower than the control at 40 days of storage, respectively ($p < 0.05$). This showed that L-Cys and GABA treatments effectively inhibited the increase in MDA content at 40 days of storage and reduced the accumulation of harmful substances that damage cells, resulting in better quality of stored loquat fruit.

2.5. Effects of L-Cys and GABA Treatments on Total Phenols, Total Flavonoids, Flavanols, and Carotenoids in Loquat Fruit after Harvest

As shown in Figure 4A, the total phenolic content measurements for loquat fruit treated with 0.05% L-Cys, 0.1% L-Cys, 0.05% GABA, and 0.025% GABA were 19.46%, 15.05%, 12.11%, and 5.41%, respectively, higher compared to the control. Thus, L-Cys and GABA treatments maintained the total phenolic content of loquat fruit and significantly inhibited its decrease ($p < 0.05$). Among treatments, 0.05% L-Cys and 0.05% GABA treatment had the best effect.

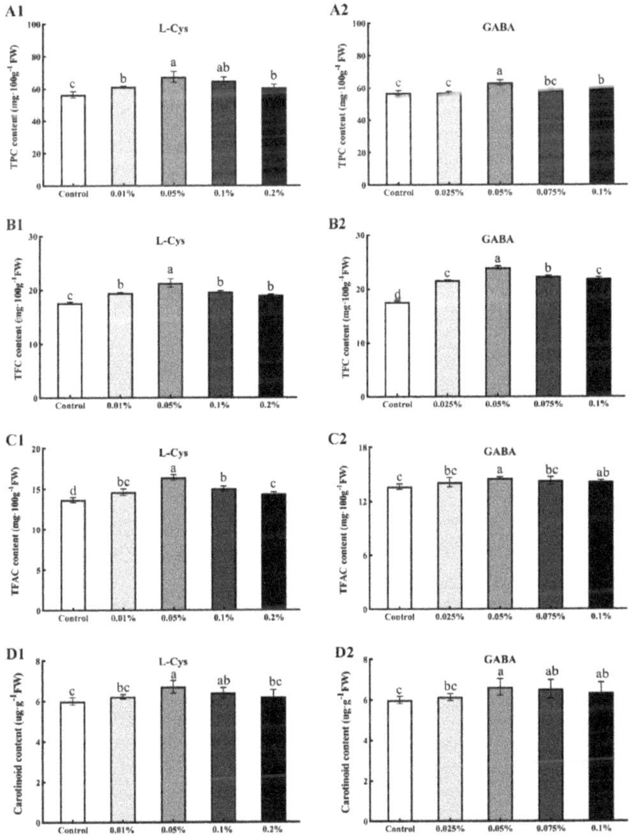

Figure 4. Effects of different concentrations of L-Cys and GABA treatments on TPC (**A**), TFC (**B**), TFAC (**C**), and carotenoid content (**D**) of loquat fruit after 40 days of storage. A1, B1, C1, D1 and A2, B2, C2, D2 indicate the effect of L-Cys and GABA, respectively. Each column represents mean ± standard deviation (n = 3), and a *t* test was performed at the 0.05 significance level ($p < 0.05$). Different letters (a, b, c, d) indicate significant differences, and the same letter indicates no significant difference.

As seen in Figure 4B, the TFC content was higher in loquat fruits treated with different concentrations of L-Cys and GABA compared to the control. The total flavonoid content of loquat fruits treated with 0.05% L-Cys and 0.05% GABA were highest at 40 days of storage, 16.67% and 33.33% higher than that in the control. L-Cys and GABA treatments significantly inhibited the decrease of total flavonoid content ($p < 0.05$).

As shown in Figure 4C, the total flavanol content measurements of loquat fruit treated with 0.05% L-Cys, 0.1% L-Cys, and 0.05% GABA were 19.71%, 9.49%, and 6.57% higher than that in the control. Of the different treatments, 0.05% L-Cys and 0.05% GABA had the best effect ($p < 0.05$).

At 40 days of storage, carotenoid content was significantly higher in loquat fruit treated with 0.05% L-Cys and 0.05% GABA than in the control ($p < 0.05$) by 11.67% and 10.5%, respectively (Figure 4D). The application of L-Cys and GABA significantly slowed the decrease of carotenoids in loquat during storage.

2.6. *Effects of L-Cys and GABA on DPPH and FRAP of Loquat Fruit after Harvest*

L-Cys and GABA treatments significantly maintained the scavenging ability of DPPH free radicals in loquat fruits after 40 days of storage (Figure 5A). The DPPH scavenging capacity in loquat fruits treated with 0.05% L-Cys and 0.05% GABA were 3.598 μmol·g^{-1}FW and 3.704 μmol·g^{-1}FW, i.e., 10.0% and 13.15% higher than the capacity of the control ($p < 0.05$).

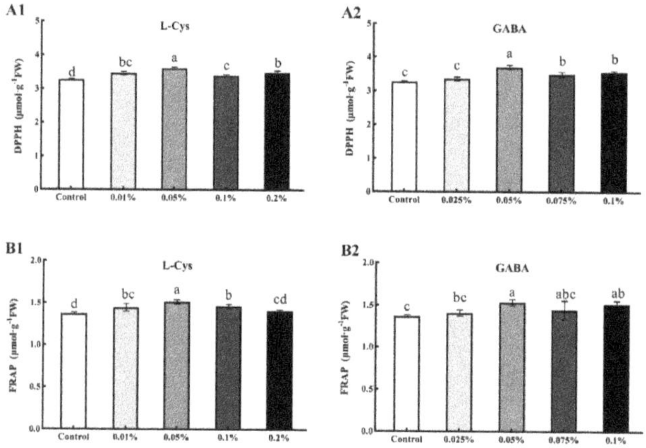

Figure 5. Effects of different concentrations of L-Cys and GABA treatments on DPPH (**A**) and FRAP (**B**) of loquat fruit after 40 days of storage. A1, B1 and A2, B2 indicate the effect of L-Cys and GABA, respectively. Each column represents mean ± standard deviation (n = 3), and a *t* test was performed at the 0.05 significance level ($p < 0.05$) Different letters (a, b, c, d) indicate significant differences, and the same letter indicates no significant difference.

As shown in Figure 5B, L-Cys and GABA treatments promoted the FRAP in loquat fruit after 40 days of storage. Treatments of 0.05% L-Cys and 0.05% GABA significantly promoted the FRAP of loquat fruit, 10.65% and 11.68% higher than that of the control ($p < 0.05$), for improved antioxidant capacity.

2.7. *Effects of L-Cys and GABA on SOD, CAT, POD, and PPO Activity of Loquat Fruit after Harvest*

SOD and CAT are antioxidant enzymes that scavenge active oxygen free radicals in fruit. As shown in Figure 6A, L-Cys and GABA treatments allowed maintenance of high SOD activity after 40 days of loquat fruit storage, with 0.05% L-Cys, 0.05% GABA, and 0.1% GABA treatments significantly higher than the control ($p < 0.05$). After 40 days of loquat fruit storage, the CAT content of the control group (Figure 6B) was significantly lower than that of other treatment groups. Treatments with 0.05% L-Cys and 0.05% GABA were 78.95% and 94.74% higher than the control group, respectively, indicating that these treatments allowed maintenance of higher CAT activity in loquat fruit.

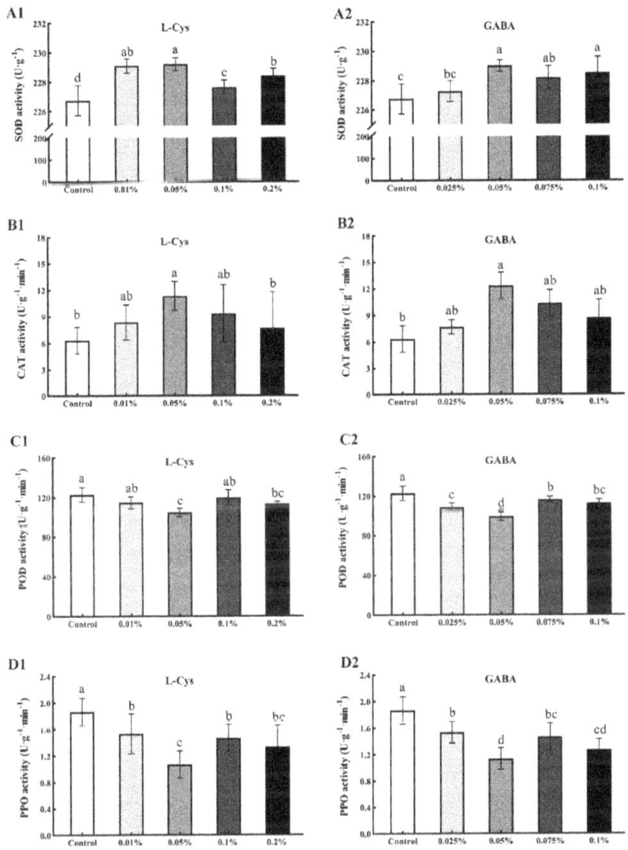

Figure 6. Effects of different concentrations of L-Cys and GABA treatments on SOD (**A**), CAT (**B**), POD (**C**) and PPO activity (**D**) of loquat fruit after 40 days of storage. A1, B1, C1, D1 and A2, B2, C2, D2 indicate the effect of L-Cys and GABA, respectively. Each column represents mean ± standard deviation (n = 3), and a *t* test was performed at the 0.05 significance level ($p < 0.05$). Different letters (a, b, c, d) indicate significant differences, and the same letter indicates no significant difference.

POD can catalyze the oxidative polymerization of phenols in fruits, resulting in browning and lignin synthesis. Treatments with 0.05% L-Cys and 0.05% GABA resulted in 17.39% and 24.21% decreases in POD activity, respectively ($p < 0.05$). The effect of L-Cys and GABA treatments to decrease the POD activity of loquat fruit reduced the incidence of fruit decay and browning.

PPO is a multifunctional enzyme widely found in fruits and vegetables, acting to catalyze the oxidation of polyphenols into quinones and cause browning of fruits and vegetables. As shown in Figure 6D, the PPO activity of treated loquat fruit was lower than that of the control group after 40 days of storage, with 0.05% L-Cys and 0.05% GABA treatment reducing the PPO activity of loquat fruit by 74.98% and 64.71%, respectively ($p < 0.05$). Thus, L-Cys and GABA treatments delayed the increase of PPO activity during storage of loquat fruit to inhibit fruit browning.

2.8. PCA

Principal component analysis (PCA) was used to explore the effects of two postharvest treatments on loquat quality. Two principal components (PCs) explained 64.88% of the total variance. PC1 and PC2 accounted for 57% and 7.88% of the total variance, respectively.

Treatments of 0.05% L-Cys and 0.05% GABA were significantly different from other groups along the PC1 axis and loaded along the negative direction of PC1. The control group was significantly separated from L-Cys and GABA-treated samples, with no significant separation between the other treated samples (Figure 7A). Loquat fruits stored for 40 days under L-Cys and GABA treatments could be distinguished by PC1. PC1 was positively correlated with some quality parameters such as weight loss rate, lignin content, respiratory rate, and PPO, and negatively correlated with fruit Vc, carotenoids, DPPH, FRAP, TFC, and CAT (Figure 7B). Overall, the results clearly indicate that 0.05% L-Cys and 0.05% GABA treatments can maintain the appearance and internal quality of loquat fruit.

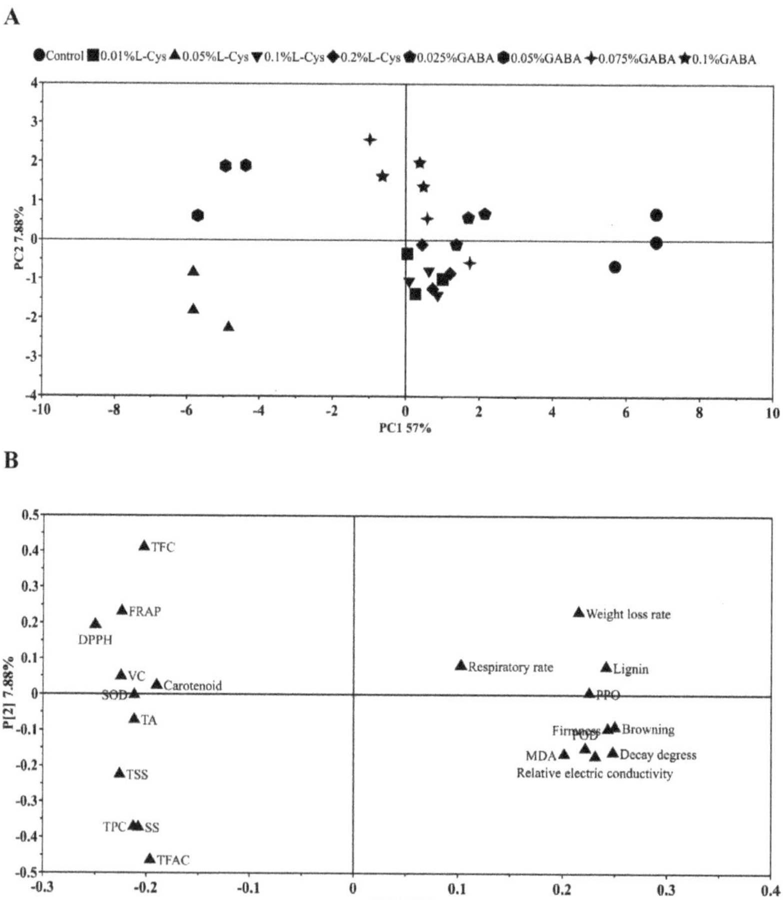

Figure 7. Principal component analysis of postharvest storage quality of loquat: (**A**) principal component score map; (**B**) PCA loading diagram.

2.9. Analysis of Relationships

Pearson correlation coefficient analysis was used to analyze the correlation between appearance quality, internal quality, phenolic substances, and antioxidant enzyme activity during loquat cold storage. Significant positive (orange) and negative (green) correlations are shown in Figure 8. The firmness was positively correlated with weight loss rate, browning index, decay index, respiration rate, relative electric conductivity, lignin content, MDA, POD, and PPO, while TPC, TFC, TFAC, DPPH, and FRAP were negatively correlated with MDA, POD, PPO, and lignin.

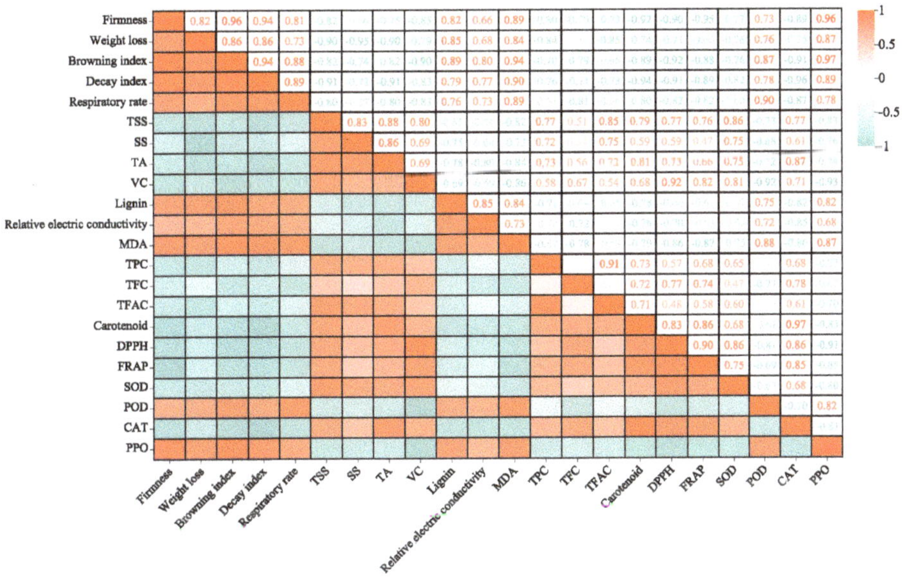

Figure 8. Correlation of fruit parameters during cold storage of loquat treated with L-Cys and GABA. The numerical value and color intensity are proportional to the correlation coefficient. Orange indicates a positive correlation, and green indicates a negative correlation ($p < 0.05$).

3. Discussion

Loquat is popular for its delicious taste and high levels of dietary antioxidants (such as carotenoids and flavonoids) and other bioactive compounds. However, due to the thin and juicy peel of loquat fruit, its post-harvest quality deteriorates rapidly. This is mainly manifested as dehydration shrinkage, decay, browning, and lignification, seriously reducing the commercial value of loquat fruit. L-Cys and GABA are important exogenous amino acids that play key roles in plant growth and stress response. This study examined the effects of L-Cys and GABA treatments on fruit quality and the antioxidant system of loquat during cold storage (Figure 9).

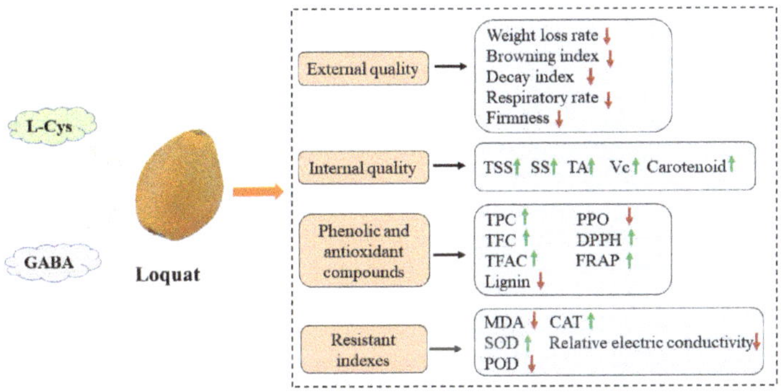

Figure 9. Changes in postharvest quality and antioxidant system of loquat fruit treated with L-Cys and GABA under cold storage conditions. The green arrow indicates an increase, and the red arrow indicates a decrease.

Weight loss rate, browning index, and decay index are important indicators reflecting the effect of fruit storage. The results showed that 0.05% L-Cys and 0.05% GABA treatments effectively reduced the weight loss rate, browning index, and decay index of loquat fruit compared with other treatments. Ali et al. [21] similarly found that 0.25% L-Cys can effectively reduce the weight loss rate and delay the browning index of litchi. Fan et al. [22] found that 1.0 mM GABA could reduce the weight loss rate of olive fruit during cold storage. Firmness is an important quality characteristic, and can defend against mechanical damage to improve the shelf life of fruits. However, during the postharvest storage of loquat fruit, the increase in firmness accompanied by the accumulation of lignin content reduces the flavor of the fruit [23]. Loquat treated with 0.05% L-Cys and 0.05% GABA exhibited a significantly delayed increase in firmness. This was consistent with previous findings [24] that the combined application of oxalic acid (OA) and modified vapor phase packaging (MAP) delayed the firmness and shelf life of loquat. Respiration rate is a good physiological index to evaluate the energy metabolism of postharvest fruit. This study found that 0.05% L-Cys and 0.05% GABA treatments significantly inhibited the respiration rate of loquat fruit after harvest, but the detailed mechanisms explaining the effects of L-Cys and GABA treatments on postharvest fruit respiration are still unclear. GABA can accumulate in *Arabidopsis* mutants lacking respiratory enzymes, indicating a key role of GABA in plant respiration [25]. Ngaffo et al. [26] found that the interaction between GABA and postharvest fruit CO_2 concentration indicated that GABA helped to reduce the respiratory activity of carambola fruit during cold storage.

Intrinsic quality is an important index that affects the taste, flavor, and nutrition of fruits, and the content of intrinsic substances in fruits gradually decreases with fruit aging [27,28]. L-Cys and GABA treatments can improve the intrinsic quality of fruits and vegetables, and in this work, loquat fruits treated with 0.05% L-Cys and 0.05% GABA maintained fruit quality better than the control treatment. Studies have found that L-Cys treatment can significantly delay the decline of soluble sugar, titratable acid, and carotenoids in litchi fruits. Other work also showed that L-Cys is a sulfur-containing amino acid that can delay fruit senescence [21,29]. Another study on olive found that 1.0 mM GABA significantly increased the soluble solids, soluble sugar, titratable acid and chlorophyll content of olive fruit [22]. Vc is related to anti-oxidation and anti-free radical [30]. This study showed that 0.05% L-Cys and 0.05% GABA treatments significantly delayed the decrease in Vc, which was consistent with the results of Sohail et al. [31].

Phenols, flavonoids, flavanols, carotenoids, and Vc in Loquat are free radical scavengers that protect important molecules in cells from oxidation. The indexes related to tissue antioxidant capacity include DPPH free radical scavenging activity and FRAP [32]. Phenolic compounds are important secondary metabolites of plants, which are essential to defend against postharvest pathogens in fruits, and have strong antioxidant activities, which can improve the fruit nutritional value [33,34]. Carotenoid is the main determinant of loquat fruit color, and its content is highest in fruit. This study found that compared with the control, 0.05% L-Cys and 0.05% GABA treatments could significantly increase the phenols and antioxidants of loquat fruit. This was similar to the conclusion of Ebrahimzadeh et al. [35] that 1% gum arabic (rich in 5.0 mM GABA) significantly increased the total phenolic content of walnut kernels and delayed the decrease in antioxidant capacity. Postharvest studies of bananas also found that GABA treatment promoted the increase in total phenol content in banana fruits during storage [19]. The DPPH free radical scavenging activity and FRAP results on the antioxidant capacity of loquat fruit were consistent with the total phenol content. Treatments of 0.05% L-Cys and 0.05% GABA treatment significantly delayed the decrease in DPPH and FRAP in loquat fruit at 40 d. Preczenhak et al. [36] showed that Cys treatment could improve the antioxidant capacity of fresh-cut red beet and lettuce.

In recent years, studies have shown that the lignin biosynthesis pathway is made up of the phenylpropanoid metabolic and lignin synthesis specific pathways. PPO and POD are key enzymes in lignin synthesis, providing structural support for plant growth, development, and defense [37,38]. PPO and POD can oxidize phenols, which are mainly involved in the browning of fruits and vegetables during cold storage. PPO is an important factor that causes enzymatic browning of fruits and vegetables, leading to spoilage [39]. POD is an antioxidant enzyme and can eliminate peroxide free radicals, ensure the integrity of the membrane system, and maintain normal metabolism [40]. Treatments of 0.05% L-Cys and 0.05% GABA significantly inhibited the activity of lignin, PPO, and POD in loquat fruit, effects that are closely related to the reduction in fruit browning rate. Sun et al. [41] showed that NO delayed the formation and inhibited the accumulation of okra lignin. Cai et al. [42] reported that low temperature storage could inhibit the activities of PPO and POD in loquat fruit, thereby reducing flesh browning. Studies have also shown that light (2000 lux) treatment significantly inhibited the PPO and POD activities of fresh-cut celery, thereby reducing the change in the browning index during storage [43].

The relative conductivity and MDA are important indicators of cell membrane permeability; the greater the value, the greater is the damage to the tissue cell membrane. In this study, L-Cys and GABA treatments reduced the relative conductivity of loquat fruit and inhibited the accumulation of MDA, which indicated that L-Cys and GABA had a positive effect on maintaining the integrity and stability of the cell membrane of postharvest loquat fruit. L-Cys and GABA reduced the chilling injury of postharvest loquat fruit by protecting the cell membrane structure. This is similar to the conclusion of Shekari et al. [44] that exogenous GABA inhibits MDA accumulation during low temperature storage of mushrooms. There are antioxidant enzyme systems in fruit and vegetable tissues. SOD and CAT are the main enzymes that scavenge reactive oxygen species in plant tissues. The coordination of these two enzymes can maintain reactive oxygen species at a low level, thereby preventing their damage to cells [45]. Loquat is a cold-sensitive fruit. When cold-sensitive fruit is stored at an unsuitable low temperature, the cell membrane first responds to low-temperature stress, reducing the activity of the oxygen-scavenging enzyme system, resulting in the disorder of active oxygen metabolism in the fruit. Excessive accumulation of reactive oxygen species in fruit tissue will accelerate membrane lipid peroxidation and eventually leads to fruit damage [46]. This study found that 0.05% L-Cys and 0.05% GABA treatments could significantly increase the activities of SOD and CAT in loquat fruit. Studies on cucumber have found that fucoidan treatment can increase SOD and CAT activity [47]. Treatment with 10 $\mu mol \cdot L^{-1}$ MeJA could increase the activities of SOD and CAT, reduce the accumulation of reactive oxygen species, and alleviate the symptoms of lignification and chilling injury caused by low temperature [48]. Overall, the results indicate that L-Cys and GABA post-harvest treatments delayed the browning and senescence of loquat fruit mainly by improving the integrity of the cell membrane and antioxidant enzyme activity during cold storage.

In summary, treatments with 0.05% L-Cys and 0.05% GABA during loquat storage resulted in an increase in the content of bioactive compounds in the fruits. The treatments with L-Cys and GABA also significantly reduced the weight loss rate, browning index, decay index, and respiratory intensity of the loquat. In addition, the treatments helped to maintain higher levels of TSS, SS, TA, Vc, and carotenoid contents; improved phenolic substances and antioxidant capacity; and delayed the increase in fruit firmness, relative electric conductivity, MDA, PPO, and POD. In general, our results show that 0.05% L-Cys and 0.05% GABA treatment are effective methods to alleviate the decay and browning of loquat fruit and maintain fruit quality during cold storage.

4. Materials and Methods

4.1. Plant Materials and Treatments

'Zaozhong No.6' loquat fruit were harvested in Sichuan Province, China (26°42′ N, 101°44′ E), at an altitude of 2080 m on 17 March 2022. Loquat fruits without disease and damage were selected and packed in a protective bag and immediately transported to the laboratory of Sichuan Agricultural University. Loquat fruits (n = 270) were randomly divided into nine groups with three replicates in each group (each replicates includes ten fruits). The nine groups of fruit were dipped in water (control), L-Cys (0.01%, 0.05%, 0.1%, 0.2%), or GABA (0.025%, 0.05%, 0.075%, 0.1%) solutions for 10 min (L-Cysteine and γ-aminobutyric acid were dissolved in distilled water). After the fruit were air-dried, they were put into PE film bags and stored at 4 ± 0.5 °C and 90% relative humidity for 35 days. The weight loss, browning index, decay index, and respiration rate were measured every 7 days. After cold storage, the fruit were stored under a simulated shelf-life environment (22 °C ± 0.5 °C, 85% relative humidity) for 5 days and then the quality was evaluated. At the same time, samples were frozen, chilled in liquid nitrogen, and stored at −80 °C.

4.2. Determination of Weight Loss Rate, Browning Index, Decay Index, and Respiratory Rate

Weight loss rate, browning index, and decay index were estimated according to the method of Cao et al. [49]. To determine the weight loss of individual loquat fruit, the fruit was weighed before storage (initial weight) and after each sampling date (final weight). The weight loss was then calculated and expressed as a percentage loss, using the equation: weight loss rate (%) = (quality before storage − quality after storage) 100/quality before storage. The browning index and decay index were calculated by grading method. The browning index is assessed by measuring the extent of the total area of brown tissue flesh of each fruit on the following scale: 0 = No browning; 1 = Slight browning (<5%); 2 = Moderate browning (5–25%); 3 = Moderate severe browning (25–50%); 4 = Severe browning (>50%). The browning index (%) was calculated as $\sum$ (browning grade × number of browning fruits at all levels) / (total number of fruits × highest grade) × 100. Decay index is assessed by measuring the extent of the total area of decayed tissue of each fruit on the following scale: 0 = No decay; 1 = Slight decay (<10%); 2 = Moderate decay (10–30%); 3 = Moderate severe decay (30–50%); 4 = Severe decay (≥50%). The decay index (%) was calculated as $\sum$ (decay grade × number of decay fruits at all levels)/(total number of fruits × highest grade) × 100. The respiration rate of loquat was measured by the method of Chen et al. [50] with slight modifications. The fruits were placed in a sealed tank for 1 h, and then the CO_2 content in the tank was measured. The results were expressed as $mg \cdot kg^{-1} \cdot h$.

4.3. Determination of Total Soluble Solids (TSS), Soluble Sugar (SS), Titratable Acid (TA) and Vitamin C (Vc) Content

The amount of total soluble solids was determined by digital refractometer, and the values were averaged from three points on the equatorial line of the fruit. The content of soluble sugar in pulp was determined by anthrone colorimetry [51]. Titratable acid in pulp was determined by sodium hydroxide titration [52]. The vitamin C content in pulp was determined by the Fe^{3+} reduction method [53].

4.4. Determination of Fruit Firmness and Lignin

The firmness was measured with a WDGY-4 fruit firmness tester. The extraction and content determination of lignin were performed according to the method of Femenia [54] The absorbance value was measured at 280 nm, and lignin content was expressed as %.

4.5. Determination of Relative Electric Conductivity and MDA

The relative electric conductivity (EC) was measured with a conductivity meter [55]. To do this, 1 g of flesh was weighed and added to a 50 mL centrifuge tube containing 20 mL of sterile water. The mixture was oscillated for 30 min, and the first conductivity was measured. The sample was then boiled for 10 min, allowed to cool to room temperature, and the final conductivity was measured.

The content of malondialdehyde was determined by thiobarbituric acid reaction [56]. First, 0.5 g tissue was mixed in 5 mL phosphate buffer (pH = 7.8) and centrifuged at 4500× g for 10 min. Then, 3 mL of 0.5% thiobarbituric acid (containing 20% TCA) was added to 2 mL of the supernatant. The mixture was incubated at 100 °C for 10 min, and immediately cooled before reading the absorbance at 532 and 600 nm, and the results were expressed as $nmol \cdot g^{-1}$.

4.6. Determination of Total Phenolic Content

Total phenolic content (TPC) was determined by the method of Zhang et al. [57]. The total phenols were extracted with 70% methanol (containing 2% formic acid), and the supernatant was mixed with Folin–Ciocalteu reagent. After 1 min, 20% saturated sodium carbonate was added and mixed well. The absorbance was measured at 760 nm with a spectrophotometer, and gallic acid was used as the standard (50–1000 $mg \cdot L^{-1}$).

4.7. Determination of Flavonoid Content

The content of total flavonoids (TFC) was determined according to the method of Zhang et al. [57]. The total flavonoids were extracted with 70% methanol (containing 2% formic acid), and the supernatant was mixed with methanol, sodium nitrite, and aluminum chloride reagent, and then mixed with sodium hydroxide after 5 min. The absorbance was measured at 510 nm by spectrophotometer. Rutin was used as the standard (20–100 $mg \cdot L^{-1}$).

4.8. Determination of Total Flavanol Content

The content of flavanols (TFAC) was determined according to the method of Zhang et al. [57]. The total flavonoids were extracted with 70% methanol (containing 2% formic acid), and the supernatant was mixed with distilled water and 1 mL 1% p-DMACA solution. After 10 min, the absorbance was measured by spectrophotometer at 640 nm, and catechin was used as the standard (6.25–200 $mg \cdot L^{-1}$).

4.9. Determination of Carotenoids

The carotenoid content of loquat fruit was measured by the 95% ethanol extraction method [58]. The loquat sample was soaked in ethanol (95%) until colorless. The result was indicated as $\mu g \cdot g^{-1}$ FW.

4.10. Determination of DPPH and FRAP

The DPPH scavenging ability was determined by the method of Brand-Williams et al. [59]. The extracted sample was added to distilled water and DPPH methanol solution, and the absorbance was measured at 517 nm by spectrophotometer after 20 min reaction in the dark.

FRAP was determined using the method of Benzie [60] with minor modifications. The extracted sample was added to distilled water and iron reduction antioxidant power (TPTZ) working solution, and reacted at 37 ° C for 10 min before quickly cooling to room temperature. The absorbance was recorded at 593 nm by spectrophotometer.

4.11. Determination of Superoxide Dismutase (SOD), Catalase (CAT), Peroxidase (POD), and Polyphenol Oxidase (PPO) Activity

Frozen samples of approximately 0.25 g were quickly weighed and added to 5 mL ice-cold phosphate buffer (containing 1 g PVP, 0.003085 g DTT, and 0.0037224 g EDTA), centrifuged at 8000 rpm for 15 min at 4 °C, and the supernatant was used as the crude enzyme extract to measure four enzyme activity indicators.

The method of Rao et al. [61] was used to measure SOD, with slight changes. The reaction mixture (3 mL) contained 50 $mmol \cdot L^{-1}$ sodium phosphate buffer (pH = 7.8), 130 $mmol \cdot L^{-1}$ methionine, 750 $mmol \cdot L^{-1}$ nitroblue tetrazolium (NBT), 100 $\mu mol \cdot L^{-1}$ EDTA, 20 $\mu mol \cdot L^{-1}$ riboflavin, and 0.1 mL enzyme extract. By recording the absorbance at 560 nm, the inhibition of NBT photoreduction by 50% at 560 nm wavelength per minute per gram of flesh was determined as an enzyme activity in units of $U \cdot g^{-1}$.

CAT activity was measured according to Aghdam et al. [62] with minor modifications. The reaction mixture contained 0.15 M (pH = 7.0) phosphate buffer and 30% H_2O_2, and was quickly added to 0.1 mL of extracted sample and shaken. The absorbance change at 240 nm wavelength was measured, and the absorbance change at 240 nm per minute was defined as 0.01 CAT activity unit. The results were expressed as $U \cdot g^{-1} \cdot min^{-1}$.

POD activity was determined using the guaiacol method [63] with slight modifications. The reaction mixture contained 0.2 M (pH = 6.0) phosphate buffer, 2% H_2O_2, 0.05 mol·L^{-1} guaiacol, and 0.5 mL enzyme solution mixed well. The change in absorbance value at 470 nm was measured, and a 0.01 change in absorbance value per minute at 470 nm was defined as one POD activity unit. The results were expressed as U·g^{-1}·min^{-1} of enzyme activity.

PPO was determined by the catechol method [64] with slight modifications. The reaction mixture contained 0.2 M (pH = 7.8) phosphate buffer and 0.05 mol·L^{-1} catechol, which was quickly added to 0.5 mL of crude enzyme extract and shaken. The absorbance change was immediately measured at 410 nm wavelength. A 0.01 increase per minute in absorbance value was defined as a POD activity unit, and the results were expressed as U·g^{-1}·min^{-1}.

4.12. Statistical Analysis

All data are expressed as mean ± standard error (SE) (n = 3) in this paper. The data analysis was performed with IBM SPSS 26.0 statistical software for analysis of variance (ANOVA). Duncan's multiple range test was used to determine the differences between treatments at a significance level of $p < 0.05$. Principal component analysis (PCA) was performed using SIMCA 14.0 software.

Author Contributions: Conceptualization, Q.D.; formal analysis, H.Z., J.P. and H.L.; investigation, M.W., Y.D. and X.T.; resources, X.L. and Q.D.; writing—original draft preparation, H.Z. and J.P.; and writing—review and editing, Y.W. All authors have read and agreed to the published version of the manuscript.

Funding: The research was supported by the Natural Science Foundation of Sichuan Province (2022NSFSC0092), the Technology Innovation Research and Development Project of Chengdu (2022-YF05-01144-SN) and the Project of cooperation between Ya'an and Sichuan Agriculture University.

Institutional Review Board Statement: Not applicable.

Informed Consent Statement: Not applicable.

Data Availability Statement: The data and materials supporting the conclusions of this study are included within the article.

Conflicts of Interest: The authors declare no conflict of interest.

References

1. Liu, Y.; Zhang, W.; Xu, C.; Li, X. Biological Activities of Extracts from Loquat (*Eriobotrya japonica* Lindl.): A Review. *Int. J. Mol. Sci.* **2016**, *17*, 1983. [CrossRef] [PubMed]
2. Goulas, V.; Minas, I.S.; Kourdoulas, P.M.; Vicente, A.R.; Manganaris, G.A. Phytochemical content, antioxidants and cell wall metabolism of two loquat (*Eriobotrya japonica*) cultivars under different storage regimes. *Food Chem.* **2014**, *155*, 227–234. [CrossRef] [PubMed]
3. Petriccione, M.; Pasquariello, M.S.; Mastrobuoni, F.; Zampella, L.; Di, P.D.; Scortichini, M. Influence of a chitosan coating on the quality and nutraceutical traits of loquat fruit during postharvest life. *Sci. Hortic.* **2015**, *197*, 287–296. [CrossRef]
4. Kahramanoğlu, İ. Preserving postharvest storage quality of fresh loquat fruits by using different bio-materials. *J. Food Sci. Technol.* **2020**, *57*, 3004–3012. [CrossRef]
5. Zhang, Y.; Jin, P.; Huang, Y.; Shan, T.; Wang, L.; Li, Y.; Zheng, Y. Effect of hot water combined with glycine betaine alleviates chilling injury in cold-stored loquat fruit. *Postharvest Biol. Technol.* **2016**, *118*, 141–147. [CrossRef]
6. Cao, S.; Zheng, Y.; Wang, K.; Rui, H.; Tang, S. Effect of 1-methylcyclopropene treatment on chilling injury, fatty acid and cell wall polysaccharide composition in loquat fruit. *J. Agric. Food Chem.* **2009**, *57*, 8439–8443. [CrossRef]
7. Wang, J.; Yuan, H.; Wu, Y.; Yu, J.; Ali, B.; Zhang, J.; Tang, Z.; Xie, J.; Lyu, J.; Liao, W. Application of 5-aminolevulinic acid promotes ripening and accumulation of primary and secondary metabolites in postharvest tomato fruit. *Front. Nutr.* **2022**, *9*, 1036843. [CrossRef]
8. Sun, J.; Chen, H.; Chen, Y.; Lin, M.; Hung, Y.C.; Jiang, Y.; Lin, H. ε-Poly-l-Lysine Enhances Fruit Disease Resistance in Postharvest Longans (*Dimocarpus longan* Lour.) by Modulating Energy Status and ATPase Activity. *Foods* **2022**, *11*, 773. [CrossRef]
9. Ainsworth, E.A.; Gillespie, K.M. Estimation of total phenolic content and other oxidation substrates in plant tissues using Folin-Ciocalteu reagent. *Nat. Protoc.* **2007**, *2*, 875–877. [CrossRef]
10. Wang, Z.; Mao, J.L.; Zhao, Y.J.; Li, C.Y.; Xiang, C.B. L-Cysteine inhibits root elongation through auxin/PLETHORA and SCR/SHR pathway in Arabidopsis thaliana. *J. Integr. Plant Biol.* **2015**, *57*, 186–197. [CrossRef]

11. Jiang, J.; Jiang, L.; Luo, H.; Yu, Z. Establishment of a statistical model for browning of fresh-cut lotus root during storage. *Postharvest Biol. Technol.* **2014**, *92*, 164–171. [CrossRef]
12. Ali, H.M.; El-Gizawy, A.M.; El-Bassiouny, R.E.; Saleh, M.A. Browning inhibition mechanisms by cysteine, ascorbic acid and citric acid, and identifying PPO-catechol-cysteine reaction products. *J. Food Sci. Technol.* **2015**, *52*, 3651–3659. [CrossRef]
13. Zhou, X.; Xiao, Y.; Meng, X.; Liu, B. Full inhibition of Whangkeumbae pear polyphenol oxidase enzymatic browning reaction by l-cysteine. *Food Chem.* **2018**, *266*, 1–8. [CrossRef]
14. Wu, X.T.; Guo, X.N.; Zhu, K.X. Inhibition of L-Cysteine on the Browning of Fresh Wet Noodles. *Foods* **2021**, *10*, 156. [CrossRef]
15. Pace, B.; Capotorto, I.; Ventura, M.; Cefola, M. Evaluation of L-Cysteine as Anti-Browning Agent in Fresh-Cut Lettuce Processing. *J. Food Process. Preserv.* **2015**, *39*, 985–993. [CrossRef]
16. Li, Y.; Liu, B.; Peng, Y.; Liu, C.; Li, C. Exogenous GABA alleviates alkaline stress in Malus hupehensis by regulating the accumulation of organic acids. *Sci. Hortic.* **2019**, *261*, 108982. [CrossRef]
17. Oh, S.J.; Kim, H.S.; Lim, S.T.; Reddy, C.K. Enhanced accumulation of gamma-aminobutyric acid in rice bran using anaerobic incubation with various additives. *Food Chem.* **2019**, *271*, 187–192. [CrossRef]
18. Shang, H.; Cao, S.; Yang, Z.; Cai, Y.; Zheng, Y. Effect of exogenous γ-aminobutyric acid treatment on proline accumulation and chilling injury in peach fruit after long-term cold storage. *J. Agric. Food Chem.* **2011**, *59*, 1264–1268. [CrossRef]
19. Wang, Y.; Luo, Z.; Huang, X.; Yang, K.; Du, R. Effect of exogenous γ-aminobutyric acid (GABA) treatment on chilling injury and antioxidant capacity in banana peel. *Sci. Hortic.* **2014**, *168*, 132–137. [CrossRef]
20. Liu, Q.; Li, X.; Jin, S.; Dong, W.; Zhang, Y.; Chen, W.; Shi, L.; Cao, S.; Yang, Z. γ-Aminobutyric acid treatment induced chilling tolerance in postharvest kiwifruit (*Actinidia chinensis cv.* Hongyang) via regulating ascorbic acid metabolism. *Food Chem.* **2023**, *404*, 134661. [CrossRef]
21. Ali, S.; Khan, A.S.; Malik, A.U.; Nawaz, A.; Shahid, M. Postharvest application of antibrowning chemicals modulates oxidative stress and delays pericarp browning of controlled atmosphere stored litchi fruit. *J. Food Biochem.* **2019**, *43*, e12746. [CrossRef] [PubMed]
22. Fan, Z.; Lin, B.; Lin, H.; Lin, M.; Chen, J.; Lin, Y. γ-Aminobutyric acid treatment reduces chilling injury and improves quality maintenance of cold-stored Chinese olive fruit. *Food Chem. X* **2022**, *13*, 100208. [CrossRef] [PubMed]
23. Huang, W.; Zhu, N.; Zhu, C.; Wu, D.; Chen, K. Morphology and cell wall composition changes in lignified cells from loquat fruit during postharvest storage-ScienceDirect. *Postharvest Biol. Technol.* **2019**, *157*, 110975. [CrossRef]
24. Öz, A.T.; Kafkas, E.; Bozdoğan, A. Combined effects of oxalic acid treatment with modified atmosphere packaging on postharvest quality of loquat fruit during storage. *Turk. J. Agric. For.* **2016**, *40*, 3. [CrossRef]
25. Ramos-Ruiz, R.; Felix, M.; Knauf-Beiter, G. The effects of GABA in plants. *Cogent Food Agric.* **2019**, *5*, 1670553. [CrossRef]
26. Ngaffo, M.F.; Duan, W.; Cisse, E.H.M.; Chen, T.; Xu, X. Alleviation of Postharvest Chilling Injury of Carambola Fruit by γ-aminobutyric Acid: Physiological, Biochemical, and Structural Characterization. *Front. Nutr.* **2021**, *8*, 752583. [CrossRef]
27. Cai, J.; Chen, T.; Zhang, Z.; Li, B.; Qin, G.; Tian, S. Metabolic Dynamics During Loquat Fruit Ripening and Postharvest Technologies. *Front. Plant Sci.* **2019**, *10*, 619. [CrossRef]
28. Liu, R.; Tian, S.; Qin, G.Z.; Wang, Y.Y. Molecular basis of 1-methylcyclopropene regulating organic acid metabolism in apple fruit during storage. *Postharvest Biol. Technol.* **2016**, *117*, 57–63. [CrossRef]
29. Ubeed, H.M.S.A.; Wills, R.B.H.; Bowyer, M.C.; Golding, J.B. Inhibition of postharvest senescence of green leafy vegetables by exogenous D-cysteine and L-cysteine as precursors of hydrogen sulphide. *J. Hortic. Sci. Biotechnol.* **2019**, *94*, 620–626. [CrossRef]
30. Padayatty, S.J.; Katz, A.; Wang, Y.; Eck, P.; Kwon, O.; Lee, J.H.; Chen, S.; Corpe, C.; Dutta, A.; Dutta, S.K.; et al. Vitamin C as an antioxidant: Evaluation of its role in disease prevention. *J. Am. Coll. Nutr.* **2003**, *22*, 18–35. [CrossRef]
31. Sohail, M.; Wills, R.B.H.; Bowyer, M.C.; Pristijono, P. Beneficial impact of exogenous arginine, cysteine and methionine on postharvest senescence of broccoli. *Food Chem.* **2021**, *338*, 128055. [CrossRef]
32. Zhang, W.; Zhao, X.; Sun, C.; Li, X.; Chen, K. Phenolic composition from different loquat (*Eriobotrya japonica* Lindl.) cultivars grown in China and their antioxidant properties. *Molecules* **2015**, *20*, 542–555. [CrossRef]
33. Vogt, T. Phenylpropanoid biosynthesis. *Mol. Plant* **2010**, *3*, 2–20. [CrossRef]
34. Skrovankova, S.; Sumczynski, D.; Mlcek, J.; Jurikova, T.; Sochor, J. Bioactive Compounds and Antioxidant Activity in Different Types of Berries. *Int. J. Mol. Sci.* **2015**, *16*, 24673–24706. [CrossRef]
35. Ebrahimzadeh, A.; Pirzad, F.; Tahanian, H.; Aghdam, M.S. Influence of Gum Arabic Enriched with GABA Coating on Oxidative Damage of Walnut Kernels. *Food Technol. Biotechnol.* **2019**, *57*, 554–560. [CrossRef]
36. Preczenhak, A.P.; Orsi, B.; Lima, G.P.P.; Tezotto-Uliana, J.V.; Minatel, I.O.; Kluge, R.A. Cysteine enhances the content of betalains and polyphenols in fresh-cut red beet. *Food Chem.* **2019**, *286*, 600–607. [CrossRef]
37. Liu, W.; Zhang, J.; Jiao, C.; Yin, X.; Fei, Z.; Wu, Q.; Chen, K. Transcriptome analysis provides insights into the regulation of metabolic processes during postharvest cold storage of loquat (*Eriobotrya japonica*) fruit. *Hortic. Res.* **2019**, *6*, 49. [CrossRef]
38. Cai, C.; Xu, C.J.; Li, X.; Ferguson, I.; Chen, K.S. Accumulation of lignin in relation to change in activities of lignification enzymes in loquat fruit flesh after harvest. *Postharvest Biol. Technol.* **2006**, *40*, 163–169. [CrossRef]
39. Moon, K.M.; Kwon, E.B.; Lee, B.; Kim, C.Y. Recent Trends in Controlling the Enzymatic Browning of Fruit and Vegetable Products. *Molecules* **2020**, *25*, 2754. [CrossRef]
40. Fang, L.; Jiang, B.; Zhang, T. Effect of combined high pressure and thermal treatment on kiwifruit peroxidase. *Food Chem.* **2008**, *109*, 802–807. [CrossRef]

41. Sun, M.; Yang, X.L.; Zhu, Z.P.; Xu, Q.Y.; Wu, K.X.; Kang, Y.J.; Wang, H.; Xiong, A.S. Comparative transcriptome analysis provides insight into nitric oxide suppressing lignin accumulation of postharvest okra (*Abelmoschus esculentus* L.) during cold storage. *Plant Physiol. Biochem.* **2021**, *167*, 49–67. [CrossRef] [PubMed]
42. Cai, C.; Xu, C.J.; Shan, L.L.; Li, X.; Zhou, C.H.; Zhang, W.S.; Ferguson, I.; Chen, K.S. Low temperature conditioning reduces postharvest chilling injury in loquat fruit. *Postharvest Biol. Technol.* **2006**, *41*, 252–259. [CrossRef]
43. Zhan, L.; Hu, J.; Lim, L.T.; Pang, L.; Li, Y.; Shao, J. Light exposure inhibiting tissue browning and improving antioxidant capacity of fresh-cut celery (*Apium graveolens* var. dulce). *Food Chem.* **2013**, *141*, 2473–2478. [CrossRef] [PubMed]
44. Shekari, A.; Hassani, R.N.; Aghdam, M.S. Exogenous application of GABA retards cap browning in Agaricus bisporus and its possible mechanism. *Postharvest Biol. Technol.* **2021**, *174*, 111434. [CrossRef]
45. Wojtaszek, G. Mechanisms for the generation of reactive oxygen species in plant defence—A broad perspective. *Physiol. Mol. Plant Pathol.* **1997**, *51*, 347–366. [CrossRef]
46. James, M.L. Chilling injury in plants. *Annu. Rev. Plant Physiol.* **2003**, *24*, 445–466. [CrossRef]
47. Lin, D.; Yan, R.; Xing, M.; Liao, S.; Chen, J.; Gan, Z. Fucoidan treatment alleviates chilling injury in cucumber by regulating ROS homeostasis and energy metabolism. *Front. Plant Sci.* **2022**, *13*, 1107687. [CrossRef]
48. Cao, S.; Zheng, Y.; Wang, K.; Jin, P.; Rui, H. Methyl jasmonate reduces chilling injury and enhances antioxidant enzyme activity in postharvest loquat fruit. *Food Chem.* **2009**, *115*, 1458–1463. [CrossRef]
49. Cao, S.F.; Wang, X.Q.; Yang, Z.F.; Ma, S.J.; Zheng, Y.H. Effects of methyl jasmonate treatment on quality and decay in cold-stored loquat fruit. *Acta Hortic.* **2007**, *750*, 425–430. [CrossRef]
50. Chen, J.; Hu, Y.; Wang, J.; Hu, H.; Cui, H. Combined effect of ozone treatment and modified atmosphere packaging on antioxidant defense system of fresh-cut green peppers. *J. Food Process. Preserv.* **2016**, *40*, 1145–1150. [CrossRef]
51. Munekata, P.E.S.; Yilmaz, B.; Pateiro, M.; Kumar, M.; Domínguez, R.; Shariati, M.A.; Hano, C.; Lorenzo, J.M. Valorization of by-products from Prunus genus fruit processing: Opportunities and applications. *Crit. Rev. Food Sci. Nutr.* **2022**, *14*, 1–16. [CrossRef]
52. Lin, Y.; Li, N.; Lin, H.; Lin, M.; Chen, Y.; Wang, H.; Ritenour, M.A.; Lin, Y. Effects of chitosan treatment on the storability and quality properties of longan fruit during storage. *Food Chem.* **2020**, *306*, 125627. [CrossRef]
53. Kampfenkel, K.; Van, M.M.; Inzé, D. Extraction and determination of ascorbate and dehydroascorbate from plant tissue. *Anal. Bioanal. Chem.* **1995**, *225*, 165–167. [CrossRef]
54. Femenia, A.; Garcia-Conesa, M.; Simal, S.; Rosselló, C. Characterisation of the cell walls of loquat (*Eriobotrya japonica* L.) fruit tissues. *Carbohydr. Polym.* **1998**, *35*, 169–177. [CrossRef]
55. Xu, Y.; Tian, Y.; Ma, R.; Liu, Q.; Zhang, J. Effect of plasma activated water on the postharvest quality of button mushrooms, Agaricus bisporus. *Food Chem.* **2016**, *197*, 436–444. [CrossRef]
56. Banin, S.O.; Razavi, F.; Rabiei, V.; Gohari, G. Postharvest application of L-cysteine to prevent enzymatic browning of "Stanley" plum fruit during cold storage. *J. Food Process. Preserv.* **2022**, *44*, e14788. [CrossRef]
57. Zhang, H.; Pu, J.; Tang, Y.; Wang, M.; Tian, K.; Wang, Y.; Luo, X.; Deng, Q. Changes in Phenolic Compounds and Antioxidant Activity during Development of 'Qiangcuili' and 'Cuihongli' Fruit. *Foods* **2022**, *11*, 3198. [CrossRef]
58. Wang, X.K. *Principles and Techniques of Plant Physiological and Biochemical Experiments*, 2nd ed.; Higher Education Press: Beijing, China, 2006.
59. Brand-Williams, W.; Cuvelier, M.E.; Berset, C. Use of a free radical method to evaluate antioxidant activity. *LWT Food Sci. Technol.* **1995**, *28*, 28–30. [CrossRef]
60. Benzie, I.S.J. Ferric Reducing/Antioxidant Power Assay: Direct Measure of Total Antioxidant Activity of Biological Fluidsand Modified Version for Simultaneous Measurement of Total Antioxidant Power and Ascorbic Acid Concentration. *Methods Enzymol.* **1999**, *229*, 15–27. [CrossRef]
61. Rao, M.V.; Paliyath, G.; Ormrod, D.P. Ultraviolet-B- and ozone-induced biochemical changes in antioxidant enzymes of Arabidopsis thaliana. *Plant Physiol.* **1996**, *110*, 125–136. [CrossRef]
62. Aghdam, M.S.; Fard, J.R. Melatonin treatment attenuates postharvest decay and maintains nutritional quality of strawberry fruits (*Fragaria×anannasa* cv. Selva) by enhancing GABA shunt activity. *Food Chem.* **2017**, *221*, 1650–1657. [CrossRef] [PubMed]
63. Siragusa, S.; Fontana, C.A.; Cappa, F.; Caputo, L.; Cocconcelli, P.S.; Gobbetti, M.; Angelis, M.D. Disruption of the gene encoding glutamate dehydrogenase affects growth, amino acids catabolism and survival of *Lactobacillus plantarum* UC1001. *Int. Dairy J.* **2011**, *21*, 59–68. [CrossRef]
64. Wang, J.; Yang, X.H.; Mujumdar, A.S.; Wang, D.; Zhao, J.H.; Fang, X.M.; Zhang, Q.; Xie, L.; Gao, Z.J.; Xiao, H.W. Effects of various blanching methods on weight loss, enzymes inactivation, phytochemical contents, antioxidant capacity, ultrastructure and drying kinetics of red bell pepper (*Capsicum annuum* L.). *LWT-Food Sci. Technol.* **2017**, *77*, 337–347. [CrossRef]

International Journal of *Molecular Sciences*

MDPI

Article

Effect of *Hibiscus sabdariffa* L. on the Metabolism of Arachidonic Acid in the Isolated Kidney of a Rat Model of Metabolic Syndrome

Israel Pérez-Torres [1,†], María Elena Soto [2,†], Linaloe Manzano-Pech [1], Eulises Díaz-Díaz [3], Raúl Martínez-Memije [4], Juan Carlos Torres-Narváez [5], Verónica Guarner-Lans [6,*] and Vicente Castrejón-Téllez [6,*]

1 Department of Cardiovascular Biomedicine, Instituto Nacional de Cardiología Ignacio Chávez, Juan Badiano 1, Sección XVI, Tlalpan, México City 14080, Mexico
2 Department of Immunology and Research Direction, Instituto Nacional de Cardiología Ignacio Chávez, Juan Badiano 1, Sección XVI, Tlalpan, México City 14080, Mexico
3 Department of Reproductive Biology, Instituto Nacional de Ciencias Médicas y Nutrición Salvador Zubirán, Vasco de Quiroga 15, Sección XVI, Tlalpan, México City 14000, Mexico
4 Department de Instrumentation Electromechanically, Instituto Nacional de Cardiología Ignacio Chávez, Tlalpan, México City 14080, Mexico
5 Department of Pharmacology, Instituto Nacional de Cardiología Ignacio Chávez, Juan Badiano 1, Sección XVI, Tlalpan, México City 14080, Mexico
6 Department of Physiology, Instituto Nacional de Cardiología Ignacio Chávez, Juan Badiano 1, Sección XVI, Tlalpan, México City 14080, Mexico
* Correspondence: gualanv@yahoo.com (V.G.-L.); vicente.castrejon@cardiologia.org.mx (V.C.-T.)
† These authors contributed equally to this work.

Abstract: The renal system is engaged in metabolic syndrome (MS) and metabolites of arachidonic acid (AA) participate in renal homeostasis and disruption of functionality. *Hibiscus sabdariffa* L (HSL) is used as a diuretic and could improve renal function. The aim of this study was to assess if treatment with HSL at 2% improves renal function in MS through the metabolites of AA. A total of 24 male Wistar rats were divided into four groups: Group 1, control (C); Group 2, MS with 30% sucrose in drinking water, Group 3, MS plus HSL infusion at 2% (MS+HSL); and Group 4, C+HSL. We evaluated the perfusion pressure changes (Δ-PP), the activities of cyclooxygenases (COXs), the percentage of AA, the expressions of PLA_2, COX2, COX1, 5-LOX, TAXS and CYP450, and the concentrations of prostaglandins in the kidney from rats with MS. There was a decrease in the Δ-PP, in the activities of COXs, and the expressions of COX2 and CYP450 ($p \leq 0.03$, respectively)as well asPGE_2, TxB_2, and LKB_4 ($p \leq 0.01$, respectively). However, the percentage of AA and expressions of PLA_2 and PGE1 ($p = 0.01$, respectively) were increased in C and MS+HSL. The HSL treatment improved the function and anatomical structure of the kidneys in the MS rats, through antioxidant molecules, and inhibited the pathways that metabolize the AA including that of PLA_2, COX2, 5-LOX, TAXS, and CYP450 while favoring the COX1 pathway. This improves the vascular resistance of renal arterioles.

Keywords: *Hibiscus sabdariffa* L.; arachidonic acid; cyclooxygenase metabolism; metabolic syndrome; kidney

1. Introduction

Metabolic syndrome (MS) is a cluster of three or more pathologies that include hypertension, dyslipidemia, obesity, overweight status, insulin resistance, hyperinsulinemia, and type2 diabetes. It is largely influenced by environmental factors such as an inadequate diet, rich in carbohydrates and/or saturated fat, and the lack of physical activity [1]. The coexistence of these pathologies in the same patient causes a long-term inflammatory state, and an oxidative background with deterioration of organs and systems, including the renal system [1].The kidney maintains a stable extracellular environment that supports the function of all cells in the body; it controls the water and ionic balance, and the excretion of sodium, chloride, potassium, calcium, magnesium, and phosphate. It also manages the acid–base status and regulates the excretion of other substances such as serum creatinine. The fundamental unit of the kidney is the nephron in which the glomerulus acts as the filtering unit or sieve, keeping proteins and cells in the bloodstream but allowing the extra fluid and wastes to pass through it.

Four enzymatic pathways are involved in the metabolism of arachidonic acid (AA) in the kidney [2]. These pathways are mediated by the cyclooxygenase isoforms (COX1 and 2), the 5-, 12-, and 15-lipoxygenase (LOX) isoforms, and cytochrome p450 (1 to 4) (CYP450). Each enzyme has several subtypes and isoforms. AA is metabolized to prostaglandins E_2 (PGE_2) with vasoconstrictor and inflammatory effects by the COX2 pathway, and to prostacyclin I_2 (PGI_2) with a vasodilator effect by the COX1 pathway [3]. AA is converted to hydroperoxyeicosatetraenoic acids (HPETES) that are then reduced to the hydroxyeicosatetraenoic acids (HETEs) and leukotrines (LTs as LTB_4, LTC_4, and LTE_4) by the LOX isoforms. These last enzymes are related to an increased entry of Ca^{2+} into cells [4]. The CYP450 family produces cis-epoxyeicosatrienoic acid (EETs) with vasoconstrictor effects [5]. In addition, AA can give rise to 8-isoprostanes, which are pro-inflammatory in a pro-oxidant environment [6]. Finally, thromboxane synthase (TXAS) gives rise to thromboxanes (TxB), with a pro-aggregation and thrombotic effect. However, AA must first be released from the phospholipids (PL) of the cell membrane with the action of the family of phospholipases (PLA_2) before entering into these routes [7].

On the other hand, natural products can improve human health. As an example, Chlorella algae, a blue-green microalgae, may restore the function of pancreatic insulin-secreting cells in diabetic rats treated with streptozotocin [8]. Also, dandelion flowers have antioxidant properties that prevent chronic inflammation and development of obesity, cancer, and numerous cardiovascular risk factors [9]. Saponins, a diverse class of natural compounds present in some plant species, have the ability to inhibit various cancers in vitro and in vivo [10]. A supplement based on lipids and enriched with micronutrients improves the birth weight, gestational age, and length and head circumference of newborn babies in pre-eclamptic women [11].In this sense, *Hibiscus sabdariffa* L. (HSL) has been used as a medicinal herb in Africa, Asia, and Latin America against hypertension, urinary blander, and kidney stones, as a diuretic and against liver disorders among other pathologies. HSL calyxes contain many chemicals with antioxidant properties that include polyphenols, flavonoids, and anthocyanins [12]. The beneficial proprieties of HSL in the kidney, as a crude extract or as an infusion, have been attributed to the vasorelaxant effect of anthocyanins probably with the inhibition of the Ca^{2+} channels and to its action on the angiotensin converting enzyme (ACE). In this sense, treatment with the HSL extract reduced the activity of ACE in plasma and decreased the concentrations of sodium without modifying potassium levels in hypertensive patients [13]. Polyphenols extracted from HSL inhibit the inflammation induced by lipopolysaccharides in vitro in macrophages by decreasing the expression of COX2 [14]. Another study demonstrated that the catechin of HSL reduced albuminuria and normalized interstitial fibrosis in the kidney in diabetic rats treated with streptozotocin [13]. Our group reported that the HSL infusion at 2% decreases oxidative stress (OS), increases the total antioxidant capacity, decreases albuminuria, and protects renal function in an MS rat model [15].

Our model of MS in Wistar rats is the result of chronic consumption of 30% commercial sucrose in drinking water for 20–24 weeks. This model courses with hypertension, hypertriglyceridemia, central obesity, OS, hyperinsulinemia, insulin resistance (IR), renal dysfunction, and damage to other organs [16]. We also reported that the metabolism of AA is altered in this model, having an increased expression/activity of COX2 and a decreased expression/activity of COX1 associated with a reduction in the total AA as a free form and in the PL [17]. However, to date, there exists no studies on the effects of HSL on the metabolic pathways that use AA and their possible contribution to improve renal function in MS, despite the findings reported in the literature. Therefore, the aim of this study was to try to understand and elucidate how treatment with HSL at 2% contributes to improve the renal function through the pathways that metabolize AA in a rat model with MS.

2. Results

2.1. Determinations of Total Flavonoids, Anthocyanins, Polyphenols, and Vitamin C

The determinations of total flavonoids, anthocyanins, polyphenols, and vitamin C in the HSL infusion were made with the techniques reported by Jagota, Jia, Lee's, and Sánchez-Rangel, respectively. The HSL infusion at 2% contained cyaniding-3-glucoside (142.06 ± 25.9 mg/L), quercetin (0.38 ± 0.02 mg/L), polyphenols (0.50 ± 0.03 mM/L), and vitamin C (0.53 ± 0.02 mg/L). In addition, the rats consumed cyanidin-3-glucoside, quercetin, and polyphenols at approximately 141.06 ± 25.97, 0.38 ± 0.03, and 0.50 ± 0.01 mg/day, respectively.

2.2. Biochemical Markers in Serum

The MS rats showed significant increases in insulin, the HOMA index, TG, intra-abdominal fat, and SBP vs. control rats ($p = 0.001$). The treatment with HSL infusion at 2% in MS rats significantly decreased insulin, the HOMA index, TG, and SBP ($p \leq 0.04$) in comparison with the MS rats. In the C group, the HSL treatment did not show significant differences in any of the biochemical parameters determined. The glucose levels did not show significant changes in any of the groups (Table 1).

Table 1. General characteristic of the rats in the experimental groups.

Variables	C	MS	MS+HSL	C+HSL
Glucose (mmol/L)	99.27 ± 5.59	161.85 ± 37.07	132.85 ± 7.29	110.14 ± 5.58
Insulin (μU/mL)	5.70 ± 0.75	20.33 ± 1.68 **	11.16 ± 1.21 *	4.89 ± 0.39
HOMA index	1.49 ± 0.12	15.08 ± 1.7 **	5.02 ± 0.32 *	1.40 ± 0.11
Triglycerides (mg/dL)	48.85 ± 4.84	187.42 ± 45.77 **	89.42 ± 4.23 *	49.57 ± 8.36
Cholesterol (mg/dL)	47.85 ± 8.05	46.71 ± 10.73	50.71 ± 1.97	43.71 ± 5.67
Intra-abdominal fat (g)	3.4 ± 0.3	11.6 ± 3.3 **	7.1 ± 0.8 *	3.8 ± 0.5
Systolic blood pressure (mmHg)	118.1 ± 5.7	141.2 ± 6.9 *	132.1 ± 4.1 *	117.63 ± 8.7

Data are mean ± SE; $n = 6$ for each group. Statistically significant at * $p < 0.04$, MS vs. MS+HSL; ** $p = 0.001$, C vs. MS. Abbreviations: C = control; MS = metabolic syndrome; MS+HSL = metabolic syndrome plus *Hibiscus sabdariffa* L.; C+HSL = control plus *Hibiscus sabdariffa* L.

2.3. Markers of the Renal Function

Table 2 represents the indicators of renal function in the experimental groups, where the MS rats showed deterioration of the renal function in comparison with the C and MS+HSL groups. This was evidenced by a significant increase in CCr, albuminuria, and urine volume of 24 h ($p \leq 0.001$). In addition, the MS rats showed an increase in water consumption in comparison with the C and MS+HSL groups ($p \leq 0.001$ and $p = 0.01$, respectively). However, the treatment with HSL infusion at 2% did not produce changes in this parameter in the C+HSL group. The weight of the right kidney in the groups did not show a significant difference.

2.4. Prostaglandins

Table 3 shows the prostaglandin values in the experimental groups where significant increases in PGE_2 and TxB_2 were present in the MS rats in comparison with C and MS+HSL rats ($p = 0.001$ and $p = 0.01$, respectively). A tendency to increase was observed in LKB_4 in MS rats in comparison with C rats ($p = 0.09$) but there was a decrease ($p = 0.01$) in comparison with the MS+HSL rats. In the same sense, the 6-keto-$PGF_{1\alpha}$ was decreased in the MS rats in comparison with C and MS+HSL rats ($p = 0.01$). The same tendency was observed in the AA percentage between MS vs. C and MS+HSL rats ($p = 0.001$ and $p = 0.01$, respectively). The HSL treatment did not modify the prostaglandin levels, or the AA percentage in the C+HSL group.

Table 2. Renal function indicators in the rats of the experimental groups.

Variables	C	MS	MS+HSL	C+HSL
Drinking water (mL/24 h)	20.52 ± 2.58	55.10 ± 5.91 **	40.83 ± 4.54 †	33.33 ± 4.77
Weight of the right kidney (g)	1.27 ± 0.03	1.44 ± 0.05	1.28 ± 0.05	1.22 ± 0.5
Urine (mL/24 h)	14.42 ± 2.77	26.00 ± 4.27 **	24.85 ± 2.95	18.71 ± 4.79
CCr (mL/min)	2.88 ± 0.37	1.36 ± 0.14 **	2.38 ± 0.42 *	2.89 ± 0.86
Albuminuria (mg/mL)	38.30 ± 4.90	82.04 ± 9.21 **	49.54 ± 7.99 †	46.74 ± 4.18

Data are mean ± SE; $n = 6$ for each group. Statistically significant at * $p \leq 0.04$, MS+HSL vs. MS; † $p = 0.01$, C vs. MS; ** $p \leq 0.003$, C vs. MS. Abbreviations: C = control; MS = metabolic syndrome; MS+HSL = metabolic syndrome plus *Hibiscus sabdariffa* L.

Table 3. Prostaglandins in kidney homogenate of the experimental groups.

Variables	C	MS	MS+HSL	+HSL
PGE_2 (pg/mL)	77.58 ± 0.02	77.77 ± 0.03 **	77.49 ± 0.02 **	77.61 ± 0.02
6-Keto-$PGF_{1\alpha}$ (pg/mL)	112.45 ± 0.75	93.63 ± 1.68 *	195.15 ± 1.21 *	182.76 ± 0.39
LKB_4 (pg/mL)	735.45 ± 4.26	805.20 ± 41.76 †	605.80 ± 59.01 *	819.15 ± 29.36
TxB_2 (pg/mL)	149.48 ± 12.23	189.53 ± 12.52 *	109.44 ± 19.31 *	154.82 ± 39.05
AA (%)	14.22 ± 1.18	9.3 ± 0. 68 **	12.67 ± 1.24 *	13.67 ± 0.73

Data are mean ± SE; $n = 6$ for each group. Statistically significant at * $p = 0.01$, MS vs. MS+HSL; ** $p = 0.001$, C vs. MS; † $p = 0.09$, NS. Abbreviations: AA = Arachidonic acid; C = Control; MS = Metabolic syndrome; MS+HSL = Metabolic syndrome plus *Hibiscus sabdariffa* L.

2.5. Delta of the Perfusion Pression

Figure 1A shows the changes in the Δ-PP, in the experimental groups. The Δ-PP was significantly increased in MS rats in comparison with C rats (10.0 ± 1.5 mmHg vs. 15.7 ± 1.6 mmHg, $p = 0.02$). The HSL treatment in MS rats resulted in a significant decrease in the Δ-PP compared to MS rats (8.4 ± 0.2 mmHg vs. 15.7 ± 1.6, $p = 0.001$). The Δ-PP in the kidney of the MS rats when the AA was perfused in the presence of NS398, SC560, baicalein, and Indo decreased by 52.2%, 54.4%, 52.9%, and 48.6%, respectively, showing a significant difference ($p \leq 0.001$) in comparison to the condition without the inhibitors. The perfusion of the AA in the presence of the inhibitors did not show differences in the percentage of the Δ-PP in C, MS+HSL, and C+HSL rats.

2.6. Coxs'activities

Figure 1B shows the activity of the COX isoforms in the microsomal fraction of the kidney homogenate in each of the experimental groups. The use of the specific inhibitors in this fraction allowed fordiscriminating between the activities of the two COX isoforms involved in the metabolism of AA. The results showed a significant increase in the MS rats when compared to the C and the MS+HSL groups ($p < 0.001$, $p = 0.01$, respectively). In the presence of NS398, which is a specific inhibitor of COX2, there was a decrease in its activity in the MS group in comparison with the same group without the inhibitor ($p = 0.03$). However, the other experimental groups did not show statistical changes. In the presence of SC560, which is a specific inhibitor of COX1, the activity of the enzyme did not show statistical changes in the MS group in comparison with the same group without the inhibitor. In the other experimental groups, the presence of the inhibitors did not show statistical changes.

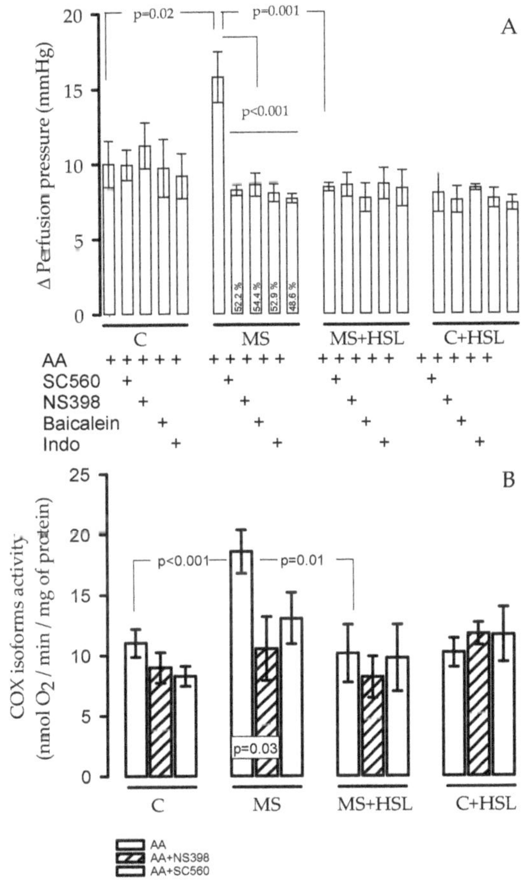

Figure 1. Panel (**A**) shows the changes in the Δ of perfusion pression (Δ-PP), by infusing 4 μg of AA, which is the substrate for the enzymes that metabolize it to eicosanoids with vasoconstrictive, prothrombotic, aggregating, or vasodilatory effects in the kidney of the rat experimental groups. The Δ-PP was significantly increased in MS rats in comparison with C and HSL rats. Panel (**B**) shows the activity of the COX isoforms in the presence of 100 μM of AA alone or with the use of the specific inhibitors, in the microsomal fraction of the kidney homogenate. The use of inhibitors in this fraction allows for discriminating between the activities of the two COX isoforms involved in the metabolism of AA. Values are expressed as mean ± SE (n = 6). Abbreviations: C = control, HSL = *Hibiscus sabdariffa* L., MS = metabolic syndrome.

2.7. Enzymatic Expressions

Figure 2A,B show an increase in the expression of 5-LOX ($p = 0.01$) and CYP450 4A ($p = 0.03$) in the kidney homogenates from the MS group in comparison with the C group. The treatment with HSL at 2% showed a tendency to decrease the expression of 5-LOX ($p = 0.09$) in MS rats. There was a statistical change in CYP450 4A ($p = 0.03$). Figure 3A,B show an increase in the expressions of PLA_2 ($p = 0.03$) and COX2 ($p = 0.01$), but a decrease in COX1 ($p = 0.02$, Figure 3C) in the MS group in comparison with the C group. The treatment with HSL at 2% showed a tendency to decrease the PLA_2 and COX2 expressions ($p = 0.05$) in MS rats, but there was an increase in the COX1 expression ($p = 0.04$).

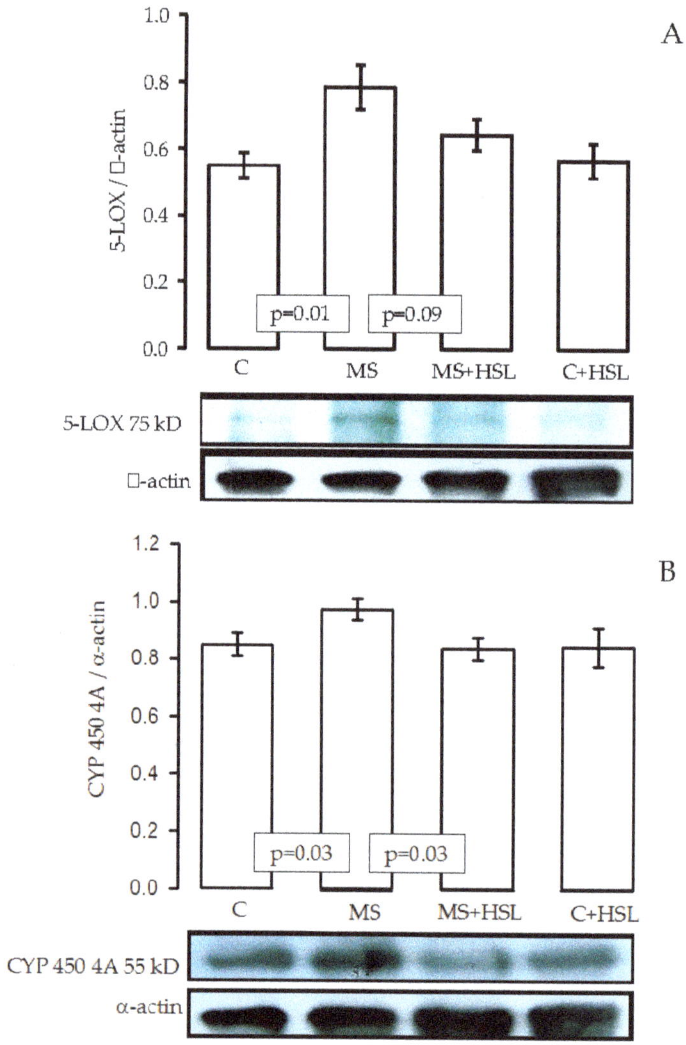

Figure 2. Panel (**A**,**B**) show the expressions of 5-LOX and CYP450 4A, respectively, in the kidney homogenates in the MS group in comparison with the C and HSL groups. Values are expressed as mean ± SE (n = 6). Abbreviations: C = control, HSL= *Hibiscus sabdariffa* L., MS = metabolic syndrome.

2.8. Histological Changes

Figure 4A–D show representative histological sections of a glomerulus in the C, MS, MS+HSL, and C+HSL groups, with the average area size of the glomerular tang expressed in arbitrary units (pixels). The densitometrical analysis showed a significant decrease in the size of the glomerulus in the MS rats when compared to the C and the MS+HSL groups ($p \leq 0.03$, panel E). The glomerular spaces and loops with their fine and delicate membrane are preserved in the C, MS+HSL, and C+HSL groups in comparison with those from MS rats, which show fibrosis, retracted glomerular tufts, and increased urinary spaces with detritus.

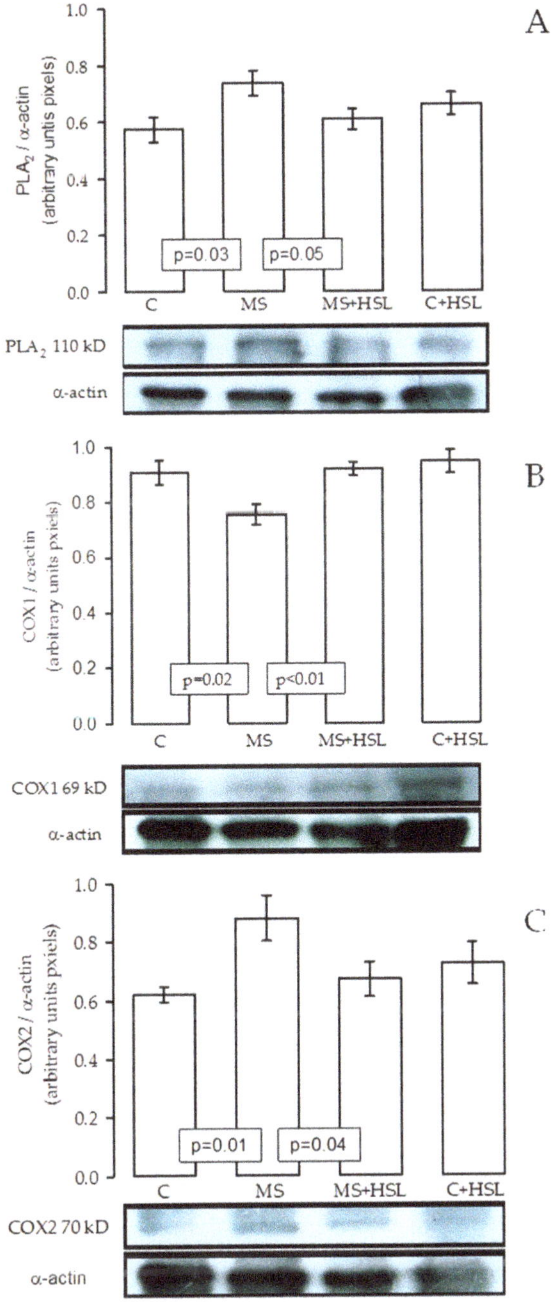

Figure 3. Panel (**A–C**) show the expressions of PLA_2, COX1, and COX2, respectively, in the kidney homogenates in the MS group in comparison with the C and HSL groups. Values are expressed as mean ± SE (n = 6). Abbreviations: C = control, HSL = *Hibiscus Sabdariffa* L., MS = metabolic syndrome.

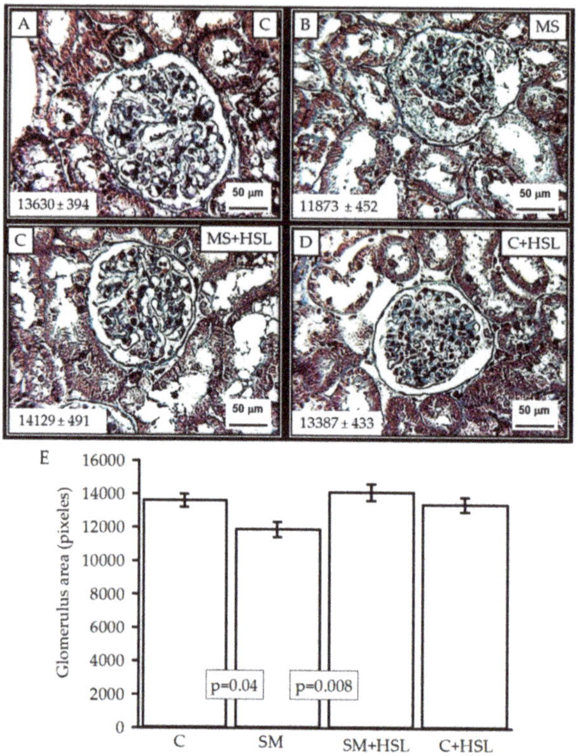

Figure 4. Representative photomicrographs of the rat renal cortex. Images show the glomerulus from Panel (**A**) C rats, Panel (**B**) MS rats, Panel (**C**) MS+HSL rats, and Panel (**D**) C+HSL rats. No abnormalities were observed under light microscopy in C, MS+HSL, and C+HSL groups, where glomerular spaces and loops with their fine and delicate membrane are preserved. However, in the glomerulus from the MS rats, there was fibrosis, retracted glomerular tufts, and increased urinary spaces with detritus. Masson's trichrome stain at 400×. Panel (**E**) shows the densitometrical analysis where a significant decrease in the size of the glomerulus in the MS rats is observed when compared to the C and the MS+HSL groups.

3. Discussion

The renal system shows different alterations in MS that result in proteinuria, infiltration of inflammatory cells, accumulation of lipids, sclerosis, and a decrease in the CCr. The production of eicosanoids through the metabolism of AA is important in the renal system, since it participates in the maintenance of the proper relaxation and contraction of the renal vasculature in the glomeruli, thin limb of the loop of Henle, connecting tubule, collecting ducts, and mesangial cells [17,18]. The traditional medicinal herb, HSL, has been used to treat different pathologies of the renal system such as urinary bladder and kidney stones, and as a diuretic [19]. Its beneficial effects have been attributed mainly to the antioxidants it contains and to its activity on the ACE. However, only a few investigations have addressed the effect of HSL on enzymes that synthesize eicosanoids including CYP450 isoforms or COX2 [20]. For this reason, the aim of this study consisted of studying how the treatment with HSL at 2% contributes to improve renal function through the pathways that metabolize AA in a rat model of MS.

AA, which is a substrate for the enzymes studied in this paper, has to be released from the PL of the cell membrane by PLA_2 before it is metabolized. This enzyme is Ca^{2+}-dependent. Our results show that the treatment with HSL resulted in a decrease in the expression of PLA_2, probably through the antioxidants it contains that are capable of modulating intracellular Ca^{2+} concentrations via the inhibition of calcium channels. As a result of the decreased expression of this enzyme, the percentage of AA in PL is increased. Another possible explanation for the increase in the AA percentage in MS rats treated with HSL is the fact that HSL contains linoleic and oleic acids [21], which are transition fatty acids that may be desaturated by the 5 and 6 desaturases, resulting in the formation of AA, which may then be reinserted in the cell membrane to increase fluidity [16]. Our MS rat model also courses with a decrease in the percentage of AA in total fatty acids and in those in the PL [17].

Our MS model in rats also has alterations in the activities or expressions of the COX isoforms and TXAS in the kidney [17]. In this sense, the treatments with HSL at 2% result in a decrease in the activity/expression of COX2 and an increase in the activity/expressions of COX1 in MS rats. This leads to a decrease in PGE_2 and an increase in PGE_1. These prostaglandins modulate the vascular tone in the afferent and efferent arterioles of the glomerulus, leading to an increase in CCr, and a decrease in albuminuria. Therefore, glomerular filtration is improved, as shown by our results. It has been previously reported that the polyphenols from HSL may reduce the expression of COX2 in LPS-treated macrophages.This enzyme also contributes to a pro-inflammatory state in hypertension. Our results confirmed these results and also showed that the treatment with HSL can favor the increase in the COX1 activity/expression. COX1 increases the synthesis of PGE1 (6-keto-$PGF_{1\alpha}$ metabolite stable of PGE1) with vasodilator, antithrombotic, and anti-inflammatory effects and contributes to the vasodilation of arterioles in the glomerulus that lowers systemic SBP [22].

In addition, there was a decreased expression of 5-LOX and CYP450 4A with the treatment with HSL at 2% in the homogenate of the kidney from MS rats. The eicosanoids produced by these enzymatic pathways such as LKB_4 are mediators of inflammation and contraction in the mesangial cells, the limb of the loop of Henle, and the collecting ducts. An increase in this prostaglandin can reduce the glomerular ultrafiltration coefficient [23]. Therefore, our results suggest that the HSL treatment may modulate the expression of these enzymes, which results in a decrease in its metabolites as previously reported [5]. In this sense, the ethanolic HSL extract causes inhibition of some CYP isoforms in vitroin human liver microsomes [24]. Another study showed that this effect may be through the protocatechuic acid and phenolic acids that are present in high concentrations in HSL [25]. Likewise, polyphenols such as quercetin, lignin, lignan, kaempferol, resveratrol, and catechins, among others, are capable of acting on the same metabolic pathways [22], and have the capacity to inhibit other enzymes of the AA metabolism such as PLA_2, 5-LOX, TXAS, and COX2 [26].

Furthermore, our results suggest that TxB_2, which is a metabolite of AA produced by TXAS, was decreased with the HSL treatment. The participation of this metabolite alongside with LKB_4 and PGE_2 may induce the transcription of laminin, fibronectin, and IV collagen, in the glomerular basement membrane with a loss of the electrical charge, and therefore result in a pro-inflammatory state, albuminuria, and sclerosis in the kidney [27,28].

The expressions of COX2 and COX1 that were modified in the kidney homogenate were reflected in the activities of the COX isoforms when using specific inhibitors for both COX2 and COX1 (NS398 and SC560, respectively). In the presence of AA, the microsomes of the kidneys from rats with MS showed that the oxygen consumption was increased, but the oxygen consumption decreased (the molecular O_2 is incorporated to AA to synthetize prostaglandins) in the presence of NS398. This suggests and supports our results of the overexpression of COX2, which participates in a larger proportion than COX1. This enzyme was not inhibited in a similar manner to COX2. However, the treatment with HSL in rats with MS did not reflect this change. This result suggests that the activities of these enzymes can be modulated using the components contained in the HSL infusion such as polyphenols, flavonoids, anthocyanins, and vitamin C, among others, resulting mainly in a decrease in COX2 activity and favoring COX1 activity.

The increase in the expression and activity of the different enzymes present in the kidney may be evaluated on the dynamics of renal function in the isolated and perfused kidney. This system evaluates renal vascular resistance in response to different drugs and metabolites resulting in changes in the Δ-PP that are reflected in glomerular filtration [28]. The results obtained here show that there is a larger vascular resistance in the kidney from the MS rats associated with the alteration in the enzymatic pathway that metabolizes AA. These changes lead to damage in the renal function. This isaccompanied by a loss of glomerular filtration rate and an increased albuminuria [5,16,17]. This was evidenced with the inhibition of the percentage of the different metabolic pathways (COX2 (52.2%), COX1 (54.4%), CYP450 (52.9%), and Indo (48.6%), with an unspecific inhibitor). However, the HSL treatment showed no significant changes in the inhibition percentage in the kidney of the rats from the MS model, but decreased the vascular resistance (reduced Δ-PP) in them. This suggests that the antioxidants present in the HSL infusion modulate the activities of the enzymes that metabolize the AA, and as a consequence, they modulate the production of the vasoconstrictor and vasodilatation eicosanoids. As an example, baicalein is a flavonoid that inhibits CYP450 [29]. Both the decrease in the vasoconstrictors and the increase in vasodilators resulting from these pathways act through the endothelium of the renal arterioles and modify the vascular tone, favoring vasodilation. Therefore, they improve the glomerular filtration rate. Our results show that the prostaglandins that are decreased using the HSL treatment and that can have an effect are PGE_2, LKB_4, and TxB_2; however, PGE_1 was increased.

The alterations of the dynamics of renal function are also accompanied by structural and anatomical modifications, which are evidenced with the histological changes in the glomeruli from MS rats, as previously reported [5,16,17]. However, treatment with HSL in MS rats contributed to the reduction in these modifications.

Last but not least, the MS rat model courses with hypertriglyceridemia, intra-abdominal fat, IR, hyperinsulinemia, and hypertension as previously reported [5,16,17]. However, the treatment with the HSL infusion decreased intra-abdominal fat, hypertriglyceridemia, hyperinsulinemia, and IR in the MS rats. These changes can be attributed to the flavonoids and polyphenols, present in the HSL infusion. These components can block lipid accumulation in the adipose tissue and the liver [30]. They may also increase the β-oxidation of fatty acids by inhibiting fatty acid synthase, cholesterol acyltransferase, 3-hydroxy-3-methylglutaryl coenzyme A reductase, and acyl-coenzyme A. These changes result in a decrease in insulin sensitivity and reduce the triglyceride levels and intra-abdominal fat [31]. This can contribute to decrease the structural and physiological abnormalities in the kidney, such as the increased tubulointerstitial damage, fibrosis, and mesangial proliferation [32,33].

Figure 5 shows the possible mechanisms by which HSL infusion at 2% may down-regulate the metabolism of AA in the kidney of the rats with MS.

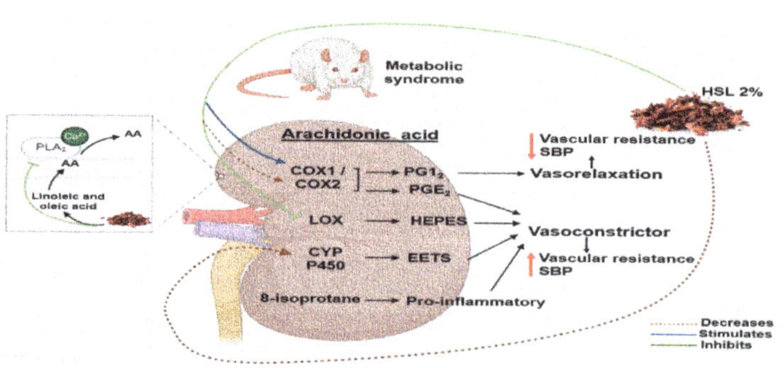

Figure 5. Summary of the possible mechanisms by which the HSL infusion at 2% acts on the enzymatic pathways that employ the AA in the kidney in an MS rat model. Uparrow = increase, down arrow = decrease.

4. Materials and Methods

The study was designed and carried out in compliance with the Laboratory Animal Care Committee of the National Institute of Cardiology Ignacio Chávez in México (INC/CICULA/011/2019) for the experiments in animals. Experiments were conducted in compliance with the Guide for the Care and use of laboratory animals of the National Institutes of Health (NIH). A total of 24 male Wistar rats were used to form 4 groups with 6 animals each for 20 weeks, and each group was treated as follows: Group 1: control (C, tap water); Group 2: MS with 30% sucrose in their drinking water, Group 3: MS with 30% sucrose plus HSL infusion (MS+HSL) at the concentration of 20 g/L (2%), and Group 4: C plus HSL at 2%. The animals were housed in ad hoc plastic boxes (Nalgene, New York, NY, USA). The animals were kept for 4 weeks under the following conditions: 12 h light/12 h dark cycle, environment temperature, and relative humidity between the ranges of 18–26 °C and 40–70%, respectively. Commercial rodent feed that contained 6% crude fiber, 4.5% crude fat, 23% crude protein, 2.5% minerals, and 8% ash (Labdiet 5008; PMI Nutrition International, Richmond, IN, USA) was provided freely. The HSL infusion at different percentages was administered orally (ad libitum) in drinking water.

4.1. Systolic Blood Pressure (SBP)

The SBP was determined at the end of the experimental period in the rats and it was measured using a tail cuff attached to a pneumatic pulse transducer method (Narco Bio-Systems Inc., Houston, TX, USA) [17]. Briefly, the rats were immobilized in an acrylic device that allowed them to breathe freely, and were rested on a plate that kept them warm. This device left the tail free to apply the tailcuff. The animals were left to relax for 15–20 min and five measurements of SBP were carried out on each rat.

4.2. Preparation of the HSL Infusion

The HSL calyces were acquired in Chilapa de Alvarez (Guerrero, México). In total, 20 g of the HSL calyces was added to a liter of boiling (90–100 °C) drinking water for 10 min and then left to cool; subsequently, 300 g of sucrose was added. The solution was filtered and stored at 4 °C for consumption by the rats of Group 3. The HSL infusion at 2% for Group 4 had no sucrose added [16,17,34].

4.3. Determinations of Polyphenols, Total Flavonoids, Total Anthocyanins, and Vitamin C in HSL Infusion

The determination of total flavonoids was conducted with the Jia method [35]. The calibration curve was made using quercetin as a standard and the absorbance was read at 510 nm. The polyphenol determination was made with aFolin–Ciocalteu reagent [36]. Determination of total anthocyanins was performed with Lee's method. The absorbance was measured at 520 nm and 700 nm and compared with a blank cell containing distilled water. The absorbance difference was performed to calculate cyaniding-3-glucoside [37]. Estimation of vitamin Cwas through the Folin–Ciocalteu reagent at 0.20 mM. The absorbance was measured at 760 nm [38].

4.4. Biochemical Variables

Commercial kits were used for the determination of some serum biochemical variables in the rats. Glucose concentration was determined with an enzymatic SERA-PAKR Plus kit (Bayer Corporation, S'ees, France). Total cholesterol (TC) and triglycerides (TG) determinations were made using commercial enzymatic kits (RANDOX Laboratories Ltd., Crumlin, County Antrim, UK). Insulin was determined using commercial radioimmunoassay kits (RIA) (Linco Research Inc., Saint Charles, MO, USA). The HOMA index of IR was calculated. HOMA−IR = insulin μU/mL × glucose mM/L/22.5.

4.5. Albuminuria and Creatinine Depuration

Albuminuria was measured using the bromocresol green reagent [39]. Serum creatinine (UCr and SCr, respectively) was measured with the Jaffe method [40] and glomerular filtration was calculated according to the following formula: clearance creatinine (CCr) = [urine creatinine]/[serum creatinine] × urinary volume (24 h)/time (1440 min).

4.6. Isolated Perfused Kidney

The rats were anesthetized with an intraperitoneal injection of sodium pentobarbital (63 mg/kg body weight). The right kidney was exposed using midline laparotomy, and the mesenteric and right renal arteries were cleared of surrounding tissue. The right renal artery was cannulated through the mesenteric artery to avoid interruption of blood flow, and the kidney was removed, suspended, and perfused at constant flow by means of a peristaltic pump (MasterFlex Easy-load II, no. 77200-50; Cole-Parmer Instrument Co, Vemon Hills, IL, USA) with a Krebs solution (in millimoles per liter)—118 of NaCl, 1.2 of NaH_2PO_4, 25 of $NaHCO_3$, 4.7 of KCl, 1.2 of $CaCl_2$, 4.2 of $MgSO_4$, and 5.5 of glucose (pH 7.4)—with a constant flow at 37 °C and oxygenated with 95% O_2/5% CO_2. The basal perfusion pressure (PP) was adjusted to 80–90 mmHg. The mean flow rate of the perfusing solution was 8 to 9 mL/min. The PP was measured with a transducer (Grass Telefactor, Grass Technologies, Astro Med, West Warwick, RI), captured, and recorded by means of a Grass model polygraph 79D and an online program (Grass PolyView). Data are expressed as changes (Δ) of PP in millimeters of mercury (mmHg). After at least 15 min of perfusion and once a stable PP had been obtained, vasoconstrictor responses to 4 µg/(mL min) of AA (Sigma Aldrich Darmstadt, Alemania) were determined in the absence and presence of 10 µmol/L of indomethacin (Indo, unspecific inhibitor of the COX pathway), 10 µmol/L of NS398 (COX2 selective inhibitor), 10 µmol/L of SC560 (COX1 selective inhibitor), or 100 µmol/L of baicalein (pathway inhibitor of CYP450). These doses were selected from published data as they seemed the most convenient after dose–response curves were obtained. Changes in the PP produced by the AA at 4 µg and the inhibitors were calculated by taking the mean of the pulsatile traces before administration and the mean of the traces at the maximal PP value after administration. Data are expressed as delta changes (Δ) of PP in mmHg. After each perfusion bolus, the kidneys were washed for a period of 20 min with a Krebs solution, to allow the organ to return to the basal PP (80–90 mmHg), and no sign of tachyphylaxis was present [16,17,41].

4.7. Kidney Homogenate

The left kidney was dissected and washed with a 0.9% saline solution. The capsule was removed, and cut in half. A half of the kidney was homogenized in a sucrose buffer (25 mM of sucrose, 10 mM of Tris, 1 mM of EDTA, and pH 7.35) with protease inhibitors (1 mM of PMSF, 2 µM of pepstatin, 2 µM of leupeptin, and 0.1% aprotinine). The homogenate was kept in ice. The kidney homogenate was centrifuged at 900× *g* for 10 min at 4 °C. The supernatant was separated and stored at −30 °C until required. Total proteins were determined with the Bradford method [42].

4.8. Microsomes

In total, 500µg of the protein of the kidney homogenate was used and suspended in 500 µL of 0.1 mol/L and centrifuged at 8000× *g* for 10 min at 4 °C. The pellet was discarded, and the supernatant was centrifuged at 44,000× *g* for 1 h at 4 °C; the resultant pellet was suspended in 250 µL of a 0.1 mol/L phosphate buffer at pH 7.35. The microsomes were used to determine the activities of the COX isoforms [17].

4.9. Cyclooxygenase Activity Assays

COXs' activities were studied by monitoring the rate of O_2 uptake using an oxymeter (YSI oxymeter model 5300A-1), which was coupled to a Clark-type electrode. To initiate the reaction, 3 mL of a 0.1 M Tris-HCl buffer, 1 mM of phenol, 85 µg of bovine hemoglobin (pH 8), and 100 μM of AA at 37 °C were added to 50 μg of kidney microsomes. Inhibition and discrimination of the catalytic activity of COX2 and 1 was performed with the addition of 10 μM of NS398, and SC560, respectively. The calibration curve was obtained with human COX2 provided by Sigma-Aldrich. A unit of cyclooxygenase activity is defined as the ability of the enzyme to catalyze oxygenation of 1 nmol of AA per minute at 37 °C.

4.10. Prostaglandins

Prostaglandins were extracted from the kidney homogenate by adding 0.5 mL of water–ethanol (1:4 vol/vol) and 10 μL of glacial acetic acid to 25 μg of protein. The mixture was well shaken and incubated at room temperature for 5 min and centrifuged at 2500× *g* for 5 min. The supernatant was applied to a Sep-Pak C18 mini-column (Millipore, Billerica, MA, USA) previously equilibrated with 2 mL of 10% ethanol. The column was then washed with 1 mL of water followed by 1 mL of hexane, and prostaglandins were eluted with 1.50 mL of ethyl acetate. The samples were dried under a nitrogen stream and re-suspended in 300 μL of phosphate-buffered saline–ethanol (2:1 *vol/vol*) [14]. The prostaglandins were determined with ELISA kits; prostaglandin E (Item No. 514531), 6-keto prostaglandin $F_{1\alpha}$ (Item No. 515211), leukotriene B_4 (Item No. 520111), and thromboxane B_2 (Item No. 501020) were obtained from Cayman Chemical (Ann Arbor, MI, USA), and were read in a visible light micro plate reader at 492/630 nm.

4.11. Extraction and Derivatization of the AA of the Phospholipids

For the extraction and derivatization of the AA of the phospholipids, 50 μg of L-α-phosphatidylcholine-di-heptadecanoyl acid as an internal standard and 1 mL of acetone were used. The mixture was shaken vigorously in a vortex for 30 s and centrifuged at 1145× *g* at room temperature for 4 min. The supernatant was removed, and the button was suspended with chloroform–methanol (2:1, *vol/vol*) with 0.002% BHT, according to the method described by Folch [43]. The AA was trans-esterified to its AA methyl esters by heating them at 90 °C for 2 h with 2 mL of methanol plus 0.002% BHT, 40 μL of H_2SO_4, and 100 μL of toluene. They were separated and identified with gas chromatography-FID in a Carlo Erba Fratovap 2300 chromatograph equipped with a capillary column packed with the stationary phase, HP-FFAP (description: 30 m length × 0.320 mm diameter × 0.25 μm film), and fitted with a flame ionization detector at 210 °C, with helium as the carrier gas at a flow rate of 1.2 mL/min. The areas under the peaks were calculated with a Shimadzu C-R6A Chromatopac integrator coupled to the gas chromatograph.

4.12. Western Blotting for COXs'Isoforms, CYP4A, 5-LOX, and PLA_2

An amount of 25 μg of protein of the kidney homogenate was run on 12% SDS-PAGE, blotted onto a polyvinylidene difluoride membrane (0.22 μm, Millipore, Billerica, MA, USA), and then blocked for 1 h at room temperature with Tris-buffer-solution–0.01%-Tween (TBS-T 0.01%) and 5% non-fat milk. The membranes were incubated overnight at 4 °C with mouse primary monoclonal antibodies: a COX2 antibody (H-3: sc-376861), COX1 antibody (H-1: sc-166573), and CYP4A1/A2/A3 antibody (clo4: sc-53247), providedfrom Santa Cruz Biotechnology, Santa Cruz, CA, USA;an Anti-5 Lipoxygenase/5-LO antibody (EP6072(2), ab169755) provided from abcam; and C-PLA2 (#SAB4502200) provided from Sigma Aldrich at a dilution of 1:5000, respectively. Then, the membranes were incubated overnight at 4 °C with a secondary antibody conjugated with horseradish peroxidase at a dilution of 1:10,000 (Santa Cruz Biotechnology, Santa Cruz, CA, USA). All of the blots were incubated with an α-actin antibody as a load control. The enzymes were detected with a chemiluminescence assay (Clarity Western ECL Substrate, Bio-Rad Laboratories, Inc., Hercules, CA, USA). Chemiluminescence that was emitted in this process was detected in X-ray films (AGFA, Ortho CP-GU, Agfa HealthCare NV, Mortsel-Belgium). Images from each film were acquired with a GS-800 densitometer (including Quantity One software from Bio-Rad Laboratories, Inc., Hercules, CA, USA). The values of the density of each band are expressed as arbitrary units (AU).

4.13. Histopathological Analysis

Tissue was processed for light microscopy according to standard techniques. A segment of the kidney was dissected, decapsulated, and washed in 0.9% NaCl for 30 s, fixed in a 10% formalin solution for 24 h, gradually dehydrated in ethanol, cleared in xylene, and embedded in paraffin. The kidney was cut into five-micrometer-thick slices with a microtome (Leica RM212RT, Wetzlar, Germany); the paraffin sections were stained with Masson's trichrome. Histological sections were analyzed at 400× magnification using a model 63,300 light microscope (Carl Zeiss, Oberkochen, Germany), equipped with a Tucsen (9 megapixels) digital camera coupled to TSview 7.3.1 software. The glomerular area was analyzed with densitometry using Sigma Scan Pro 5 Image Analysis software (Systat Software Inc., San Jose, CA, USA). The density values are expressed as pixel units.

4.14. Statistical Analysis

The Sigma Plot program (SigmaPlot® version 14.5, Jandel Corporation) was used for the statistical analysis. The data are presented as the mean ± standard error. Statistical significance was determined with Tukey's one-way ANOVA and a post hoc test. A $p \leq 0.05$ was considered as significant.

5. Conclusions

The treatment with the HSL infusion at 2% improves the physiological renal function and anatomic structure of the kidneys in the rats with MS, through the antioxidant molecules that it possesses. These compounds inhibit pathways that metabolize the AA such as PLA_2, COX2, 5-LOX, TAXS, and CYP450, but favor the COX1 pathway. This improves the vascular resistance of the renal arterioles, which increases the glomerular filtration rate, through the synthesis of prostaglandins.

6. Study Limitations

This study is a basic investigation where an HSL infusion at 2% was provided to a rat model that shows several characteristics of metabolic syndrome. The model was generated with consumption of 30% sucrose. Although the results in this study are clear, they may not be predictive for humans. In this work, we did not quantify the consumption of food for rodents, but in previous works already reported by our group, there is a significant difference in this regard between MS and C rats [44]. We did not assess the physical activity of the experimental animals either. On the other hand, we previously evaluated the increase in the activity of some antioxidant enzymes in the plasma of patients with Marfan syndrome before and after consumption, for 2 months, of an infusion of HSL at 2% with very promising results. Therefore, the current dosage has been previously implemented for human use [34]. In addition, in this previous study, none of the patients showed displeasure for the infusion since it had excellent palatability. Furthermore, infusions of HSL at a similar dose are commonly consumed with food in different countries, including Mexico. Therefore, the results from the present study support that the 2% HLS infusion could be used as an adjuvant therapy in the pathologies that comprise MS.

Author Contributions: I.P.-T. and M.E.S.: Contributed to conceptualization, wrote the manuscript, and performed some experiments; L.M.-P.: Performed some experiments and created the graphical abstract and Figure 5; E.D.-D. and J.C.T.-N.: Biochemical variable performance; R.M.-M.: Software for araquidonic acid quantification; V.G.-L. and V.C.-T.: Contributed to conceptualization, and revised and edited the paper. All authors have read and agreed to the published version of the manuscript.

Funding: This study was supported by funds from authorized direct expenditure of the Basic Research subdirection of the Instituto Nacional de Cardiologia, Ignacio Chávez.

Institutional Review Board Statement: The study was designed and carried out in compliance with the Laboratory Animal Care Committee of the National Institute of Cardiology Ignacio Chávez in México (INC/CICULA/011/2019) for the experiments in animals. Experiments were conducted in compliance with the Guide for the Care and use of laboratory animals of the National Institutes of Health (NIH).

Informed Consent Statement: Not applicable.

Data Availability Statement: The data in our study are available from the corresponding authors upon reasonable request.

Acknowledgments: We thank Benito Chávez Rentería for histology technical support.

Conflicts of Interest: The authors declare no conflict of interest.

References

1. Rubio-Ruiz, M.E.; Hafidi, M.E.; Pérez-Torres, I.; Baños, G.; Guarner, V. Medicinal agents and metabolic syndrome. *Curr. Med. Chem.* **2013**, *20*, 2626–2640. [CrossRef] [PubMed]
2. Imig, J.D. Eicosanoids and renal vascular function in diseases. *Clin. Sci.* **2006**, *111*, 21–34. [CrossRef] [PubMed]
3. Hao, C.M.; Breyer, M.D. Roles of lipid mediators in kidney injury. *Semin. Nephrol.* **2007**, *27*, 338–351. [CrossRef] [PubMed]
4. Chakrabarti, S.K.; Cole, B.K.; Wen, Y.; Keller, S.R.; Nadler, J.L. 12/15-lipoxygenase products induce inflammation and impair insulin signaling in 3T3-L1 adipocytes. *Obesity* **2009**, *17*, 1657–1663. [CrossRef] [PubMed]
5. Zúñiga-Muñoz, A.M.; Guarner-Lans, V.; Soria Castro, E.; Diaz-Díaz, E.; Torrico-Lavayen, R.; Tena-Betancourt, E.; Pérez-Torres, I. 17β Estradiol modulates perfusion pressure and expression of 5-LOX and CYP450 4A in the isolated kidney of metabolic syndrome female rats. *Int. J. Endocrinol.* **2015**, *2015*, 149408. [CrossRef]
6. Mesaros, C.; Lee, S.H.; Blair, I.A. Targeted quantitative analysis of eicosanoid lipids in biological samples using liquid chromatography-tandem mass spectrometry. *J. Chromatogr. B Analyt. Technol. Biomed. Life. Sci.* **2009**, *877*, 2736–2745. [CrossRef]
7. Hu, Y.; Li, T.T.; Zhou, W.; Lu, T.T.; Li, F.F.; Ding, B.; Liu, B.L.; Xie, X.J.; Ma, J.H. Lipoprotein-associated phospholipase A2 is a risk factor for diabetic kidney disease. *Diabetes Res. Clin. Pract.* **2019**, *150*, 194–201. [CrossRef]
8. Amin, A.; Lotfy, M.; Mahmoud-Ghoneim, D.; Adeghate, E.; Al-Akhras, M.A.; Al-Saadi, M.; Al-Rahmoun, A.; Hameed, R. Pancreas-protective effects of chlorella in STZ-induced diabetic animal model: Insights into the mechanism. *J. Diab.Mel.* **2011**, *1*, 36–45. [CrossRef]
9. Hamza, A.A.; Mohamed, G.M.; Mohamed, F.; Amin, L.A. Dandelion prevents liver fibrosis, inflammatory response, and oxidative stress in rats. *J. Basic App. Zool.* **2020**, *81*, 43. [CrossRef]
10. Bouabdallah, S.; Al-Maktoum, A.; Amin, A. Steroidal saponins: Naturally occurring compounds as inhibitors of the hallmarks of cancer. *Cancers* **2023**, *15*, 3900. [CrossRef]
11. Mohammad, N.S.; Nazli, R.; Zafar, H.; Fatima, S. Effects of lipid based Multiple Micronutrients Supplement on the birth outcome of underweight pre-eclamptic women: A randomized clinical trial. *Pak. J. Med. Sci.* **2022**, *38*, 219–226. [PubMed]
12. Riaz, G.; Chopra, R. A review on phytochemistry and therapeutic uses of *Hibiscussabdariffa* L. *Biomed. Pharmacother.* **2018**, *102*, 575–586. [CrossRef] [PubMed]
13. Herrera-Arellano, A.; Flores-Romero, S.; Chávez-Soto, M.A.; Tortoriello, J. Effectiveness and tolerability of a standardized extract from *Hibiscus sabdariffa* in patients with mild to moderate hypertension: A controlled and randomized clinical trial. *Phytomedicine* **2004**, *11*, 375–382. [CrossRef] [PubMed]
14. Kao, E.S.; Hsu, J.D.; Wang, C.J.; Yang, S.H.; Cheng, S.Y.; Lee, H.J. Polyphenols extracted from *Hibiscus sabdariffa* L. inhibited lipopolysaccharide-induced inflammation by improving antioxidative conditions and regulating cyclooxygenase-2 expression. *Biosci. Biotechnol. Biochem.* **2009**, *73*, 385–390. [CrossRef]
15. Hase, M.; Babazono, T.; Karibe, S.; Kinae, N.; Iwamoto, Y. Renoprotective effects of tea catechin in streptozotocin- induced diabetic rats. *Int. Urol. Nephrol.* **2006**, *38*, 693–699. [CrossRef]
16. Rodríguez-Fierros, F.L.; Guarner-Lans, V.; Soto, M.E.; Manzano-Pech, L.; Díaz-Díaz, E.; Soria-Castro, E.; Rubio-Ruiz, M.E.; Jiménez-Trejo, F.; Pérez-Torres, I. Modulation of renal function in a metabolic syndrome rat model by antioxidants in *Hibiscus sabdariffa* L. *Molecules* **2021**, *26*, 2074. [CrossRef]
17. Pérez-Torres, I.; El Hafidi, M.; Pavón, N.; Infante, O.; Avila-Casado, M.C.; Baños, G. Effect of gonadectomy on the metabolism of arachidonic acid in isolated kidney of a rat model of metabolic syndrome. *Metabolism* **2010**, *59*, 414–423. [CrossRef]
18. Currie, M.G.; Needleman, P. Renal arachidonic acid metabolism. *Annu. Rev. Physiol.* **1984**, *46*, 327–341. [CrossRef]
19. Zheoat, A.M.; Gray, A.I.; Igoli, J.O.; Ferro, V.A.; Drummond, R.M. Hibiscus acid from Hibiscus sabdariffa (Malvaceae) has a vasorelaxant effect on the rat aorta. *Fitoterapia* **2019**, *134*, 5–13. [CrossRef]
20. Johnson, S.S.; Oyelola, F.T.; Ari, T.; Juho, H. In vitro inhibitory activities of the extract of *Hibiscus sabdariffa* L. (family Malvaceae) on selected cytochrome P450 isoforms. *Afr. J. Tradit. Complement. Altern. Med.* **2013**, *10*, 533–540. [CrossRef]
21. Nzikou, J.M.; Bouanga-Kalou, G.; Matos, L.; Ganongo-Po, F.B.; Mboungou-Mboussi, P.S.; Moutoula, F.E.; Panyoo-Akdowa, E.; Silou, T.H.; Desobry, S. Characteristics and nutritional evaluation of seed oil from roselle (*Hibiscus sabdariffa* L.) in congo-Brazzaville. *Curr. Res. J. Sci.* **2011**, *3*, 141–146.

22. Zhen, J.; Villani, T.S.; Guo, Y.; Qi, Y.; Chin, K.; Pan, M.H.; Ho, C.T.; Simon, J.E.; Wu, Q. Phytochemistry, antioxidant capacity, total phenolic content and anti-inflammatory activity of Hibiscus sabdariffa leaves. *Food Chem.* **2016**, *190*, 673–680. [CrossRef] [PubMed]
23. Dunn, M. The role of arachidonic acid metabolites in renal homeostasis. Non-steroidal anti-inflammatory drugs renal function and biochemical, histological and clinical effects and drug interactions. *Drugs* **1987**, *33*, 56–66. [CrossRef] [PubMed]
24. Ochan, P.C.; D'Mello, P. Antioxidant and antihyperlipidemic activity of Hibiscus sabdariffa Linn. leaves and calyces extracts in rats. *Indian. J. Exp Biol.* **2009**, *47*, 276–282.
25. Krajka-Kuźniak, V.; Szaefer, H.; Baer-Dubowska, W. Modulation of cytochrome P450 and phaseIIenzymes by protocatechuicacid in mouse liver and kidney. *Toxicology* **2005**, *216*, 24–31. [CrossRef] [PubMed]
26. Costa, V.; Costa, M.; Videira, R.A.; Branquinho, P.; Paiva-Martins, A.F. Anti-Inflammatory activity of olive oil polyphenols-the role of oleacein and Its metabolites. *Biomedicines* **2022**, *10*, 2990. [CrossRef] [PubMed]
27. Nakano, Y.; Nakatani, Y.; Takami, M.; Taniyama, Y.; Arima, S. Diverse associations between oxidative stress and thromboxane A_2 in hypertensive glomerular injury. *Hypertens. Res.* **2019**, *42*, 450–458. [CrossRef]
28. Hong, J.; Smith, T.J.; Ho, C.T.; August, D.A.; Yang, C.S. Effects of purified green and black tea polyphenols on cyclooxygenase- and lipoxygenase-dependent metabolism of arachidonic acid in human colon mucosa and colon tumor tissues. *Biochem. Pharmacol.* **2001**, *62*, 1175–1183. [CrossRef]
29. Tublin, M.; Tessier, F.; Murphy, M. Correlation between renal vascular resistente, pulse pressure, and the resistive index in isolated perfusedrabbit kidneys. *Radiology* **1999**, *23*, 258–264. [CrossRef]
30. Li, C.; Kim, M.; Choi, H.; Choi, J. Effects of baicalein on the pharmacokinetics of tamoxifen and its main metabolite, 4-hydroxytamoxifen, in rats: Possible role of cytochrome P450 3A4 and P-glycoprotein inhibition by baicalein. *Arch. Pharm. Res.* **2011**, *34*, 1965–1972. [CrossRef]
31. Ojulari, O.V.; Gi, L.S.; Nam, J.O. Beneficial effects of natural bioactive compounds from *Hibiscus sabdariffa* L on obesity. *Molecules* **2019**, *24*, 210. [CrossRef] [PubMed]
32. Sasaki, R.; Nishimura, N.; Hoshino, H.; Isa, Y.; Kadowaki, M.; Ichi, T.; Tanaka, A.; Shin, N.; Fukuda, I.; Ashida, H.; et al. Cyanidin 3-glucoside ameliorates hyperglycemia and insulin sensitivity due to downregulation of retinol binding protein 4 expression in diabetic mice. *Biochem. Pharmacol.* **2007**, *74*, 1619–1627. [CrossRef] [PubMed]
33. Hashimoto, S.; Maoka, T.; Kawata, T.; Mochizuki, T.; Koike, T.; Shigematsu, T. Roles of insulin receptor substrates (IRS) in renal function and renal hemodynamics. *PLoS ONE* **2020**, *15*, e0242332. [CrossRef]
34. Soto, M.E.; Zuñiga-Muñoz, A.; Guarner-Lans, V.; Duran-Hernández, E.J.; Pérez-Torres, I. Infusion of *Hibiscus sabdariffa* L. Modulates oxidative stress in patients with Marfan syndrome. *Mediators. Inflamm.* **2016**, *2016*, 8625203. [CrossRef] [PubMed]
35. Jia, Z.; Tang, M.; Wu, J. The determination of flavonoid contents in mulberry and their scavenging effects on superoxide radicals. *Food. Chem.* **1999**, *64*, 555–559.
36. Sánchez-Rangel, J.C.; Benavides, J.; Heredia, J.B.; Cisneros-Zevallos, L.; Jacobo-Velázquez, D.A. The Folin-Ciocalteu assay revisited: Improvement of its specificity for total phenolic content determination. *Anal. Method* **2013**, *5*, 5990–5999. [CrossRef]
37. Lee, J.; Durst, R.; Wrolstad, R. Determination of total monomeric anthocyanin pigment content of fruit juices, beverages, natural colorants, and wines by the Ph differential method: Collaborative study. *J. AOAC Int.* **2005**, *88*, 1269–1278. [CrossRef]
38. Jagota, S.K.; Dani, H.M. A new colorimetric technique for the estimation of vitamin C using Folin phenol reagent. *Anal. Biochem.* **1982**, *127*, 178–182. [CrossRef]
39. BenGershôm, E. Screening for albuminuria: A case for estimation of albumin in urine. *Clin. Chem.* **1975**, *21*, 1795–1798. [CrossRef]
40. Jaffe, M. Ueber den niederschlag welchen pikrinsäure in normalen harn erzeugt und über eine neue reaction des kreatinins. *Z. Physiol. Chem.* **1886**, *10*, 391–400.
41. Pérez-Torres, I.; Ibarra, B.; Soria-Castro, E.; Torrico-Lavayen, R.; Pavón, N.; Diaz-Diaz, E.; Flores, P.L.; Infante, O.; Baños, G. Effect of glycine on the cyclooxygenase pathway of the kidney arachidonic acid metabolism in a rat model of metabolic syndrome. *Can. J. Physiol. Pharmacol.* **2011**, *89*, 899–910. [CrossRef] [PubMed]
42. Bradford, M.M. A rapid and sensitive method for the quantitation of microgram quantities of protein utilizing the principle of protein-dye binding. *Anal. Biochem.* **1976**, *72*, 248–254. [CrossRef] [PubMed]
43. Folch, J.; Lees, M.; Sloane-Stanley, G.H. A simple method for the isolation and purification of total lipids from animal tissues. *J. Biol. Chem.* **1957**, *226*, 497–509. [CrossRef] [PubMed]
44. Pérez-Torres, I.; Gutiérrez-Alvarez, Y.; Guarner-Lans, V.; Díaz-Díaz, E.; Manzano-Pech, L.; Caballero-Chacón, S.D.C. Intra-Abdominal fat adipocyte hypertrophy through a progressive alteration of lipolysis and lipogenesis in metabolic syndrome rats. *Nutrients* **2010**, *11*, 1529. [CrossRef] [PubMed]

International Journal of
Molecular Sciences

MDPI

Article

Impact of Treatment with Antioxidants as an Adjuvant to Standard Therapy in Patients with Septic Shock: Analysis of the Correlation between Cytokine Storm and Oxidative Stress and Therapeutic Effects

Israel Pérez-Torres [1,†], Alfredo Aisa-Álvarez [2,†], Sergio Casarez-Alvarado [3], Gabriela Borrayo [4], Ricardo Márquez-Velasco [3], Verónica Guarner-Lans [5], Linaloe Manzano-Pech [1], Randall Cruz-Soto [3], Omar Gonzalez-Marcos [2], Giovanny Fuentevilla-Álvarez [5], Ricardo Gamboa [5], Huitizilihuitl Saucedo-Orozco [6], Juvenal Franco-Granillo [2] and María Elena Soto [3,7,8,*]

[1] Cardiovascular Biomedicine Department, Instituto Nacional de Cardiología Ignacio Chávez, Juan Badiano No. 1, Col. Sección XVI, Mexico City 14380, Mexico; pertorisr@yahoo.com.mx (I.P.-T.); loe_mana@hotmail.com (L.M.-P.)
[2] Critical Care Department, American British Cowdray (ABC) Medical Center, PAI ABC Sur 136 No. 116, Col. las Américas, Mexico City 01120, Mexico; alfredoaisaa@gmail.com (A.A.-Á.); ogmarcos35@gmail.com (O.G.-M.); jfranco@abchospital.com (J.F.-G.)
[3] Immunology Department, Instituto Nacional de Cardiología Ignacio Chávez, Juan Badiano No. 1, Col. Sección XVI, Mexico City 14380, Mexico; secazalv@gmail.com (S.C.-A.); marquezric@hotmail.com (R.M.-V.); randallcruz44@hotmail.com (R.C.-S.)
[4] Instituto Mexicano del Seguro Social, Dirección de Prestaciones Médicas Coordinación de Innovación en Salud, Ciudad de México 06700, Mexico; gabriela.borrayo@imss.gob.mx
[5] Physiology Department, Instituto Nacional de Cardiología Ignacio Chávez, Juan Badiano No. 1, Col. Sección XVI, Mexico City 14380, Mexico; gualanv@yahoo.com (V.G.-L.); fuentevilla_alvarez@hotmail.com (G.F.-Á.); rgamboaa_2000@yahoo.com (R.G.)
[6] Cardioneumology Department, Specialty Hospital, National Medical Center "La Raza", Mexico City 02990, Mexico; huitzilihuitls@outlook.com
[7] Research Direction Instituto Nacional de Cardiología Ignacio Chávez, Juan Badiano No. 1, Col. Sección XVI, Mexico City 14380, Mexico
[8] Cardiovascular Line in American British Cowdray (ABC) Medical Center, PAI ABC Sur 136 No. 116, Col. Las Américas, Mexico City 01120, Mexico
* Correspondence: elena.soto@cardiologia.org.mx; Tel.: +52-553888-0897
† These authors contributed equally to this work.

Abstract: Cellular homeostasis is lost or becomes dysfunctional during septic shock due to the activation of the inflammatory response and the deregulation of oxidative stress. Antioxidant therapy administered alongside standard treatment could restore this lost homeostasis. We included 131 patients with septic shock who were treated with standard treatment and vitamin C (Vit C), vitamin E (Vit E), N-acetylcysteine (NAC), or melatonin (MT), in a randomized trial. Organ damage quantified by Sequential Organ Failure Assessment (SOFA) score, and we determined levels of Interleukins (IL) IL1β, Tumor necrosis factor alpha (TNFα), IL-6, monocyte chemoattractant protein-1 (MCP-1), Transforming growth factor B (TGFβ), IL-4, IL-10, IL-12, and Interferon-γ (IFNγ). The SOFA score decreased in patients treated with Vit C, NAC, and MT. Patients treated with MT had statistically significantly reduced of IL-6, IL-8, MCP-1, and IL-10 levels. Lipid peroxidation, Nitrates and nitrites (NO_3^- and NO_2^-), glutathione reductase, and superoxide dismutase decreased after treatment with Vit C, Vit E, NAC, and MT. The levels of thiols recovered with the use of Vit E, and all patients treated with antioxidants maintained their selenium levels, in contrast with controls ($p = 0.04$). The findings regarding oxidative stress markers and cytokines after treatment with antioxidants allow us to consider to future the combined use of antioxidants in a randomized clinical trial with a larger sample to demonstrate the reproducibility of these beneficial effects.

Keywords: antioxidants; septic shock; melatonin; cytokines

1. Introduction

Sepsis and septic shock are global health problems and their prevalence generates a high economic cost; 51% of patients are admitted to intensive care units (ICUs), and 17.3% are treated in intermediate care (IC) or coronary units (UCCs) [1]. Approximately 80% of patients have a risk of developing multiple organ failure (MOF) [2].

In patients with sepsis and septic shock, dysfunction and loss of cellular and metabolic homeostasis are conditioned by the patient's response to the infection and are associated with a high mortality [3–5]. Since viruses, bacteria, and fungi cause infections leading to these conditions, the immune response must be rapid and coordinated; however, when the immune response is dysregulated or excessively triggered, a state of hyperinflammation occurs [6–8]. Several damage mechanisms interact, and it is vital to control the inflammatory response and the pro-oxidant state during treatment [9], because their deregulation leads to endothelial dysfunction, capillary permeability alterations, and variations in the coagulation system [10].

Immunoregulation by cytokines is vital; interferons mediate natural immunity, activate mononuclear phagocytes and neutrophils, and increase the molecules of the major histocompatibility complex. During septic shock, the participation of cytokines, such as IL-2 and IL-4, regulate the activation, proliferation, and differentiation of leukocytes [11–13]. Chemokines promote the adhesion of monocytes and neutrophils [14], and interferon-gamma (IFN γ) acts as a potent activator of mononuclear phagocytes and neutrophils to increase the expression of MHC class I and class II molecules [15]. TNF-α activates mononuclear neutrophils, and eosinophils, it acts on hepatocytes to induce the synthesis of acute-phase proteins and constitutes a central mediator of responses of the host against Gram-negative bacteria [16].

Similarly, IL-6 mediates natural immunity since it induces the synthesis of acute-phase proteins in the liver and is a proliferation factor for activated B cells. Its levels increase in patients with infection and are associated with mortality in patients with sepsis [17–19]. Likewise, elevated IL-6, IL-8, and TNF-α concentration are associated with septic shock and multiple organ failure [20–25]. The measurement of IL-8 identifies serious infections in patients with neutropenia, disseminated intravascular coagulation (DIC), lactic acidosis, severe hypoxemia, and high risk of mortality [26,27].

The endothelium actively participates in the pathophysiology and damage associated with sepsis, leading to systemic inflammation [28]. Interleukin 1 (IL-1) acts on the vascular endothelium and mononuclear phagocytes, activating endothelial functions, and it can lead to loss of anticoagulant properties and vaso permeability [20]. In perfusion models, elevated levels of Il-2 are associated with capillary leak syndrome where it activates neutrophils adhered to endothelial cells leading to the production of oxygen radicals [20]. Similarly, intercellular adhesion molecule 1 (ICAM-1) promotes diapedesis and participates in endothelial damage. High levels are present in patients with septic shock [28].

The deregulation of the host response and multiorgan failure due to infection is attributed to the interaction of various mechanisms, where oxidative stress (OS) participates. Mitochondrial dysfunction plays a vital role in the pathogenesis of septic shock [29]. Modulation of mitochondrial function could be a possible therapeutic strategy in the management of sepsis in the future because mitochondrial dysfunction leads to MOF in septic patients [30].

Some cytokines regulate inflammation, such as interleukin 10 (IL-10), which inhibits the activity and production of TNF-α, IL-1, IL-2, and chemokines [20,21]. A change in the ratio of IL-6 to IL-10 levels might be a marker to evaluate the systemic inflammatory response [31].

Immune cells express pattern recognition receptors (PRRs) that rapidly initiate host defense responses upon detecting tissue damage or microbial infection. Intracellular damage-associated molecular proteins (DAMPs) mediate immune recognition of damaged tissue. Toll-like receptors (TLRs), a subfamily of PRRs, have emerged as crucial receptors for DAMP recognition and initiation of the inflammatory response. During sepsis, immune response activation occurs due to the release of high levels of DAMPs originating from invading microorganisms and damaged host tissue, leading to overstimulation of immune cells. This unbalanced response is known as a cytokine storm and transforms its function from fighting excess infection into damage due to inflammation [32].

Furthermore, the stimuli produced by the infection lead to cell apoptosis through the intracellular or extracellular pathway (intrinsic or extrinsic pathway). Intrinsically, proapoptotic members of the Bcl family regulate the exit of cytochrome c from mitochondria, such as Bid and Bax. Other Bcl members (Bcl-XL and Bcl-2) are anti-apoptotic and are critical in signaling pathways leading to apoptosis or cell survival. Endoplasmic reticulum stress is another mechanism of apoptosis triggered by intracellular stimuli and mediated by caspase 12. In the extrinsic way, activation occurs due to binding a ligand, a polypeptide of the TNF family, with its receptor. This leads to activating initiator caspases 2, 8, and 10. Once activated, the caspases are cleaved and activate other effector caspases (3, 6, and 7), which enhance the cleavage of cytoskeletal and nuclear proteins, leading to apoptosis cells. The immune system's responsiveness is impaired during sepsis, and increased apoptosis of immune cells may explain the development of immune dysfunction. Therefore, it is necessary to eliminate a more significant number of apoptotic cells. Despite critical efforts in septic shock research, no therapy significantly modifies its outcome, and new treatments have led to several experimental studies focusing on the apoptotic process [33].

Cytokines [34] and reactive oxygen species (ROS) [35] play a catastrophic role in sepsis. Cell respiration, protein folding, or by-products of metabolism mainly produce some reactive oxygen species (ROS); others are mainly generated by NADPH oxidase [36]. Also, leukocytes are attracted to the affected sites after infection and release cytokines and ROS [37].

Recently, antagonistic therapies anticytokine therapy targeting (tumor necrosis factor alpha [TNF-alpha], interleukin-1 [IL-1]) and anti-endotoxin strategies have been proposed as important therapeutic targets in sepsis. They use antibodies against endotoxins or endotoxin receptor/carrier molecules (anti-CD14 or anti-LPS binding proteins) [38]. These therapies are based on the presence of an exponential increase in many cytokines in inflammatory conditions where a cytokine storm is present. Recent studies demonstrate the effectiveness and usefulness of an early start of comprehensive therapeutic management in sepsis [39].

On the other hand, many experimental studies of in vitro models of sepsis using mouse hepatocytes AML12 have been published. These cells were treated with lipopolysaccharide (LPS) to induce sepsis by hepatocyte injury. In cells pretreated with melatonin, the effects on oxidative stress, inflammation, mitophagy, mitochondrial biogenesis, and adenosine triphosphate (ATP) levels were reduced, and the mitochondrial quality was improved. This finding demonstrated that the antioxidant melatonin had potential benefic therapeutic effects in sepsis induced in the liver by injury, promoting mitophagy and stimulating the biogenesis of mitochondrial activity. These findings justify research to explore the precise effects, the underlying mechanisms, and the effectiveness of melatonin in the clinical setting [40].

In animal models of sepsis, there were significant anti-inflammatory, antioxidant, and anti-apoptotic effects with the use of Sulforaphane, which inhibits the TLR4/NF-κB signaling pathway. There was a reduction in the cardiac damage caused by sepsis [41]. This finding supports the idea that the use of antioxidant therapy together with standard therapy in sepsis might be beneficial.

During septic shock, the cytokine storm is caused by the invasion of pathogens and a sustained high level of cytokines, which alter homeostasis and cause potentially fatal organ dysfunction [42,43].

Hyperinflammation is a prooxidant state with increased expression of reactive oxygen species (ROS). [Superoxide (O_2^-), Hydroxyl radical (OH), Hydroperoxyl radical (OOH), Peroxyl radical (ROO)], RNS [Nitric oxide (NO), Nitrogen dioxide (NO_2) radical] are present in the circulating immune cells and in the affected organs, [44,45]. Elevated levels of intracellular superoxide are produced by NADPH oxidase, cyclooxygenase-2 (COX-2), and xanthine oxidase in the mitochondria during sepsis. SOD neutralizes superoxides under physiological conditions, but not during sepsis. In mouse models, the inhibition of NOX-2 prevents organ damage induced by sepsis. The inhibition of COX-2 also leads to a decrease in peroxynitrite in the experimental model of sepsis [46–48]. These findings suggest the importance of limiting excessive ROS generation with antioxidants rather than their effect on hyperinflammation to prevent sepsis. The clinical value of this strategy could be better if tested in the clinical environment.

Clinical trials to control only inflammation in septic shock have not had overwhelming success for many years. However, new immunomodulatory therapies and cytokine blockers have shown great success in severe cases of COVID-19 [49]. Standard management in intensive care units solves many fundamental conditions associated with septic shock, such as the cardiometabolic conditions and hemodynamic imbalances. Many studies in animal and experimental models use antioxidant therapy in septic shock; however, few studies demonstrate that the use of antioxidant therapy added to standard management in human patients with septic shock reduces organ dysfunction, improves clinical conditions, and reduces oxidative stress [50]. For this reason, our objective was to analyze whether adding antioxidant therapy in a randomized fashion to the standard treatment applied to patients with septic shock helps control cytokine levels and oxidant stress markers, and improves organ dysfunction.

2. Results

2.1. Population Studied

A total of 131 patients were included in this study, of which 61 (47%) were male and 70 (53%) female. The median age was 68 (58–78) years. Table 1 shows the demographic characteristics of the patients according to their assigned antioxidant treatment.

Table 1. Demographic characteristics of patients in each group of treatment.

	Vit C *n* = 25	Vit E *n* = 27	NAC *n* = 24	MT. *n* = 26	Control *n* = 29	*p*
Age	62 (58–78)	70 (51–77)	68.5 (58.5–78)	62.5 (58–69)	75 (65–81)	0.10
Weight, kg	70 (65–80)	71 (60–82)	69.5 (56.5–80)	66 (60–78)	70 (61–80)	0.80
Height	1.7 (1.5–1.7)	1.7 (1.6–1.7)	1.7 (1.6–1.8)	1.7 (1.6–1.71)	1.7 (1.6–1.7)	0.39
Body mass index	24.9 (23–30.1)	24.6 (22.8–29)	23 (20.7–6.1)	25.35 (21–28)	25 (23.4–28.6)	0.41
Male/Female	10 (40)/15 (60)	17 (62.96)/10 (37.04)	14 (58.33)/10 (41.67)	13 (50)/13 (50)	16 (55.17)/13 (44.83)	0.53 §
SAPSII	39.44 ± 14.10	45.70 ± 16.37	42.41 ± 19.84	40.88 ± 16.80	47.20 ± 7.11	0.60 ¥
APACHEII	14 (12–19)	20 (15–24)	15.5 (11–20.5)	16.5 (10–21)	17 (15–25)	0.15
SOFA	8 (6–9)	9 (7–11)	8 (4–10)	8 (6–9)	9 (7–11)	0.42
NUTRIC	4.16 ± 2.21	4.81 ± 1.64	4.08 ± 1.83	3.88 ± 1.79	5.10 ± 1.54	0.41 ¥
Diabetes mellitus	7 (28.00)	5 (18.52)	5 (20.83)	6 (23.08)	8 (27.59)	0.90 *
Arterial hypertension	10 (40.00)	11 (40.74)	12 (50.00)	8 (30.77)	15 (51.72)	0.53 §
COPD	1 (4.00)	5 (18.52)	4 (16.67)	2 (7.69)	0 (0.00)	0.05 *
Smoking	17 (68.00)	12 (44.44)	9 (37.50)	15 (57.69)	14 (48.28)	0.22 §
Cancer	6 (24.00)	11 (40.74)	8 (33.33)	8 (30.77)	14 (48.28)	0.39 §
Cirrhosis	2 (8.00)	2 (7.41)	1 (4.17)	1 (3.85)	4 (13.79)	0.71 *
CKD	2 (8.00)	3 (11.11)	4 (16.67)	3 (11.54)	3 (10.34)	0.92 *
Hypothyroidism	4 (16.00)	4 (14.81)	2 (8.33)	6 (23.08)	7 (24.14)	0.56 *
CVD	3 (12.00)	0 (0.00)	1 (4.17)	2 (7.69)	3 (10.34)	0.41 *
Heart stroke	1 (4.00)	0 (0.00)	3 (12.50)	2 (7.69)	2 (6.90)	0.43 *
Atrial fibrillation	3 (12.00)	2 (7.41)	3 (12.50)	5 (19.23)	4 (13.79)	0.79 *
DVT	0 (0.00)	0 (0.00)	1 (4.17)	0 (0.00)	2 (6.90)	0.39 *
PE	0 (0.00)	0 (0.00)	2 (8.33)	1 (3.85)	2 (6.90)	0.42 *

Vit C: vitamin C; Vit E: vitamin E; NAC: N-acetylcysteine; MT: melatonin; TX: treatment; COPD: chronic obstructive pulmonary disease; CKD: chronic kidney disease; CVD: cerebral vascular dis-ease; DVT: Deep venous thrombosis; PE: Pulmonary embolism. Values are expressed as median (p25–p75); Statistical test: Kruskal–Wallis test, ¥ one-way ANOVA, § chi-squared test, and * Fisher's exact test.

According to the different conditions that patients may present when entering an Intensive Care Unit, such as the reason for admission, place of access, diagnosis, and location of the initial infection, we compare these variables between the antioxidant treatment groups to evaluate if they were homogeneous. We show no differences, which reveals the homogeneity of these clinical conditions (Table 2).

Table 2. Demographic characteristics of patients according to treatment group.

Characteristics	Vit C *n* = 25	Vit E *n* = 27	NAC *n* = 24	MT. *n* = 26	Control *n* = 29	*p*
			Reason for admission, *n* (%)			
Surgical	7 (28.00)	6 (22.22)	5 (20.83)	4 (15.38)	13 (44.83)	0.14
Non-surgical	18 (72.00)	21 (77.78)	19 (79.17)	22 (84.62)	16 (55.17)	
Place of admission, *n* (%)						
Emergency	14 (58.33)	15 (55.56)	13 (54.17)	18 (69.23)	14 (48.28)	
Operating room	4 (16.67)	5 (18.52)	3 (12.50)	3 (11.54)	7 (24.14)	0.76
Hospitalization	4 (16.67)	7 (25.93)	8 (33.33)	5 (19.23)	7 (24.14)	
Diagnoses at admission, *n* (%)						
Cardiovascular	0 (0.00)	0 (0.00)	0 (0.00)	2 (7.69)	2 (6.90)	
Respiratory	6 (25.00)	6 (22.22)	7 (29.17)	4 (15.38)	5 (17.24)	
Gastrointestinal	9 (37.50)	5 (18.52)	4 (16.67)	5 (19.23)	6 (20.69)	
Neurological	1 (4.17)	1 (3.70)	1 (4.17)	2 (7.69)	2 (6.90)	
Sepsis	6 (25.00)	1 (3.70)	7 (29.17)	9 (34.62)	12 (41.38)	0.82
Trauma	0 (0.00)	1 (3.70)	1 (4.17)	0 (0.00)	0 (0.00)	
Metabolic	1 (4.17)	1 (3.70)	0 (0.00)	0 (0.00)	1 (3.45)	
Hematologic	0 (0.00)	2 (7.41)	0 (0.00)	1 (3.85)	1 (3.45)	
Renal/Genitourinary	0 (0.00)	1 (3.70)	3 (12.50)	3 (11.54)	0 (0.00)	
Site of infection, *n* (%)						
Pulmonary	7 (29.17)	11 (40.74)	9 (39.13)	11 (42.31)	10 (34.48)	
Gastrointestinal	10 (41.67)	8 (29.63)	5 (21.74)	5 (19.23)	11 (37.93)	
Nefrourinary	3 (12.50)	3 (11.11)	6 (26.09)	6 (23.08)	3 (10.34)	
CNS	0 (0.00)	2 (7.41)	0 (0.00)	0 (0.00)	1 (3.45)	0.83
Skin and soft tissues	2 (8.33)	2 (7.41)	2 (8.70)	2 (7.69)	2 (6.90)	
Endocarditis	0 (0.00)	0 (0.00)	0 (0.00)	0 (0.00)	1 (3.45)	
Gastrointestinal	0 (0.00)	1 (3.70)	0 (0.00)	2 (7.69)	1 (3.45)	

Vit C: vitamin C; Vit E: vitamin E; NAC: N-acetylcysteine; MT: melatonin.

2.2. SOFA Score Assessment

The SOFA score was assessed in patients during the five days of antioxidant treatment to identify the impact of the addition of antioxidant therapy on organ damage. Supplementary Material S1 shows the reduction in the score. Statistical analysis using repeated measures (which evaluates changes over time) showed that there was a statistically significant reduction in the score in patients treated with Vit C from 8 to 3.5, (p = 0.001) between day 0 and day 5. The reduction with NAC was from 7 to 4, (p = 0.003) and with melatonin from 8 to 2 (p= 0.001). The patients who received Vit E and the control group also showed a decrease in the score; however, the difference was not statistically significant, and the final SOFA score remained high [50].

In patients treated with antioxidants, selenium levels were maintained at the same level as at admission; however, in the control group, they decreased (Figure 1A–C).

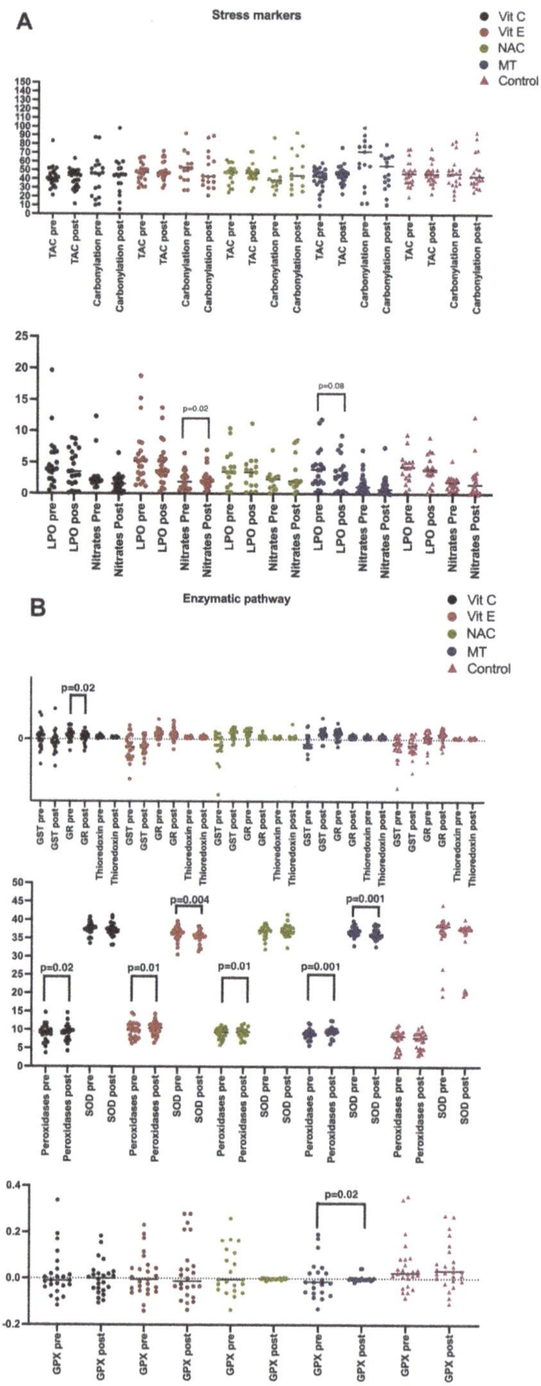

Figure 1. *Cont.*

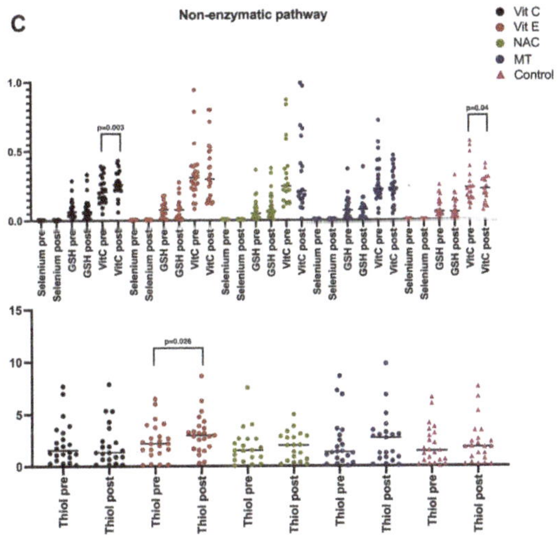

Figure 1. (**A**) Values of oxidative stress markers before and after antioxidant therapy. Groups: Vit C (n = 25), Vit E (n = 27), NAC (n = 24), MT (n = 26), and control (n = 29). Vit C: vitamin C; LPO: lipoperoxidation (nmol MDA/mL); TAC: total antioxidant capacity (TroloxnM/mL); Thiol (μM/mL); Nitrates and Nitrites (HNO_3^-/NO_2^- nmol/mL of serum). Carbonylation (ngcarbonyl/mL); (**B**) Enzymatic pathway before and after antioxidant therapy. Peroxidases, (U/L) SOD: superoxide dismutase (U/mgL); (**C**) Enzymatic pathway before and after antioxidant therapy. GPX: glutathione peroxidase (μmol of NADPH/min/mL); GSH: glutathione (μM/mL of serum); GST: (μM/mg/protein); GR: glutathione reductase (U/min/mL Thioredoxin (μM/mg/protein).

2.3. Lipoperoxidation Levels: Enzymatic and Non-Enzymatic Antioxidant Pathways

Lipid peroxidation and oxidative stress markers were increased at the beginning of the treatment in patients with septic shock, and a decrease was observed in all patients who received antioxidant therapy. However, only MT showed a statistically significant change. NO_3^- and NO_2^- levels were elevated before treatment, and there was a statistically significant decrease in patients treated with Vit C. The antioxidant capacity of the patients showed low levels before treatment, and the levels increased with the use of Vit E, NAC, and MT; nevertheless, the difference was not statistically significant.

Regarding the enzymatic antioxidant pathway, we evaluated glutathione and found increased levels in patients treated with NAC $p = 0.05$. Glutathione peroxidase showed changes with the use of MT $p = 0.02$. SOD significantly decreased with the use of Vit E and MT $p = 0.004$ and $p = 0.001$, respectively. Glutathione reductase showed elevated levels before treatment and were decreased with the use of Vit C $p = 0.02$. Thioredoxins did not increase in the groups treated with antioxidants and the control group. All patients who received antioxidants had increased peroxidases, and the differences were statistically significant. The level decreased in the control group, but there was no statistical difference.

The levels of non-enzymatic antioxidants showed that there was an increase in thiols in patients treated with Vit E. All patients had low levels of Vit C but the only group that showed an increase was the one treated with Vit C $p = 0.003$.

2.4. Cytokine Quantification

IL-1β, IL-2, IL4, IL-6, IL-8, IL-10, IL-12p70, IL-17A, TNF-α, INFγ, IP-10, MCP-1, and TGFβ-1 were quantified using the ELISA method. Table 3 shows the levels of proinflammatory and anti-inflammatory cytokines before and after treatment with the antioxidant therapy. Paired-sample t-tests showed significant differences between the pretreatment and posttreatment values. With the use of Vit C, IL-6 decreased ($p = 0.006$), while Vit E increased IL-2 and IL-12 ($p = 0.01$ for both), as well as IFNγ ($p = 0.03$). NAC decreased TNFα ($p = 0.004$)

and increased IL-12 ($p = 0.04$), while MT increased MCP-1 ($p = 0.01$) and decreased IL-6, IL-8, and IL-4 ($p = 0.001$, $p = 0.001$, and $p = 0.04$, respectively). In the control group, there was an increase in MCP-1 ($p = 0.002$) and a decrease in IL-6 and IL-8 ($p = 0.002$ and $p = 0.001$, respectively). We also found that all antioxidants increased TGF-β; Vit C before treatment 44.7 (8.8–414.2) after 46.9 (0–532.2) $p = 0.80$, MT 45.09 (8.8–626.3) to 51.4 (8.8–359.6) $p = 0.53$, NAC 27.6 (8.8–411.6) to 52.9 (8.8–789.7) $p = 0.08$, control 45.1 (8.8–1033) to 75.8 (8.8–356-9) 0.67 except Vit E 63 (8.8–452.8) to 52.9 (0–826.7). The changes are shown in Figure 2.

Table 3. Levels of cytokines in patients from the different treatment groups.

	Proinflammatory			**Anti-Inflammatory**		
	Before	**After**		**Before**	**After**	
Group	IL-1β	IL-1β	*p*	IL-4	IL-4	*p*
Vit C	9.8 (8.12–5700)	8.1 (6.4–178.79)	0.16	3.3 (0.39–328)	2.2 (0.39–65.2)	0.53
Vit E	8.1 (8.1–360)	8.1 (8.1–170.7)	0.66	5.35 (0.39–275.5)	13.7 (0.39–484.9)	0.21
NAC	10.8 (8.1–139.5)	8.1 (8.1–172.7)	0.27	4.05 (0.39–4262.7)	6.4 (0.39–844.4)	0.39
MT	8.1 (8.1–691)	8.1 (7.1–1050)	0.34	3.8 (0.39–184)	2.09 (0.39–117)	0.04
Control	8.1 (8.1–4110)	8.2 (8.1–1337)	0.28	3.7 (0.39–1307.7)	1.5 (0.39–1000)	0.52
	IL-6	IL-6		IL-10	IL-10	*p*
Vit C	124.89 (22.35–546.49)	59.82 (14.93–148.1)	0.006	73.3 (2.97–1094.30)	66 (2.97–2762.5)	0.56
Vit E	106.345 (5.64–512.885)	16.965 (0.94–233.24)	0.07	106.3 (2.9–1322)	74 (2.9–362.9)	0.17
NAC	127.48 (31.27–997.595)	69.85 (3.805–231.895)	0.90	56.5 (4.6–1007.97)	54.7 (7.34–1370.7)	0.88
MT	344.87 (15.13–564.88)	23.25 (7.45–238.45)	0.001	70.5 (6.7–1815)	57.4 (4.1–5.6)	0.11
Control	236.24 (22.92–2197.01)	35.97 (6.91–112.95)	0.002	107 (2.97–1683)	55.1 (1.9–3960)	0.08
	IL-8	IL-8	*p*	TGF-β	TGF-β	*p*
Vit C	12.6 (1.45–83.6)	6.1 (1.45–46.34)	0.14	44.7 (8.8–414.2)	46.9 (0–535.2)	0.80
Vit E	19.05 (1.45–241)	8.6 (1.45–269)	0.16	63 (8.8–452.8)	52.9 (0–826.7)	0.41
NAC	14.8 (1.45–283)	9.5 (1.84–269.9)	0.73	27.6 (8.8–411.6)	52.9 (8.8–789.7)	0.08
MT	13.8 (1.45–227.17)	8.9 (1.45–367)	0.001	45.09 (8.8–616.3)	51.4 (8.8–359.06)	0.53
Control	21.9 (1.45–688.8)	7.6 (1.45–100)	0.001	45.1 (8.8–1033)	75.8 (8.8–356.9)	0.67
	Proinflammatory			Regulatory cytokines		
	TNF-α	TNF-α		IL-2	IL-2	*p*
Vit C	0.80 (0.41–99.6)	0.65 (0.41–18.9)	0.26	1.4 (0.45–100.4)	0.45 (0.45–10.6)	0.36
Vit E	0.74 (0.41–119)	0.8 (0.41–144.9)	0.85	0.56 (0.30–51.4)	1.13 (0.45–102.7)	0.01
NAC	1.06 (0.41–76.6)	0.1 (0.41–144.9)	0.04	0.58 (0.45–51.4)	0.66 (0.45–102.7)	0.15
MT	0.64 (0.41–20.4)	0.80 (0.41–49.6)	0.84	0.45 (0.45–8.9)	0.45 (0.45–18.4)	0.34
Control	0.41 (0.41–32.6)	0.54 (0.41–65)	0.79	0.51 (0.45–3.80)	0.45 (0.45–15.3)	0.54
	IP-10	IP-10	*p*	IL-12	IL-12	*p*
Vit C	73.3 (2.97–1094.3)	66.05 (2.97–2762.53)	0.57	1.2 (0.85–21.8)	1.06 (0.85–32.1)	0.12
Vit E	106.3 (2.97–1322.2)	74.5 (2.9–362.9)	0.17	1.49 (0.85–176.4)	1.45 (0.85–222.7)	0.01
NAC	56.5 (4.68–1007.9)	54.7 (7.34–1370.7)	0.88	1.49 (0.85–62.36)	1.4 (0.85–187.02)	0.04
MT	69.6 (6.7–1001.7)	57.4 (4.1–565.8)	0.11	1.1 (0.85–27.8)	0.92 (0.85–49.1)	0.11
Control	107.6 (2.9–1683.8)	55.1 (1.9–3960.96)	0.08	1.2 (0.85–46.7)	1.2 (0.85–59.4)	0.75
	MCP-1	MCP-1	*p*	IFN-γ	IFN-γ	*p*
Vit C	324 (15.1–4321)	216 (5.3–1641)	0.11	3.08 (3.05–31.40)	3.05 (3.05–23.01)	0.46
Vit E	189.1 (1.6–2700)	119 (1.6–2866)	0.45	3.15 (3.05–20.77)	3.05 (3.05–96.16)	0.03
NAC	257.6 (78.08–2954.7)	152.8 (23.5–2569.9)	0.44	3.49 (3.05–23.98)	3.93 (3.05–36.71)	0.15
MT	295 (23.2–3479)	210.7 (31.2–2504.4)	0.01	3.05 (3.05–18.18)	3.05 (3.05–14.09)	0.53
Control	316 (38–4110)	185. (11.03–1000)	0.002	3.2 (3.05–67.44)	3.05 (3.05–33.24)	0.21

Vit C: vitamin C. Vit E: vitamin E. NAC: N-acetylcysteine. MT: melatonin. MCP-1: chemotactic protein from monocytes 1: range test signaled by Wilcoxon. Values are expressed in pg/mL.

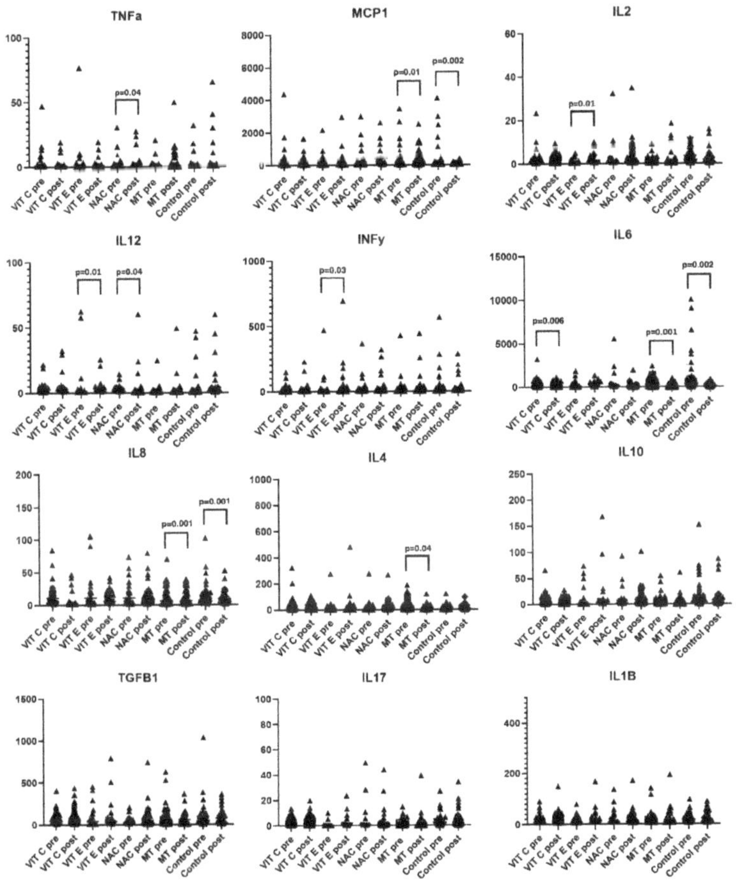

Figure 2. Proinflammatory interleukins show decreases with a significant delta when evaluating the pretreatment and posttreatment levels. With Vit C, there is a decrease in IL-6 (p = 0.006); with Vit E there is an increase in IL-2, IL-12, and IFNγ (p = 0.01, p = 0.01, and p = 0.03, respectively); with MT, there is an increase in MCP-1 and a decrease in IL-6, IL-8, and IL-4 (p = 0.001, p = 0.001, p = 0.01, and p = 0.04, respectively); and with NAC, there is a decrease in TNFα (p= 0.04) and an increase in IL-12 (p = 0.04). Individuals without antioxidant therapy had decreased levels of IL-6 (p = 0.002), IL-8 (p = 0.001), and MCP-1 (p = 0.002).

2.5. Canonical Correlation

A canonical correlation analysis was conducted. It included clinical variables such as SOFA score, PCT PCR laboratory, and oxidant stress biomarkers, and considered explanatory variables (u1, u2, u3) and effect variables (v1, v2, v3).

The first evaluation of the canonical correlation carried out was with laboratory parameters and biomarkers. We included the explanatory variables and those of possible effect (cytocines). We found that in patients with septic shock, there is a high correlation of 0.95 when there are high levels of procalcitonin and carbonylation and low levels of total antioxidant capacity (TAC), glutathione, and vitamin C. This correlates with low levels of IL4 and elevated levels of IP10, IL1B, MCP1, IL-6, IL-17, and TNFα.

In the second evaluation of the canonical correlation, we found a correlation of 0.86 when there are high levels of LPO, carbonylation, and low levels of GSH, selenium, and thiols; this correlates with low levels of IL-4, il12p70, and IFNy and high levels of IL1B, mcp1, and IL-6. There was a high correlation of 0.82 (although lower than the two previous correlations) which shows us that high levels of the SOFA score, CRP, and CRP, and low levels of GSH and GSH peroxidase correlate with low levels of IL-4, TGFB1, and Il12p70 and high levels of IL1B and IL8 (Table 4).

Table 4. Canonical correlation analysis between cases and controls with septic shock pretreatment.

Variable	Level	Variable	Coef.	Std. Err.	t	*p*-Value	IC95%		Corr.
u1	High	PCT	0.0172182	0.0042318	4.07	0.0001	0.0085873	0.0258491	0.9529
		Carbonyl	−0.0068545	0.0030781	−2.23	0.033	−0.0131324	−0.0005766	
	Low	TAC	−0.0002907	0.0000967	−3.01	0.005	−0.000488	−0.0000935	
		GSH	5.295229	1.522535	3.48	0.002	2.189998	8.400459	
		Vit C	6.144225	0.6809717	9.02	0.001	4.755374	7.533075	
v1	Low	IL4	−0.0073968	0.0025368	−2.92	0.007	−0.0125706	−0.002223	
	High	IP10	0.0009865	0.0001994	4.95	0.001	0.0005798	0.0013932	
		IL1B	0.0002558	0.000123	2.08	0.046	5.01×10^{-6}	0.0005067	
		MCP 1	−0.0006084	0.0001506	−4.04	0.001	−0.0009155	−0.0003012	
		IL6	0.0003399	0.0000436	7.8	0.001	0.0002511	0.0004288	
		IL17a	−0.0098983	0.0025994	−3.81	0.001	−0.0151998	−0.0045967	
		TNFα	0.049159	0.0144382	3.4	0.002	0.0197122	0.0786059	
u2	High	LPO	−0.1764843	0.0744333	−2.37	0.024	−0.328292	−0.0246766	0.8643
		Carbonyl	0.0164766	0.0056307	2.93	0.006	0.0049928	0.0279604	
	Low	GSH	−11.67688	2.785091	−4.19	0.001	−17.35711	−5.996647	
		Selenium	492.1174	139.0012	3.54	0.001	208.6226	775.6122	
		Thiol	−0.1522415	0.0385513	−3.95	0.001	−0.2308675	−0.0736155	
v2	Low	il4	0.0124842	0.0046404	2.69	0.011	0.00302	0.0219484	
		IL12p70	0.0367106	0.0155704	2.36	0.025	0.0049544	0.0684667	
		IFNy	−0.0140323	0.0035662	−3.93	0.001	−0.0213055	−0.006759	
		TGFB1	0.0031482	0.00154	2.04	0.049	7.40×10^{-6}	0.006289	
	High	IL1B	−0.0007156	0.000225	−3.18	0.003	−0.0011744	−0.0002568	
		MCP1	0.0013014	0.0002755	4.72	0.001	0.0007396	0.0018632	
		IL6a	0.0002323	0.0000797	2.92	0.007	0.0000698	0.0003947	
u3	High	SOFA	−0.3846204	0.1060993	−3.63	0.001	−0.6010113	−0.1682295	0.8218
		PCT	−0.0219292	0.009223	−2.38	0.024	−0.0407397	−0.0031188	
		PCR	0.0441247	0.0167844	2.63	0.013	0.0098926	0.0783568	
	Low	GSH	6.547262	3.318249	1.97	0.057	−0.2203513	13.31488	
		GSHpx	−6.29239	1.747836	−3.6	0.001	−9.857125	−2.727655	
v3	Low	IL4	0.0070789	0.0055288	1.28	0.21	−0.004197	0.0183549	
		TGFB1	0.0038936	0.0018348	2.12	0.042	0.0001515	0.0076356	
		IL12p70	−0.0400625	0.0185511	−2.16	0.039	−0.0778978	−0.0022272	
	High	IL1B	−0.0007372	0.000268	−2.75	0.01	−0.0012838	−0.0001905	
		IL8	0.0300677	0.0085392	3.52	0.001	0.0126518	0.0474836	

The canonical correlation revealed a significantly high correlation between the first (0.9529), the second (0.8643), and third (0.8218). Lambda test with an F = 0.001; the Lawley–Hotteling trace and Loy's largest root yielded the same statistical significance.

2.6. Protein–Protein Interaction Network

To present the participation of different mechanisms of damage in sepsis, we show the protein–protein interactions of the cytokines evaluated in this work; a diagram was created based on the Kyoto Encyclopedia of Genes and Genomes (KEGG), as well as the different signaling pathways that could be detected (Supplementary Material S2). The quantified interleukins might be involved in the presence of viral or bacterial infections that can lead to the development of septic shock. The interaction network was generated using the ShinyGO 0.76 server [51]. Also, we show in Figure 3 that in sepsis, the intrinsic and extrinsic pathways participate in the different mechanisms involved in damage to tissues and organs. And, we also explain our work on a hypothesis of inflammatory damage and oxidative stress through a graphical abstract.

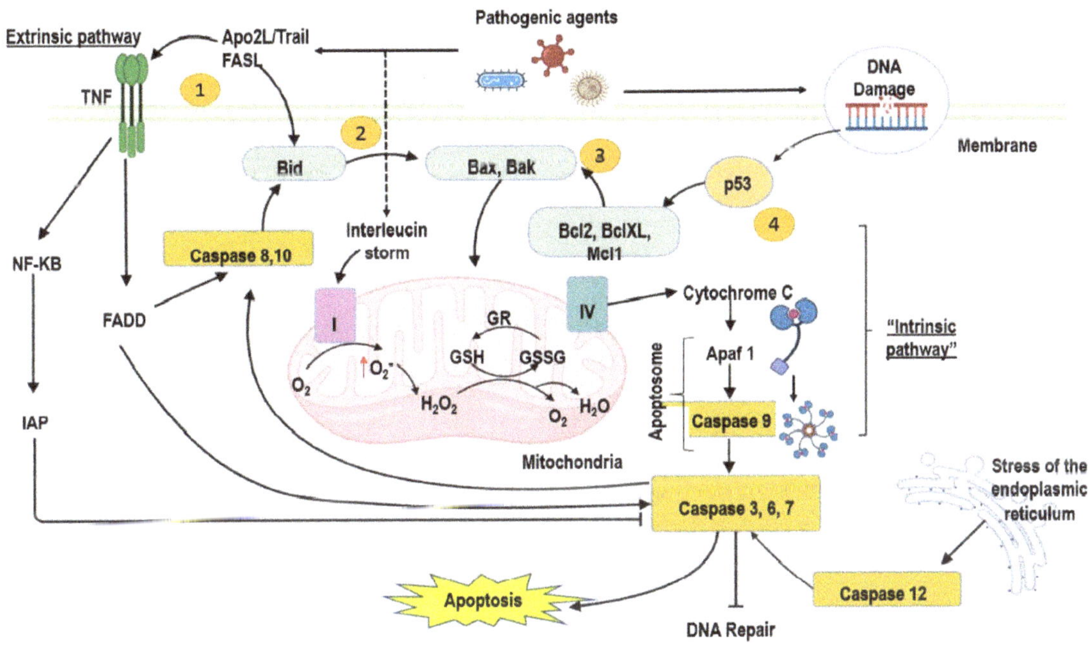

Figure 3. Intrinsic and extrinsic pathway for induction of apoptosis. Intrinsic and extrinsic path way for induction of apoptosis. (1) The extrinsic pathway of apoptosis can be activated by TNF-dependent Apo2L/TRAIL receptors. (2) In some extional cases, Apo2L/TRAIL can be intrinsically pathway activated by Bid. (3) The intrinsic pathway can be activated by DNA or microtubule damage, involving the interaction of some proteins such as BAX and BAK. (4) Apoptosis can also be promoted by activation of p53. Abbreviations: Apaf-1 = apoptotic protease-activating factor 1, Apo2L/Trail = Recombinant human apoptosis ligand 2/inducing tumor necrosis factor-related apoptosis ligand 2, FADD = the fas-associated death domain protein, GR = glutathione reductase, GSH = glutathione, GSSG = oxidized glutathione, H_2O_2 = hydrogen peroxide, IAP = apoptosis inhibitor, NF-KB = kappa nuclear factor, O_2^- = superoxide anion, TNF = tumor necrosis factor.

3. Discussion

The control of the activation of pathophysiological mechanisms such as oxidative stress, and innate and adaptive inflammatory response in septic shock is vital [52], Deterioration and organ damage result from deregulation of physiological functions that lead to proteolysis [53,54] activation of the complement system, the coagulation pathway, the fibrinolytic system, lipid pathways, oxidative stress, and cytokine production [43].

On the other hand, the participation of pathogen-derived molecular patterns (PAMPs) and damage-associated molecular patterns (DAMPs), recognized by specific receptors on the cell surface, initiate the transcription cascade of inflammatory molecules due to the translocation of the nuclear factor NF-kβ in activated cells.

Imbalance requires early therapeutic management [55–57] since the cytokine storm leads to a potentially fatal result [42,57].

Cytokines activate and inhibit cellular functions by negatively regulating effector mechanisms and restoring homeostasis through the release of cytokines IL-10 and transforming growth factor-β. They also regulate protein deposition in the extracellular matrix and angiogenesis [58].

The effects of the use of antioxidant therapy in patients with septic shock and its effect on cytokines were among the objectives of this study since we have previously reported reduced biomarkers of oxidant stress with the use of this therapy. Furthermore, there was better disease control in patients with COVID-19. Although the results in previous works are good, experimental and clinical studies are still required to confirm this hypothesis [59].

In this sense, cytokines involved in innate immunity are attractive targets for therapeutic intervention [60]. They are essential in diseases whose pathogenesis is the result of a defective regulation of the cytokine network; therefore, cytokine-targeted therapies with monoclonal antibodies, soluble receptors, or small molecule inhibitors promise new possibilities for the treatment of patients resistant to standard drug regimens. The proposal to add antioxidants to standard therapy in septic shock is based on the reduction in oxidative stress and the increase in antioxidant capacity, and our group was one of the first to carry out studies in humans employing this therapy [50,60–62].

In this study, we support the hypothesis raised by us and other researchers that the function of cytokines changes with the use of antioxidant therapy. We found an increase in the level of IL-2 after treatment with Vit E, NAC, and MT. Only Vit E showed a statistically significant difference. This finding coincides with a previous one in children undergoing dialysis treatment, where they combined NAC with Vit E and demonstrated regulation of the cellular redox state and modulation of the cytokine profile. That study also found a reduction in organ damage at the level of the kidney in patients with septic shock [63]. IL-2 stimulates the proliferation of NK cells, increases their cytolytic function, and acts on B cells as a proliferation factor. Its increase is relevant since it acts as a stimulus for synthesizing antibodies and has a regulatory function.

The combined use of Vit E and NAC to stimulate the production of IL4 had a regulatory effect on oxidative stress, inflammation, and organ failure in experimental studies with an animal model [64]. In patients with septic shock, we found that, in addition to the attenuation of oxidative stress, there was an improvement in organic damage, and IL-4 levels increased in patients who received vitamin E. However, the delta did not reach statistical significance ($p = 0.06$).

NAC increased IL-10, which has regulatory properties. The increase in IL-10 levels due to the use of antioxidants is an important finding since there is multiorgan dysfunction that involves uncontrolled production of reactive oxygen species (ROS) and a cytokine storm in septic shock. At the same time, exhaustion of antioxidants that contribute to the progression to septic shock occurs. Macrophages secrete proinflammatory and anti-inflammatory mediators during the infectious process. The binding of lipopolysaccharide (LPS) to Toll-like receptor 4 (TLR4) releases TNF-α, which initiates proinflammatory events through tumor necrosis factor receptor 1 (TNFR1) signaling, which is an effect of IL-10 [65]. In humans with septic shock, the effect achieved with MT may be related to improving IL-10 levels.

IL-12 has a regulatory function and there was no association between its levels and severity of the disease in a study of children with septic shock [66]. In another study, IL12 levels were not modified in patients in a septic state after surgery [67]. The increase or decrease in IL-12 levels during septic shock could be measured longitudinally to determine its relevance in this condition. Unlike other studies, we found that Il-12 levels increased with Vit E and NAC therapy $p = 0.001$. This finding is relevant because this cytokine promotes cellular immunity by inducing the maturation of T cells into Th1 cells and stimulating the secretion of gamma interferon (IFN-γ), which activates natural killer cells, macrophages, and cytotoxic T cells. Therefore, its deficiency leads to deterioration of immune functions, favoring the spread of the infection.

The increase in IL-12 found in this study has particular importance given that the most important function of IL-12 is the induction of gamma interferon (IFN), which is a mediator in viral, fungal, bacterial, and parasitic resistance. The induction of IL-12 triggered by microbial products is a potent stimulus for IFNy, which leads to the quick control of the infection by these agents. However, if the response is not controlled, the synthesis of IL-12 can result in an excessive activation of the immune system which causes damage to the host tissue and even death. This occurs in autoimmune diseases and/or septic shock.

The Induction of IL-12 during acute infection is crucial to initiate the synthesis of INFγ by NK cells and T lymphocytes, and mediators of resistance against viruses, bacteria, fungi, and parasites. In several animal models, IL-12 has been used as immunotherapy to protect cellular immunity and to control or reject inappropriate responses to pathogenic infections in organisms with immune disorders. In any case, increasing its concentration can be beneficial. However, the excess IL-12 response appears to be responsible for some problems observed during severe microbial infections. Under these conditions, IL-12 antagonists have attenuated immunopathology and prevented death. The toxic effects of the proposed therapy with IL-12 were observed in mice, and therefore its use as therapy in humans is controversial. A study with different doses is still necessary, or its combined use with antimicrobials with other molecules or with antioxidants [68].

In this study, IL-6 showed elevated levels before antioxidant therapy. Levels decreased in patients treated with Vit C, MT, and NAC. However, this effect was similar in the control group. This finding coincides with that from a systematic cohort study of patients with sepsis, in which the therapy was only with antibiotics and considered Il6 as a possible predictor of response to this therapy. The study also documented the substantial interindividual variability in the induction of IL-6 during sepsis since it depends on the type of infection, type of bacteria, and origin of the organ involved. All patients in our study received antibiotics. Therefore, the change and difference observed with antioxidants based on this cytokine will require an appropriate methodological strategy to discern whether the antioxidant effect creates synergy with the impact or is only related to the antibiotic [69].

Another regulatory cytokine involved in controlling the infectious process is TGF-β since it inhibits viral replication and cell proliferation. We found that all our patients had low levels of TGF-β before treatment, and levels increased in the group treated with antioxidants and in the control group. However, none showed a significant statistical difference; only the group treated with NAC had a clear trend of $p = 0.08$. Other studies have reported that its increase requires nutritional products such as sesame seeds and that it has synergistic properties with antioxidants such as vitamin E [70].

There is evidence for the complex pathogenesis of septic shock and LPS-induced re-actions. In recent years, mediators such as cytokines have aroused great interest for their role in inflammatory responses induced by LPS, TNFα, and IL-1 [71–74].

The relevance of TGFB1 in septic shock has been studied in mouse macrophages, and it inhibits JNK activity stimulated by lipopolysaccharide (LPS). It might have a possible inhibitory mechanism on TGF-beta in the signaling of inflammatory responses induced by LPS [75]. We found increased TGF-beta levels using all the antioxidants studied (Vit C, Vit E, NAC). In the control group, the levels decreased; it is precisely in this group of patients where SOFA remained elevated.

Several studies have demonstrated the involvement of proinflammatory cytokines in septic shock, and patients receiving standard treatment improve in intensive care units [76,77]; However, this response to treatment depends on the number of organs affected since admission [78] or if it was associated with surgical intervention, site of infection [79,80], burn [81,82], endocarditis [83,84], necrotizing fasciitis [85,86], meningitis [87,88], and septic arthritis [89–91]. The active participation of cytokines correlates with severity and prognosis [76].

High amounts of IL-1 and TNF-α are released during systemic inflammation, resulting in hypotension and shock. However, IL-1 is frequently undetectable and, when present, has little predictive value [92]. This study confirms this observation since the cytokines' IL-1β and TNFα levels did not show relevant changes after treatment. In this regard, we consider that with our findings, we show that there are favorable changes in some cytokines with standard comprehensive management and antioxidant therapy. Comprehensive management probably requires exploring the synergistic therapeutic link of these pathways since there is information that with the use of some specific biologicals, the beneficial effect has not been fully proven in humans with infectious processes. During the COVID-19 pandemic, the use of anakinra, an IL-1 antagonist, or IL-6 antagonists were not wholly conclusive [93].

On the other hand, results using antioxidants at the experimental level have shown that some have regulatory capacity on oxidative stress. MT has radical scavenging properties and protects cell membrane lipids, cytosol proteins, nuclear proteins, and mitochondrial DNA. In our research, it reduced LPO, similar to Galley's findings [94]. Also, the beneficial effect of MT has been demonstrated more widely in experimental cells, plants, and animals, although its mechanism of action remains unknown. The effects of MT could be related to its detoxifying capacity, thus protecting molecules from the destructive effects of OS in various conditions, such as ischemia/reperfusion (cerebrovascular accident and myocardial infarction), ionizing radiation, and drug toxicity. In sepsis, the protective effects of MT are associated with the inhibition of apoptosis and the reduction in OS [95]. MT has been shown to increase total antioxidant capacity (TAC) in experimental studies, equally with the associated use of Vit E and NAC [95–97]. On the other hand, the use of MT decreased the SOFA score, confirming the same effect found in multiorgan lesions induced by sepsis [98,99].

After release, plasma melatonin is rapidly distributed and transported to the mitochondria, acting as an antioxidant. In mammals, MT is an agonist of MT1 and MT2 receptors [100,101], and has pharmacological and physiological anti-inflammatory effects through receptor-dependent and non-receptor-dependent pathways [102]. It reduces levels of lipid peroxidation [103]. Also, we found an increase in GR activity in patients before treatment, which may lead to a decrease in GSH concentrations and an increase in ROS. GR activity decreased in patients treated with vitamin C, and GST activity remained unchanged in patients treated with Vit C and other antioxidants, but in the control they decreased. These findings on the activity of these enzymes with antioxidant treatment suggest that they could contribute to the reduced levels of LPO, and the SOFA score that evaluates MOF.

GPx uses selenoprotein P as a cofactor and decreases during septic shock [104]. We found low levels of Se in patients with septic shock and MOF. Although Se levels did not increase with the administration of antioxidant therapy, they were maintained, contrasting with the decreased levels found in untreated patients, who also showed no improvement in the SOFA score. The exact mechanism is not yet understood; however, our results suggest that the antioxidant therapy can maintain Se levels in septic shock patients, leading to improved GPx activity.

One of the inducers of sepsis is endotoxin or lipopolysaccharide (LPS), which comes from Gram-negative bacteria and leads to sepsis and multiorgan dysfunction [105]. LPS increases heart rate and decreases systemic vascular resistance, leading to hypotension. LPS also induces the synthesis of cytokines and elevates neutrophils, free radicals, and proteolytic enzymes that lead to endothelial dysfunction [106].

The glutathione peroxidase enzyme metabolizes intermediates formed by aerobic metabolism, such as hydrogen peroxide; this reaction occurs in the cytosol and mitochondria, resulting in oxidized glutathione, which is reduced by glutathione reductase. GR employs NADPH as a cofactor, helping to recycle an oxidation–reduction [107,108].

Glutathione is a tripeptide composed of three amino acids: glutamic acid, glycine, and cysteine. It plays a vital role in septic shock since it protects against damage caused by free radicals. Like glutathione, Vit E is a fat-soluble antioxidant [109]. It reacts with free radicals generated in the lipid phase and protects the lipids in the membranes. Vit C has radical scavenging activity, stabilizes membrane structures, oxidizes vitamin E, and has a synergistic mechanism [110].

In this study, Vit C and Vit E decreased LPO. Vit E and MT increased the total antioxidant capacity. NAC, a direct precursor of glutathione, increased glutathione levels and improved multiorgan failure (MOF), in accordance with previous research [111].

The effect of various antioxidants, such as NAC, MT, vitamins A, C, and E, enzyme cofactors (selenium and zinc), endogenous compounds (ubiquinone, α-lipoic acid, bilirubin, albumin, ferritin), and quercetin, can inhibit reactive nitrogen, oxygen species (ROS), and

RNS [61]. The effect NAC has on improving antioxidant capacity is due to the replenishment of glutathione (GSH) and the sequestration of ROS. It also improves hemodynamic variables, cardiac index, oxygenation, and static lung compliance [62], and reduces the levels of IL-6 and ICAM-1 [60].

Innate cytokines are attractive targets for therapeutic intervention [63]. They are essential in diseases whose pathogenesis is the result of a defective regulation of the cytokine network; therefore, cytokine-targeted therapies with monoclonal antibodies, soluble receptors, or small molecule inhibitors promise new possibilities for the treatment of patients resistant to standard drug regimens [64,65]. The proposal to add antioxidants to standard therapy in septic shock has been previously proposed for its benefits in terms of reduction in oxidative stress and an increase in antioxidant capacity. Our group was one of the first to propose this [50,61].

In septic shock, interleukins are the first to increase, reaching a maximum peak at 2 h, when procalcitonin increases, and IL-6 and IL-8 levels decrease [112]. This aspect limits the interleukins' determination during clinical practice; therefore, having little use in emergencies [112]. Moreover, as the deterioration of the organ progresses, determining cytokines, considering their lifespan, could add more information to clinical treatment. C-reactive protein (CRP) and procalcitonin (PCT) could be of help in the guidance of management. Determining oxidative stress markers is not routinely feasible today; however, this study and the experimental ones that show the participation of oxidative stress in the progression of damage in patients with septic shock allow us to suggest using antioxidant therapy associated with standard management.

In sepsis, organ dysfunction is potentially fatal, and it is caused by deregulation of the host response to infection. The pathological process is complex, and several cytokines participate in the damage mechanism, which, when dysfunctional, lead to an intertwined storm whose outcome worsens function and causes structural damage at the cellular, tissue, and organ level. The uncontrolled inflammation cascade requires regulation of both the cytokines involved and other intertwined mechanisms, such as the deregulation of oxidative stress, which requires simultaneous treatment and modulation to repair tissues and improve organ function.

Standard therapy in sepsis includes hydroelectrolytic hemodynamic management and antibiotic therapy that essentially control the initial inflammatory pathophysiological damage. However, deregulation of oxidative stress must also participate and is necessary to improve organ dysfunction and interact in the modulation of some cytokines.

Although there are limitations with using oxidative stress markers in emergency conditions, in this research we show that there is deregulation of oxidative stress in septic shock, which is related to cytokine storm, and that the complex interaction produced by treatment with antioxidants can improve organic dysfunction.

A limitation of this study is that the sample size would have to be larger to confirm the statistical power for all cytokines.

4. Methods and Materials

This study was a prospective, longitudinal, aleatorized, and blind clinical trial with a cohort of patients who were treated between April 2018 and January 2022. This dataset corresponds to the same population used in another study of our group and the study was completed with a new study of cytokines, which were not previously analyzed [50].

4.1. Study Population

Patients of any gender, over 18 years of age who were admitted to the intensive care unit of the ABC Observatory Medical Center and Santa Fe campus with a diagnosis of septic shock, according to the Third International Consensus on Sepsis and Septic Shock, were included [73]. The characterization of septic shock included refractory hypotension requiring the use of vasopressors despite resuscitation with sufficient fluids (20 mL/kg colloid or 40 mL/kg crystalloid) to maintain blood pressure $\geq$ 65 mmHg and lactate > 2 mmol/L.

Patients were required to give a signed consent or an informed assent of the patient or responsible caregiver. Patients with a history of having signed an advance directive, chronic use of steroids and/or statins in the last six months, or with recent use of antioxidant treatment before the moment of septic shock, and who had any contraindication for the use of Vit C, Vit E, NAC, or MT, as well as pregnant women, were excluded and patients who, having already been included, withdrew their informed consent were removed from the study.

An advance directive is a legal instrument through which a competent person determines in writing their will regarding the treatment that they would and would not like to receive in a situation where they cannot express their will.

In standard therapy, patients receive various therapies; in general, they receive antibiotics, blood products, sedation and analgesia, steroids, and intravenous solutions. Hemodynamic support is monitored, and goals related to lactate control, glycemic control, cultures, mechanical ventilator withdrawal test, stress ulcer prophylaxis, and nutritional care are performed [72].

4.2. Sample Size

The sample size was calculated considering the mean difference between low levels of ascorbic acid and improvement in treatment with antioxidants. The test suggested the inclusion of 11 patients in each group for a desired power of 80% and an alpha error of <0.05 based on a previous study [74]. To achieve greater statistical power, the sample size was increased to >24 patients per group, and a power of 97% was reached with an alpha error of 0.03.

4.3. Randomization

Computerized electronic selection was used to divide the patients into blocks; Randomization was 1:1 in balanced blocks. There were 6 blocks, with an approximate number of 25 patients per block. Personnel who were not involved in the study participated in the blinding and placed the indicated therapy in identical opaque envelopes numbered from 1 to 125, and these were applied consecutively. Group 1 received Vit C, Group 2 received Vit E, Group 3 received NAC, Group 4 received MT, and Group 5 was the control group with antioxidant treatment. Of 3745 patients admitted to the ICU during this study, only 131 had septic shock. Once the diagnosis of septic shock was established (inclusion criterion), we applied treatment with antioxidants to patients according to randomization. We show the number of patients who entered each group and the type of treatment they received. Additionally, we provide the conditions that occurred during follow-up (intervention discontinuation, adverse effects, death) in each group and show the number of patients finally included in the analysis (Supplementary Material S3).

4.4. Masking and Drug Administration

The random assignment sequence for the administration of antioxidants was generated at the coordination center using a computer-generated randomization program. The pharmacy maintained the blinding. Blinding was also carried out from the study's beginning until the results' analysis. A pharmaceutical professional maintains surveillance and care of the form of application of each of the antioxidants according to the type of absorption and interactions with other medications and the most appropriate way to administer them under the conditions that each patient requires.

All antioxidants were administered orally or through a nasogastric tube for 5 days in addition to standard therapy. NAC was administered in 1200 mg tablets every 12 h. In addition, patients were given 50 mg of MT in 5 mg tablets once a day and vitamin C in 1 g effervescent tablets every 6 h. Vit E capsules of 400 IU were administered every 8 h. The preparation and administration of these drugs is presented in Supplementary Material S4. The doses of antioxidants were chosen according to what is reported in the literature [113–116].

The pharmacotherapeutic follow-up of the antioxidants used in the clinical trial had strict surveillance of the administration of medications by the CMABC Pharmacy Department to establish that antioxidants were administered as much as possible under the same conditions, both in the preparation and preparation administration hours (Supplementary Material S4).

4.5. Data Collection Method

Upon admission to the intensive care unit, a complete medical history and physical examination of the patients were performed to obtain the demographic data of the patients. APACHE II and SAPS II prognostic scores were calculated upon admission to the intensive care unit, and the SOFA score for organ dysfunction (neurological, respiratory, hemodynamic, hepatic, and hematological) was performed. Laboratory measurements were taken upon admission to the ICU and for every day of hospital stay, which included complete blood count, blood chemistry, serum electrolytes (sodium, potassium, chloride, calcium, and magnesium), liver function tests, C-reactive protein, procalcitonin, and venous and arterial blood gases.

The SOFA score was evaluated using 6 important systems, including respiration (PaO_2/FiO_2), coagulation (platelet count), liver (bilirubin), cardiovascular (mean arterial pressure), central nervous system (Glasgow Coma Scale, GCS), and renal system (creatinine and/or urine output) [117]. A detailed definition of the SOFA criteria including the relevant thresholds is shown in Supplementary Material S5.

4.6. Standard Therapy in the ICU

The patients were treated according to the recommendation of the International Guidelines for the Management of Sepsis and Septic Shock. This management includes fluid replacement, hemodynamic support, lactate goals, control of the infectious focus, antibiotic cultures, use of blood products, sedation and test for withdrawal from mechanical ventilation, use of steroids, glycemic control, prophylaxis of stress ulcers, and nutrition [77].

4.7. Sample Collection and Storage

Blood samples were obtained from each patient that entered the draw, before initiation of the treatment, and 48 h after its administration. The blood samples were centrifuged for 20 min at 936× *g* and 4 °C. The plasma of the samples was placed in 3 or 4 aliquots and stored at −30 °C.

4.8. Oxidative Stress Markers in Plasma

4.8.1. NO_3^-/NO_2^- Ratio

The NO_3^- was reduced to NO_2^- by the nitrate reductase enzyme reaction. A quantity of 100 μL of plasma previously deproteinized with 0.5 N, NaOH and 10%, $ZnSO_4$ was mixed, and the supernatant was incubated for 30 min at 37 °C in presence of nitrate reductase (5 units). At the end of the incubation period, 200 μL of sulfanilamide 1% and 200 μL of N-naphthyl-ethyldiamine 0.1% were added and the total volume was adjusted to 1 mL. The absorbance was measured at 540 [118].

4.8.2. GSH Concentration

A total of 100 μL of plasma previously deproteinized with 20% trichloroacetic acid (vol/vol) and centrifugated to 10,000× *g* for 5 min was added to 800 μL of phosphate buffer 50 mM, pH 7.3, plus 100 μL of 5, 50-dithiobis-2-nitrobenzoic acid 1 M. The mixture was incubated at room temperature for 5 min and absorbance was read at 412 nm [119].

4.8.3. Evaluation Total Antioxidant Capacity (TAC)

Briefly, 100 μL of plasma were suspended in 1.5 mL of a reaction mixture prepared as follows: 300 mM acetate buffer pH 3.6, 20 mM hexahydrate of ferric chloride, and 10 mM of 2,4,6-Tris-2-pyridil-s-triazine dissolved in 40 mM HCl. These reactives were added in a relation of 10:1:1 v/v, respectively. After mixing, samples were incubated at 37 °C for 15 min in the dark. The absorbance was measured at 593 nm [120].

4.8.4. Lipid Peroxidation (LPO)

A quantity of 50 μL CH_3-OH with 4% butylated hydroxytoluene plus phosphate buffer pH 7.4 was added to 100 μL of plasma. It was incubated and centrifuged at 4000 rpm in room temperature for 2 min. Then, the n-butanol phase was extracted, and absorbance was measured at 532 nm [118].

4.8.5. Carbonylation Protein Concentration

Briefly, 100 μL of plasma were added to 500 μL of HCl 2.5 N in parallel with another sample with 500 μL of 2, 4-dinitrophenylhydrazine (DNPH) and incubated. At the end of the incubation period, they were centrifuged at 15,000× g for 5 min. The supernatant was discarded. Two washings were performed. The mixture was incubated again at 37 °C for 30 min. Absorbance was read in a spectrophotometer at 370 nm, using bi-distilled water as blank and a molar absorption coefficient of 22,000 M^{-1} cm^{-1} [118].

4.8.6. Determination of Selenium (Se)

Selenium (Se) determination was performed using 200 μL of serum according to the method described by Soto et al., and the absorbance was read at 600 nm [118].

4.8.7. Thiols

This test consists of Ellman's reagent reaction with a thiol group, commonly a thiolate, producing thiol-nitrobenzoate through potassium hydride. The technique used was previously described by Erel and Neselioglu [121], with some modifications carried out in our laboratory as previously reported [122]. A quantity of 50 μL of serum was used for the determination and the absorbance was measured at 415 nm. The calibration curve was obtained with solution GSSG 1 mg/1 mL and the absorbance was measured at 415 nm.

4.8.8. GPx Activity

A total of 100 μL of serum was suspended in 1.6 mL of 50 mM phosphate buffer (KH_2PO_4, pH 7.3), 0.2 mM NADPH, 1 mM GSH, and 1 IU/mL glutathione reductase. The mixture was incubated for 3 min at 37 °C; then, 100 μL of 0.25 mM H_2O_2 was added to start the reaction and the absorbance was monitored for 7 min at 340 nm [123]. The units are expressed in μmol NADPH oxidized/min/mL in serum with an extinction coefficient of 6220 M^{-1} cm^{-1} at 340 nm of NADPH [124].

4.8.9. GST Activity

A total of 100 μL of serum was added to 700 μL of phosphate buffer (KH_2PO_4, 0.1 M, pH 6.5) with 100 μL of 0.1 mM GSH and 100 μL of 0.1 mM 1-chloro-2,4-dinitrobenzene (CDNB). The sample was incubated and monitored for 7 min at 37 °C and read at 340 nm. GST activity was expressed in units of GS-DNB μmo/min/mL of serum with an extinction coefficient of 14,150 M^{-1} cm^{-1}.

4.8.10. Extracellular Superoxide Dismutase (ecSOD)

ecSOD activity was determined via electrophoresis using native 10% polyacrylamide gels. Electrophoresis was carried out at 120 V for 4 h, as previously described by Pérez-Torres et al. [39]. In brief, 100 L of serum were used; the gel was incubated in 2.45 mM NBT solution for 20 min. The liquid was discarded, and then the gel was incubated in a TEMED solution with 36 mM potassium phosphate (pH 7.8) and 0.028 mM riboflavin. The gel was exposed to a UV lamp for 10 min and washed with distilled water to stop the reaction. A standardized curve was obtained using a serial dilution (2.5, 5, 10, 15, 30, and 60 ng) with SOD from bovine erythrocytes (Sigma Aldrich Chemical SA de RL de C.V., Toluca, México). SOD activity was calculated.

4.8.11. Vitamin C Levels

Briefly, 100 μL of 20% trichloroacetic acid were added to 100 μL of plasma and centrifuged at 5000 rpm for 5 min. Then, 200 μL of Folin–Ciocalteu reagent 0.20 mM was added to the supernatant. The mixture was incubated for 10 min. The absorbance was measured at 760 nm [103,125].

4.9. Cytokine Measurement

Serum cytokine levels were determined using a flow cytometry bead-based LEGENDplex™ HU Essential Immune Response Panel (BioLegend, San Diego, CA, USA), a 13-plex human panel to analyze the following cytokines and chemokines: IL-1β, IL-2, IL-4, IL-6, IL-8, IL-10, IL-12p70, IL-17A CCL2 (MCP-1) CXCL10 (IP-10), IFN-γ, TGF-β1, and TNF-α. The assay was performed following the manufacturer's protocol; briefly, samples were diluted 2-fold with assay buffer and 50 mL of sample or standard was mixed with 25 μL of previously vortexed mixed beads in a V-bottom plate, then incubated for 2 h, centrifuged at 250 rpm for 5 min, and washed two times with 200 μL wash buffer, 25 μL of detection antibody were added and incubated as above, and without washing, 25 μL of SA-PE were added and incubated for 30 min, and afterward, centrifugation and washing steps were repeated, adding a final 150 μL of wash buffer to each well before being individually transferred to microtubes for the acquisition in a BD FACS AriaTM Fusion (BD Biosciences, San Jose, CA, USA). Analytes concentrations were calculated using the LEGENDplex Data Analysis SoftwareCatalog 740932 Version 9. Software GraphPad Prism version 9.4.0.

4.10. Statistical Analysis

Continuous variables are expressed as mean ± standard deviation or median with the minimum and maximum values. Categorical variables, such as frequencies and percentages, are reported. Normality distribution was evaluated using the Shapiro–France test. Non-parametric tests (Mann–Whitney) or Student's tests, according to the Gaussian distribution, were performed to detect significant independent variables.

Some variables were standardized and Bonferroni correction was adjusted for multiple comparisons. For the paired analyses (before–after), the Friedman or Wilcoxon test was used with the signed-rank test according to the distribution of the data. Pearson's Chi-square (χ^2) test or Fisher's exact test were used to compare proportions between two groups. For the multivariate analysis with the confounding factors, binary logistic regression was used. For time and group analyses, repeated-samples analysis and panel data tests of different models (pooled model, model for longitudinal data, marginal approximation model, and multilevel model) were performed. A general adjustment was made to select by random age using propensity matching between treated and untreated patients.

We perform a canonical correlation analysis (CC) to analyze the correlation of two subsets of variables. In the first set, we include the x expressed as U (p) (X1, X2, ... XP) in which the variables that they made up were (IL-1β, IL-2, IL4, IL-6, IL-8, IL-10, IL-12p70, IL-17A, TNF-α, INFγ, IP-10, MCP-1, and TGFβ-1). In the second set of variables, expressed by V(q) (Y1, Y2, ... Yq). We included the SOFA score, C-reactive protein, procalcitonin, lipoperoxidation, selenium, thiols, vitamin C, carbonylation, glutathione, Gpx, GST, GR, SOD, and antioxidant capacity (CAT). We include more than two variables on both sides of the equation in this type of analysis. The main objective of CC is to identify the linear combinations and the value of the response of that correlation.

$$\mathrm{U1}\ \frac{1}{4}\ \mathrm{a11X1}\ þ\ \mathrm{a12X2}\ þ\ldots\ldots þ\ \mathrm{a1pXp}$$

$$\mathrm{U2}\ \frac{1}{4}\ \mathrm{a21X1}\ þ\ \mathrm{a22X2}\ þ\ldots\ldots þ\ \mathrm{a2pXp}$$

$$\mathrm{Ur}\ \frac{1}{4}\ \mathrm{ar1X1}\ þ\ \mathrm{ar2X2}\ þ\ldots\ldots þ\ \mathrm{arpXp}$$

And with the predictive variable, it is expressed as follows:

$$V1 \frac{1}{4} b11Y1 \text{ þ } b12Y2 \text{ þ} \ldots\ldots \text{ þ } b1pYp$$

$$V2 \frac{1}{4} b21Y1 \text{ þ } b22Y2 \text{ þ} \ldots\ldots \text{ þ } b2pYp$$

$$Vr \frac{1}{4} br1Y1 \text{ þ } br2Y2 \text{ þ} \ldots\ldots \text{ þ } brpYp$$

The largest correlation was identified between U1 and V1; the second largest correlation was between U2 and V2, as long as there was no correlation between U1 and U2, nor between V1 and V2; and the third largest correlation was between U3 and V3 and, as described above. There could be no alteration between U3, and U1 and U2, nor between V3, and V1 and V2. These combinations are known as canonical variables. Differences were considered statistically significant when the *p*-value was <0.05. Statistical analyses were performed using the STATA V.16 Software and Sigma Software Plot 14 (Jendel Corporation, 1986–2017, New York, NY, USA).

4.11. Ethical Aspects

A signed informed consent form was obtained from each participant in accordance with the Declaration of Helsinki, as amended at the Congress in Tokyo, Japan. This research was approved by the Ethics, Biosafety and Research Committees of the National Institute of Cardiology (INCICh registry number: PT 10-0-76) and the ABC Campus Observatory Medical Center, approval number ABC-18-19; Registry of Trial: Clinical-Trials.gov Identifier: NCT 03557229.

5. Conclusions

Patients treated with antioxidants such as Vit C, Vit E, NAC, or MT, in addition to standard therapy, show reduced levels of proinflammatory cytokines and improved regulatory function in Il-2, IL-12, and IFN. There is improvement of the antioxidant capacity and reduction in biomarkers of oxidative stress with evidence of decreased organ damage measured by SOFA score. Combining antioxidants such as Vit C, MT, and NAC associated with standard therapy is a potential perspective in investigation through randomized clinical trials since it could improve other clinical conditions and pathophysiological mechanisms that converge in septic shock. Regarding the damage mechanisms involved, there is no evidence of improvement with the use of anticytokine biologicals and anti-apoptosis therapy, which evidences the need of research studies to follow for the comprehensive management of sepsis.

Supplementary Materials: The supporting information can be downloaded at: https://www.mdpi.com/article/10.3390/ijms242316610/s1. References [126–133] are cited in the supplementary materials.

Author Contributions: Conception and design of the research: M.E.S., A.A.-Á. and G.B.; acquisition of data: A.A.-Á., R.C.-S., I.P.-T., O.G.-M. and J.F.-G.; participation in patient enrolment, patient UCI, procedures, and sample collection: A.A.-Á., R.C.-S., R.M.-V., S.C.-A., L.M.-P. and J.F.-G.; statistical analysis and interpretation of data: M.E.S., A.A.-Á., H.S.-O. and R.G.; writing of the manuscript: M.E.S., A.A.-Á., I.P.-T. and G.F.-Á.; critical revision of the manuscript for intellectual content: M.E.S., I.P.-T., G.F.-Á. and V.G.-L. All authors have read and agreed to the published version of the manuscript.

Funding: This work was supported by Centro Medico A.B.C. Observatorio Project (ABC number 18-19) and Instituto Nacional de Cardiologia Ignacio Chavez.

Institutional Review Board Statement: The study was conducted in accordance with the Declaration of Helsinki and approved by the Instituto Nacional de Cardiologia Ignacio Chavez, registration number INCICh: PT 10-0-76, and Centro Medico A.B.C. Campus Observatory, approval number ABC-18-19. Trial Registration: ClinicalTrials.gov Identifier: NCT 03557229.

101. Dubocovich, M.L. Molecular pharmacology regulation and function of mammalian melatonin receptors. *Front. Biosci.* **2003**, *8*, 1089. [CrossRef]
102. Tarocco, A.; Caroccia, N.; Morciano, G.; Wieckowski, M.R.; Ancora, G.; Garani, G.; Pinton, P. Melatonin as a master regulator of cell death and inflammation: Molecular mechanisms and clinical implications for newborn care. *Cell Death Dis.* **2019**, *10*, 317. [CrossRef]
103. Aisa-Alvarez, A.; Soto, M.E.; Guarner-Lans, V.; Camarena-Alejo, G.; Franco-Granillo, J.; Martínez-Rodríguez, E.A.; Ávila, R.G.; Pech, L.M.; Pérez-Torres, I. Usefulness of Antioxidants as Adjuvant Therapy for Septic Shock: A Randomized Clinical Trial. *Medicina* **2020**, *56*, 619. [CrossRef]
104. Forceville, X.; Mostert, V.; Pierantoni, A.; Vitoux, D.; Le Toumelin, P.; Plouvier, E.; Dehoux, M.; Thuillier, F.; Combes, A. Selenoprotein P, Rather than Glutathione Peroxidase, as a Potential Marker of Septic Shock and Related Syndromes. *Eur. Surg. Res.* **2009**, *43*, 338–347. [CrossRef]
105. Manthous, C.A.; Hall, J.B.; Samsel, R.W. Endotoxin in Human Disease. *Chest* **1993**, *104*, 1872–1881. [CrossRef]
106. Stephens, R.; Mythen, M. Endotoxin immunization. *Intensive Care Med.* **2000**, *26*, S129–S136. [CrossRef]
107. Jena, A.B.; Samal, R.R.; Bhol, N.K.; Duttaroy, A.K. Cellular Red-Ox system in health and disease: The latest update. *Biomed. Pharmacother.* **2023**, *162*, 114606. [CrossRef]
108. Wicha, P.; Tocharus, J.; Janyou, A.; Jittiwat, J.; Changtam, C.; Suksamrarn, A.; Tocharus, C. Hexahydrocurcumin protects against cerebral ischemia/reperfusion injury, attenuates inflammation, and improves antioxidant defenses in a rat stroke model. *PLoS ONE* **2017**, *12*, e0189211. [CrossRef]
109. Pita Rodríguez, G. Funciones de la Vitamina E en la nutrición humana/Role of vitamin E in human nutrition. *Rev. Cuba. Aliment. Nutr.* **1997**, *11*, 46–57.
110. Tahir, M.; Foley, B.; Pate, G.; Crean, P.; Moore, D.; McCarroll, N.; Walsh, M. Impact of vitamin E and C supplementation on serum adhesion molecules in chronic degenerative aortic stenosis: A randomized controlled trial. *Am. Heart J.* **2005**, *150*, 302–306. [CrossRef]
111. Haileselassie, B.; Joshi, A.U.; Minhas, P.S.; Mukherjee, R.; Andreasson, K.I.; Mochly-Rosen, D. Mitochondrial dysfunction mediated through dynamin-related protein 1 (Drp1) propagates impairment in blood brain barrier in septic encephalopathy. *J. Neuroinflamm.* **2020**, *17*, 36. [CrossRef]
112. Riedel, S.; Carroll, K.C. Laboratory Detection of Sepsis. *Clin. Lab. Med.* **2013**, *33*, 413–437. [CrossRef]
113. Menon, V.; Mohamed, Z.U.; Prasannan, P.; Moni, M.; Edathadathil, F.; Prasanna, P.; Menon, A.; Nair, S.; Greeshma, C.; Sathyapalan, D.T.; et al. Vitamin C Therapy for Routine Care in Septic Shock (ViCTOR) Trial: Effect of Intravenous Vitamin C, Thiamine, and Hydrocortisone Administration on Inpatient Mortality among Patients with Septic Shock. *Indian J. Crit. Care Med.* **2020**, *24*, 653–661. [CrossRef]
114. Fowler, A.A.; Truwit, J.D.; Hite, R.D.; Morris, P.E.; DeWilde, C.; Priday, A.; Fisher, B.; Thacker, L.R., 2nd; Natarajan, R.; Brophy, D.F.; et al. Effect of Vitamin C Infusion on Organ Failure and Biomarkers of Inflammation and Vascular Injury in Patients With Sepsis and Severe Acute Respiratory Failure. *JAMA* **2019**, *322*, 1261. [CrossRef]
115. Lowes, D.A.; Almawash, A.M.; Webster, N.R.; Reid, V.L.; Galley, H.F. Melatonin and structurally similar compounds have differing effects on inflammation and mitochondrial function in endothelial cells under conditions mimicking sepsis. *Br. J. Anaesth.* **2011**, *107*, 193–201. [CrossRef]
116. Howe, K.P.; Clochesy, J.M.; Goldstein, L.S.; Owen, H. Mechanical Ventilation Antioxidant Trial. *Am. J. Crit. Care* **2015**, *24*, 440–445. [CrossRef]
117. Khwannimit, B. A comparison of three organ dysfunction scores: MODS, SOFA and LOD for predicting ICU mortality in critically ill patients. *J. Med. Assoc. Thail.* **2007**, *90*, 1074–1081.
118. Soto, M.E.; Manzano-Pech, L.G.; Guarner-Lans, V.; Díaz-Galindo, J.A.; Vásquez, X.; Castrejón-Tellez, V.; Gamboa, R.; Huesca, C.; Fuentevilla-Alvárez, G.; Pérez-Torres, I. Oxidant/Antioxidant Profile in the Thoracic Aneurysm of Patients with the Loeys-Dietz Syndrome. *Oxid. Med. Cell Longev.* **2020**, *2020*, 5392454. [CrossRef]
119. Zúñiga-Muñoz, A.M.; Pérez-Torres, I.; Guarner-Lans, V.; Núñez-Garrido, E.; Velázquez Espejel, R.; Huesca-Gómez, C.; Gamboa-Ávila, R.; Soto, M.E. Glutathione system participation in thoracic aneurysms from patients with Marfan syndrome. *Vasa* **2017**, *46*, 177–186. [CrossRef]
120. Benzie, I.F.F.; Strain, J.J. The Ferric Reducing Ability of Plasma (FRAP) as a Measure of "Antioxidant Power": The FRAP Assay. *Anal. Biochem.* **1996**, *239*, 70–76. [CrossRef]
121. Erel, O.; Neselioglu, S. A novel and automated assay for thiol/disulphide homeostasis. *Clin. Biochem.* **2014**, *47*, 326–332. [CrossRef]
122. Soto, M.E.; Guarner-Lans, V.; Díaz-Díaz, E.; Manzano-Pech, L.; Palacios-Chavarría, A.; Valdez-Vázquez, R.R.; Aisa-Álvarez, A.; Saucedo-Orozco, H.; Pérez-Torres, I. Hyperglycemia and Loss of Redox Homeostasis in COVID-19 Patients. *Cells* **2022**, *11*, 932. [CrossRef]
123. Flohé, L.; Günzler, W.A. *[12] Assays of Glutathione Peroxidase*; Academic Press: Cambridge, MA, USA, 1984; pp. 114–120. [CrossRef]
124. Soto, M.E.; Soria-Castro, E.; Lans, V.G.; Ontiveros, E.M.; Mejía, B.I.H.; Hernandez, H.J.M.; García, R.B.; Herrera, V.; Pérez-Torres, I. Analysis of Oxidative Stress Enzymes and Structural and Functional Proteins on Human Aortic Tissue from Different Aortopathies. *Oxid. Med. Cell Longev.* **2014**, *2014*, 1–13. [CrossRef]

125. Jagota, S.K.; Dani, H.M. A new colorimetric technique for the estimation of vitamin C using Folin phenol reagent. *Anal. Biochem.* **1982**, *127*, 178–182. [CrossRef]
126. LYSOMUCIL—PLM. Available online: https://www.medicamentosplm.com/Home/productos/lysomucil_tabletas_efervescentes/42/101/36846/218 (accessed on 26 June 2020).
127. REDOXON—PLM. Available online: https://www.medicamentosplm.com/Home/productos/redoxon_tabletas_masticables/21/101/9525/220 (accessed on 26 June 2020).
128. ETERNAL CÁPSULAS 400 UI de México. Available online: https://www.vademecum.es/mexico/medicamento/1280368/eternal+capsulas+400+ui (accessed on 2 September 2023).
129. BENEDORM—PLM. Available online: https://www.medicamentosplm.com/Home/productos/benedorm_tabletas_sublinguales/187/101/10598/224 (accessed on 26 June 2020).
130. Johnson, C.E.; Cober, M.P.; Thome, T.; Rouse, E. Stability of an extemporaneous alcohol-free melatonin suspension. *Am. J. Health Pharm.* **2011**, *68*, 420–423. [CrossRef]
131. National Center for Biotechnology Information. *Vitamin E | C29H50O2*; National Center for Biotechnology Information: Bethesda, MD, USA, 2022.
132. Galli, F.; Azzi, A.; Birringer, M.; Cook-Mills, J.M.; Eggersdorfer, M.; Frank, J.; Cruciani, G.; Lorkowski, S.; Özer, N.K. Vitamin E: Emerging aspects and new directions. *Free Radic. Biol. Med.* **2017**, *102*, 16–36. [CrossRef]
133. Outline, C.; Concepts, A. Distribution in Foods. 2012; pp. 181–211. Available online: https://health2016.globalchange.gov/food-safety-nutrition-and-distribution (accessed on 9 October 2023).

International Journal of
Molecular Sciences

MDPI

Article

Effect of Natural Variation and Rootstock on Fruit Quality and Volatile Organic Compounds of '*Kiyomi tangor*' (*Citrus reticulata* Blanco) Citrus

Tie Wang [†], Zhendong Zheng [†], Lijun Deng, Weijia Li, Ya Yuan, Mingfei Zhang, Guochao Sun, Siya He, Jun Wang, Zhihui Wang * and Bo Xiong

College of Horticulture, Sichuan Agricultural University, Chengdu 611130, China;
wangtie@stu.sicau.edu.cn (T.W.); zhengzhendong@stu.sicau.edu.cn (Z.Z.); 2020205017@stu.sicau.edu.cn (L.D.);
liweijia@stu.sicau.edu.cn (W.L.); yuanya@stu.sicau.edu.cn (Y.Y.); zhang_mingfei@sicau.edu.cn (M.Z.);
71203@sicau.edu.cn (G.S.); hesiya@sicau.edu.cn (S.H.); 13560@sicau.edu.cn (J.W.);
xiongbo1221@sicau.edu.cn (B.X.)

* Correspondence: wangzhihui318@sicau.edu.cn

† These authors contributed equally to this work.

Abstract: In this study, we compared the fruit quality and color of 'Kiyomi' (WT) and its mutant (MT) grafted on Ziyang xiangcheng (Cj) (WT/Cj, MT/Cj), and the MT grafted on Trifoliate orange (Pt) (MT/Pt). The differences in sugar, organic acid, flavonoids, phenols, and volatile substances of the three materials were also analyzed by high performance liquid chromatography (HPLC) and headspace solid-phase microextraction-gas chromatography-mass spectrometry (HS-SPME-GC–MS). The results showed significant differences in the appearance of WT/Cj, MT/Cj, and MT/Pt. MT/Pt, compared to WT/Cj, MT/Cj, had lower sugar, acid, phenol and flavonoid contents in the pulp. However, MT/Pt pulp was higher in vitamin C (V_C), and the peel had significantly higher total phenol and flavonoid contents. In terms of pulp, WT/Cj had the greatest diversity of volatile organic compounds (VOCs). 4-methyl-1-pentanol was significantly higher in MT/Cj pulp, while MT/Pt pulp had a unique octanoic acid, methyl ester. VOCs were more diverse in the peels of the three materials. β-Myrcene and valencen were significantly higher in MT/Cj peels. In contrast, 16 unique VOCs were detected in MT/Pt, and D-limonene content was significantly higher than in WT/Cj and MT/Cj. The results suggest Trifoliate orange is a suitable rootstock for MT.

Keywords: citrus; mutation; rootstock; volatile organic compounds; fruit quality

1. Introduction

Citrus belongs to the *Rutaceae* family and is the world's largest type of fruit. Citrus fruits are rich in bioactive compounds such as vitamin C (V_C), phenols, carotenoids, folic acid, and dietary fiber, which have beneficial effects on human health by preventing cancer, heart disease, and obesity [1–3]. China ranks first among citrus-producing countries in the world, with a planting area of approximately 3 million ha and a yield of approximately 51 million tons in 2021. The abundant germplasm resources play an important role in promoting citrus breeding and the development of the citrus industry.

With the diversification of popular consumption habits, there are new requirements for the quality and variety of citrus. Therefore, the selection and breeding of excellent new citrus varieties has been the focus of industrial development. Citrus is a perennial fruit tree and is polyembryonic, making traditional breeding methods time-consuming and inefficient [4]. In recent years, natural bud mutation has become a major resource for citrus and other fruit tree breeding [5]. Bud variation is the result of somatic mutation, leading to distinctly different phenotypes, also known as bud sports [6]. Previous researchers have reported that bud mutants account for 70% of citrus varieties, such as sweet orange, satsuma mandarin and clementine mandarin [7,8]. Some excellent mutants gain customers' preference by improving the fruit quality of existing varieties. It has been demonstrated that distinct carotenoid and flavonoid accumulation in a spontaneous mutant of 'Ponkan' (*Citrus reticulata* Blanco) resulted in yellowish fruit and enhanced postharvest resistance [9]. The mutation in 'Fengjiewancheng' orange affected the sugar and acid metabolism, which ripened one month later than 'Fengjie 72-1' navel orange [10]. A better understanding of natural bud mutations is critical to maintain the excellent traits of the cultivars.

Commercial production of citrus is widely dependent on rootstock grafting. The interaction between rootstock and scion affects fruit characteristics. Previous studies on this effect have been extensively documented in many species, including apple [11], pear [12], Japanese plum [13], peach [14], and mango [15]. In the case of citrus, the rootstock affected the external color, peel thickness, juice content, sugars and acids, phenol compounds, and volatile compounds of the fruit [16]. Grafting 'Or' and 'Odem' mandarins, 'Valencia' oranges and 'Redson' hybrid on sour orange and Volkamer lemon resulted in total soluble solids and acidity levels of all species being lower in fruit on Volkamer than on sour orange [17]. 'Lane Late' sweet orange grafted on six different citrus rootstocks showed significant differences in antioxidant activity and related compounds [18].

The perceived flavor of citrus fruits is determined by a combination of taste and aroma. The sweet and sour flavors are mainly derived from the sugar and acid in the juice capsule. And the aroma of the fruit is provided by a mixture of dozens of volatiles. *'Kiyomi tangor'* (*Citrus unshiu* × *Citrus sinensis* cv. Kiyomi) is the first orange hybrid bred using the Satsuma mandarin (*Citrus unshiu* Marcov.forma miyagawawase) and the Trovita orange (*Citrus sinensis* Osbeck) in Japan [19]. The late ripening characteristic of *'Kiyomi tangor'* fruits holds great promise for their market prospects [20]. The *'Kiyomi tangor'* citrus mutant was discovered in Meishan, Sichuan Province during our field survey. However, little information is known about this mutant. In this work, we compared the differences between 'Kiyomi' and its natural mutant at the levels of fruit morphological characteristics, fruit quality, antioxidant capacity, and volatile metabolites. In addition, we found significant differences between the mutant grafted on Trifoliate orange (*Poncirus trifoliata* [L.] Raf.) and Ziyang xiangcheng (*Citrus junos* Sieb. ex Tanaka). Therefore, this study explored the effects on 'Kiyomi' fruit from two dimensions: bud variation and rootstock. This result will provide germplasm resources and a theoretical basis for improving the quality of 'Kiyomi' fruits.

2. Results

2.1. Fruit Variation and Color Differences

Figure 1a shows the appearance characteristics of the fruit after WT mutation. Compared to WT, the MT peel had a distinctly protuberant character, and the trait was more pronounced especially on grafted Pt rootstocks. In addition, there were significant differences in peel color. Combining the color parameters, we found that MT grafted on Pt had significantly higher L*, a*, and C values, indicating higher brightness, redder color, and higher color purity of MT/Pt peel. In contrast, MTs grafted on Cj rootstocks possessed significantly lower a* values, indicating lower MT/Cj peel color redness [21]. Thus, this result indicates that the mutation gives the MT peel a distinctly different raised character from that of the WT, and that the grafting on different rootstocks exhibits different performance, with the raised character of the MT being more distinct and redder when Pt is used as the rootstock.

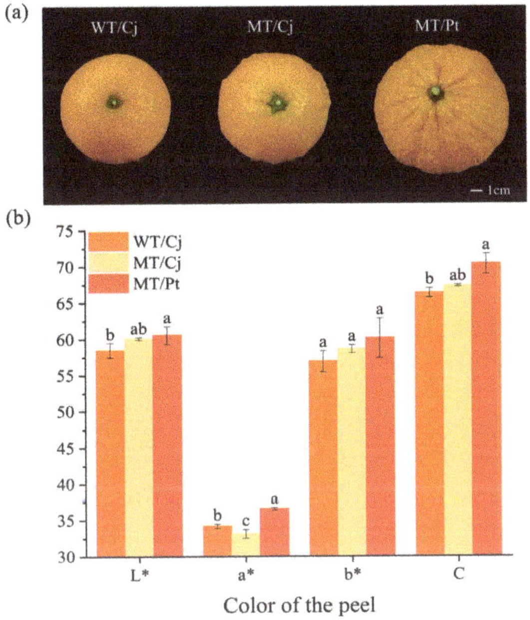

Figure 1. Fruit shape and color differences between '*Kiyomi tangor*' and varieties: (**a**) fruit shape (**b**) color of the peel. Different letters above bars indicate significant differences at the $p < 0.05$ level according to Duncan's test.

2.2. Standard Quality Parameters

The effects of different rootstocks on standard quality parameters of MT fruit are shown in Table 1. Compared to WT/Cj and MT/Cj, MT/Pt fruits possessed significantly greater fruit weight, vertical/transverse diameter, and significantly higher V_C. WT/Cj had significantly higher TSS and TSS/TA. The differences in fruit shape index and TA were not significant among the treatments. These data indicated that both mutation and rootstock treatments had an effect on standard quality parameters of MT fruit.

Table 1. The comparison of fruit quality of '*Kiyomi tangor*' variation grafted on Pt and Cj rootstocks.

Treatments	Fruit Weight (g)	Vertical Diameter (mm)	Transverse Diameter (mm)	Fruit Shape Index	TSS (g/100 g FW)	TA (g/100 g FW)	TSS/TA Ratio	V_C (g/100 mL FW)
WT/Cj	254.13 ± 4.50 b	72.19 ± 1.93 b	81.82 ± 0.21 b	0.88 ± 0.03 a	10.53 ± 0.25 a	0.82 ± 0.07 a	12.91 ± 1.12 a	28.61 ± 0.47 b
MT/Cj	194.67 ± 16.97 c	70.18 ± 1.49 b	75.31 ± 2.28 c	0.93 ± 0.02 a	7.00 ± 0.10 c	0.80 ± 0.02 a	8.75 ± 0.30 c	23.80 ± 1.23 c
MT/Pt	357.03 ± 11.62 a	81.83 ± 3.34 a	92.34 ± 2.09 a	0.89 ± 0.05 a	8.30 ± 0.17 b	0.81 ± 0.02 a	10.25 ± 0.13 b	35.78 ± 0.49 a

TSS, total soluble solid; TA, titratable acidity. Different letters after the numbers indicate significant differences at the $p < 0.05$ level according to Duncan's test.

2.3. Sugar and Acid Components

Among the sugar components, we found that MT/Cj fruits had significantly lower soluble sugar components content and total content. MT/Pt fruits were not significantly different from WT/Cj treatment in sugar content except for the lower sucrose content (Figure 2a). In the comparison of acid components, we found that Cj rootstock conferred significantly higher levels of oxalic acid, citric acid, and total acid to MT fruits. MT/Pt had significantly higher malic acid and acetic acid content compared to WT/Cj and MT/Cj (Figure 2b).

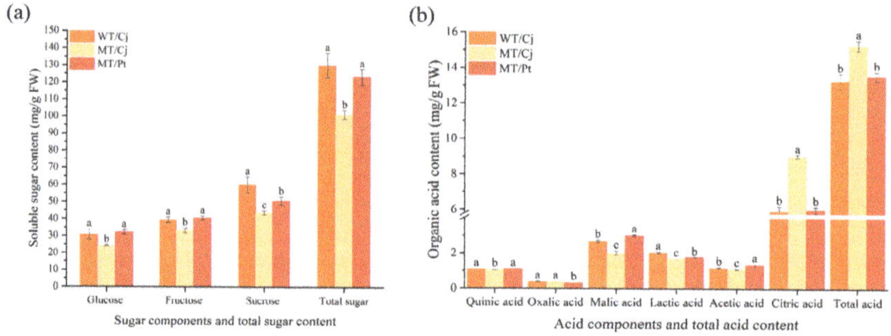

Figure 2. Differences in WT and MT sugar and acid components of different treatments: (**a**) soluble sugar (**b**) organic acids. Different letters above bars indicate significant differences at the $p < 0.05$ level according to Duncan's test.

2.4. Phenols Content

2.4.1. Total Flavonoids and Phenols Content

As shown in Figure 3a, WT/Pt peel possessed significantly higher total flavonoids and total phenols content, while the differences between WT/Cj and MT/Cj peel were not significant. In the pulp, total flavonoids content was significantly higher in WT/Cj and lowest in MT/Cj. The difference between WT/Cj and MT/Pt total phenols content was not significant, but MT/Cj content was significantly lower (Figure 3b).

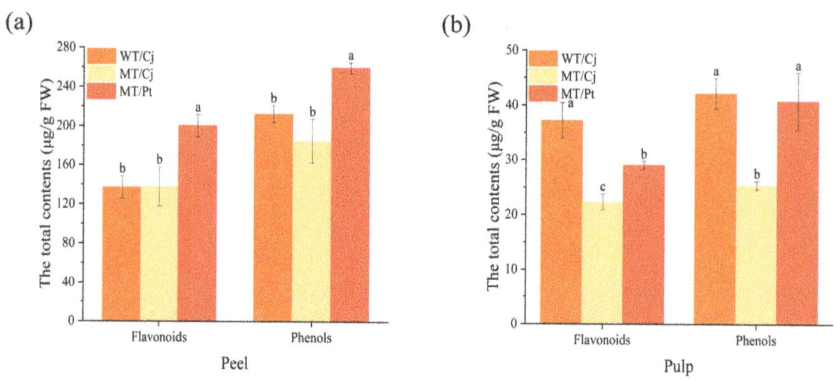

Figure 3. Total flavonoids, phenols in (**a**) peels and (**b**) pulp. Different letters above bars indicate significant differences at the $p < 0.05$ level according to Duncan's test.

2.4.2. The Contents of Phenol Compounds in Pulp

The importance of phenols in human health is well-known, and their potential protective effects are widely recognized [22,23]. Since the pulp is the main edible part of citrus fruit, we conducted an in-depth study using High Performance Liquid Chromatography (HPLC) in order to analyze the reasons for the differences in total pulp phenols between treatments. As shown in Figure 4, a total of eight phenols were detected in the pulp of different treatments. The highest content was sinapic acid and the lowest was *p*-hydroxybenzoic acid. In addition, all WT/Cj components were significantly highest between treatments, except for gallic acid and rutin. In MT fruits, there were effects of different rootstocks on gallic acid, naringin, and rutin contents, and non-significant differences in the other five phenols. Overall, mutation reduced fruit phenol accumulation and different rootstocks also influenced fruit phenol accumulation.

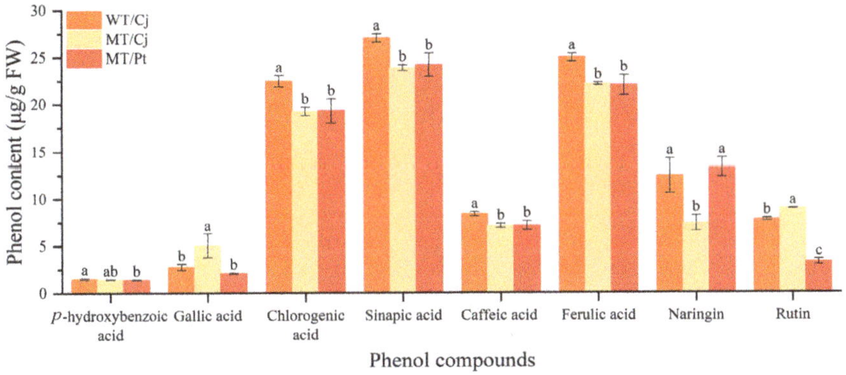

Figure 4. The contents of phenol compounds in pulp. Neohesperidin and naringenin were not detected. Different letters above bars indicate significant differences at the $p < 0.05$ level according to Duncan's test.

2.5. Principal Components Analysis (PCA) and Correlation Analysis

PCA was used to study the overall variation in the relationship between various quality parameters between treatments [24]. In the present work, the results of the experiment showed significant differences between the treatments (Figure 5). The differences between MT/Cj and WT/Cj treatments with the same rootstock were smaller but significantly larger than MT/Pt. The combined results showed that both mutation and rootstock had an effect on each quality index, but the rootstock effect was significantly more pronounced. In the peel, a* and C showed a significant positive correlation with total flavonoids, indicating that total flavonoids had a significant effect on peel color. In the pulp, we found a positive correlation between TSS and all sugar components, especially sucrose.

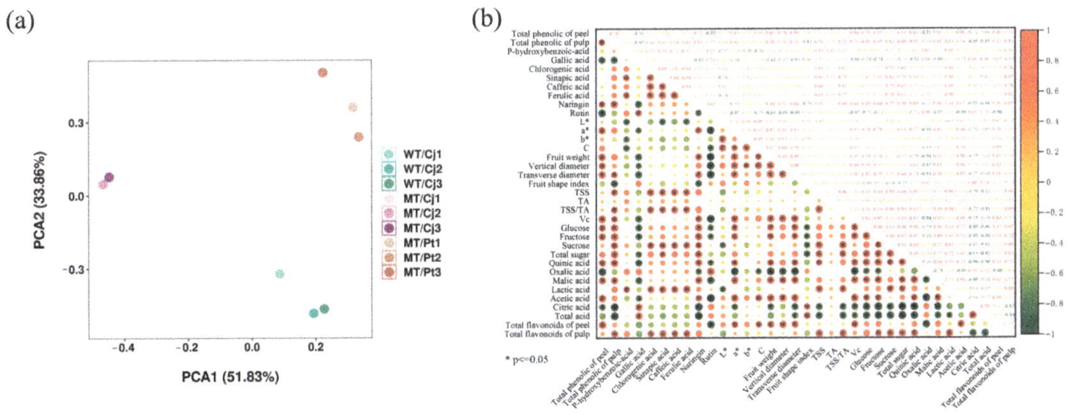

Figure 5. PCA and correlation analysis of the quality parameters of the different treatments: (**a**) PCA analysis. (**b**) correlation analysis.

2.6. VOCs

2.6.1. Identification and Quantification of the VOCs in the Pulp of Different Treatments

As shown in Figure 6a,b, the WT/Cj treatment had the highest number of pulp VOCs species and higher content, especially D-limonene (Table S1) and β-Myrcene. The pulp 4-methyl-1-pentanol content was significantly higher in MT/Cj treatment compared to WT/Cj, while MT/Pt had higher octanal. Furthermore, although the MT/Pt treatment pulp had the fewest types and lower levels of VOCs, it contained the unique octanoic acid, methyl ester. PCA analysis showed small material differences between WT/Cj and MT/Cj treatments, but both differed significantly from MT/Pt treatments (Figure 6c).

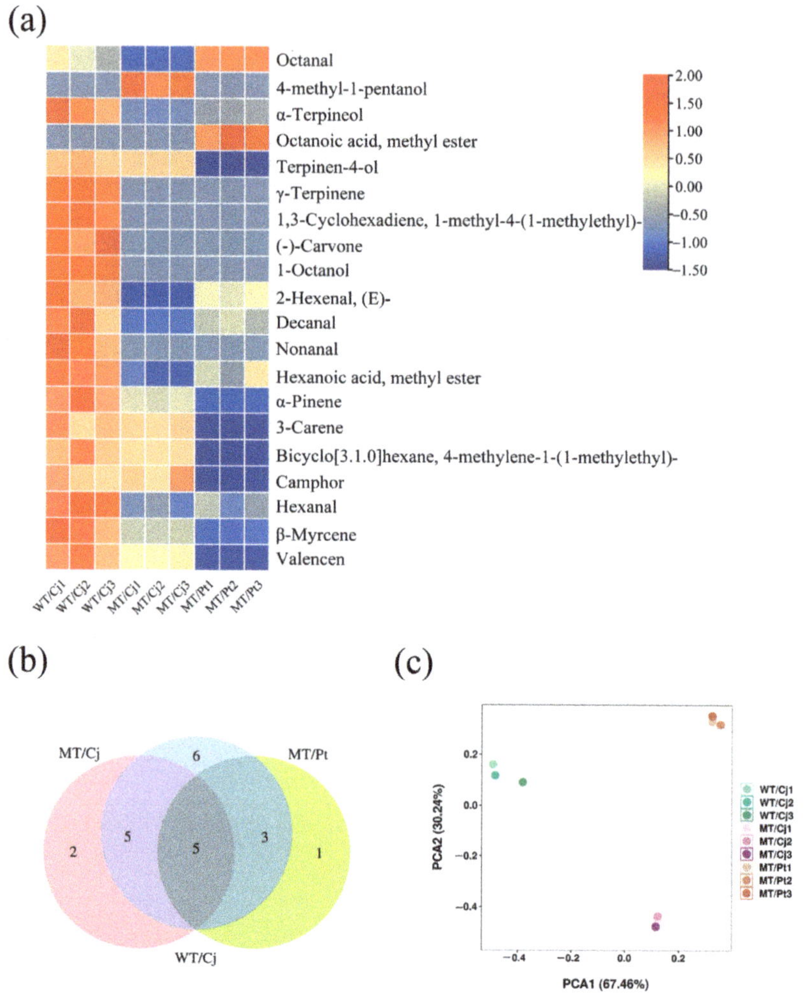

Figure 6. Analysis of the VOCs in the pulp of different treatments: (**a**) heatmap (**b**) Venn diagram (**c**) PCA analysis. D-limonene is not shown in the heatmap.

2.6.2. Identification and Quantification of the VOCs in the Peel of Different Treatments

Compared to the pulp, the peel VOCs were significantly more diverse, with 50, 51, and 67 VOCs in the WT/Cj, MT/Cj, and MT/Pt treatments, respectively (Figures 7 and 8). Among them, MT/Pt treatment peel contained unique 16 VOCs such as *p*-mentha-1,8-dien-7-yl acetate, 2,6-dimethyl-1,3,5,7-octatetraene, E, E-, and d-carvone. Overall, the MT/Pt treatment peel had significantly higher VOC content, especially D-limonene, which was 60.65% and 27.06% higher than WT/Cj and MT/Cj, respectively (Table S2). However, MT/Cj contained significantly higher β-Myrcene and valencen. PCA analysis showed large material differences between treatments, especially MT/Pt to WT/Cj, which may imply that rootstocks have a similar function to mutation, both altering the type and content of peel VOCs.

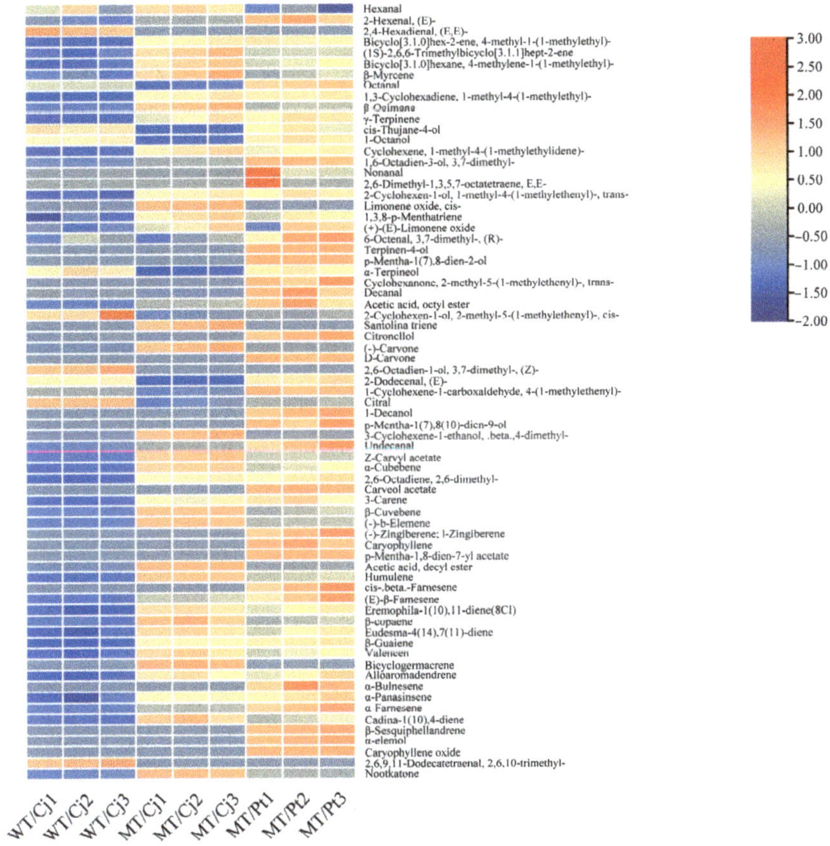

Figure 7. Heatmap of the VOCs in the peel of different treatments. D-limonene is not shown in the heatmap.

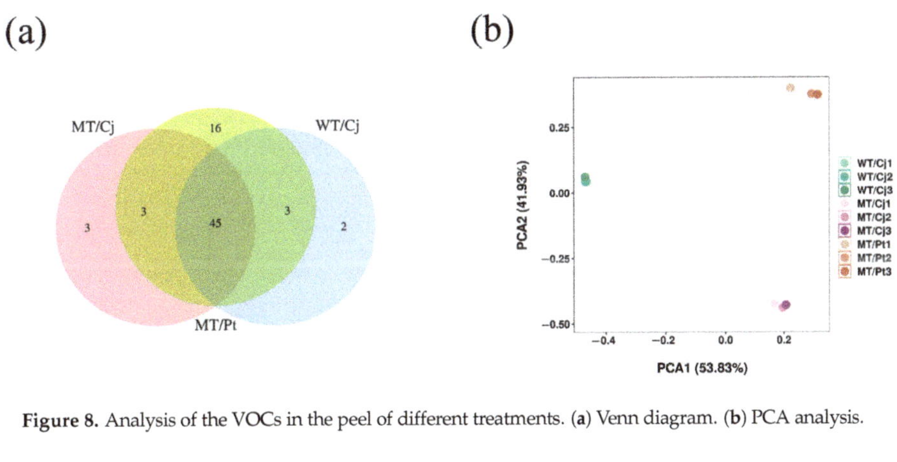

Figure 8. Analysis of the VOCs in the peel of different treatments. (**a**) Venn diagram. (**b**) PCA analysis.

3. Discussion

3.1. Natural Variation and Rootstocks Effect the General Morphological Characteristics and Nutrient Content of 'Kiyomi tangor' Citrus

Citrus is the most produced fruit in the world and provides important nutrients to people [25]. Although cultivars are numerous, they are mainly derived from spontaneous mutations [26]. Our results showed that the mutation conferred a protuberant character to the MT peel. In addition, this trait can become more pronounced by grafting onto Pt rootstocks (Figure 1a). Currently, this unique trait makes MT fruits popular in the market, but the underlying mechanism still needs further study. The more vivid the color of the citrus fruit, the more attention it gets [27]. In the present study, mutation caused a decrease in MT peel redness, but by grafting Pt rootstocks, MT peel redness increased significantly, as evidenced by an increase in a* values. This indicates that the peel color is not only genetically related, but also regulated by the rootstock. This is consistent with the results of previous studies [28,29].

Fruit size and weight are very important economic traits in fruit breeding research [30]. In this study, our results have found significantly smaller weight and size of MT fruits, but significantly larger MT/Pt fruits (Table 1). It is hypothesized that the variation in this trait may be the result of environment, developmental instability, and plasticity [31]. That this variation produces unexpected results is worthy of further study. Previous studies in orange and grape found that the mutants had significantly higher TSS/TA values compared to the wild type, indicating that they would ripen slightly earlier than the parental plants [32,33]. Contrary to this, our results have found that MT fruit TSS/TA was significantly lower, presumably maturing later in MT than WT. In general, the main soluble sugars in most citrus are glucose, fructose, and sucrose, but mainly sucrose. Organic acids, on the other hand, are dominated by citric acid [34,35]. Consistent with this, our results show a similar pattern (Figure 2). Overall, MT/Cj fruits showed low sugar and high acid characteristics, but MT/Pt was not significantly different from WT/Cj. It indicated that rootstocks play an important role in the regulation of sugar-acid metabolism in MT fruits.

3.2. Natural Variation and Rootstock Effect the Antioxidant Capacity of 'Kiyomi tangor' Citrus Peel and Pulp

Plant polyphenols are phenylpropanoids, including phenol acids, stilbene, curcumin, and flavonoids. These compounds have a wide range of biopharmacological characteristics, such as being antioxidant, anti-inflammatory, and antiviral, and have promising applications [36]. Antioxidant capacity is the link and bridge between multiple biological activities. Citrus flavonoids play a crucial role in the regulation of oxidative stress for health benefits [37]. In this study, compared with WT/Cj, the difference in total flavonoid content of MT/Cj peel was not significant, but MT/Pt peel became significantly higher, presumably the rootstock promoted the accumulation of total flavonoids in MT peel. In contrast, in the pulp, we observed a significantly lower total flavonoid content of MT pulp, speculating that the mutation may have reduced the accumulation of total flavonoids in the pulp. However, it is noteworthy that rootstock (Pt) had a significant effect on the accumulation of total flavonoids in MT pulp (Figure 3). Similar results have been observed in grapes [38] and apples [39]. Among the flavonoid components, we detected only naringin and rutin; neohesperidin and naringenin were not detected (Figure 4). This is in agreement with our previous results [40]. In addition, we also found that the content of flavonoid components and total flavonoid content varied in the same trend between treatments, indicating that our results are reliable.

Citrus peel is a good source of phenols because it is rich in phenol acids [41]. In this study, the total phenols content of MT/Cj and WT/Cj peels were not significantly different, but the total phenol content of MT/Cj pulp was significantly lower. In addition, the MT/Pt peel and pulp total phenol content were significantly higher compared to MT/Cj (Figure 3). This result is consistent with the trend in total flavonoids, which further suggests that mutation may reduce the accumulation of total flavonoids and total phenols in MT pulp, but Pt rootstock has the opposite effect. In the phenol acids, the common phenol acids include caffeic, ferulic, *p*-coumaric, sinapic acids, and so on [42,43]. As shown in Figure 4, we detected a total of 6 phenol acids in the pulp of the fruit. The phenol acid content of the different treatments showed the same pattern, with sinapic acid being the most abundant and *p*-hydroxybenzoic acid being the least abundant. This is consistent with the previous findings in navel orange [44]. In addition, previous authors noted that sinapic acid also increased significantly in Thai fruits during ripening, and they also demonstrated a strong positive correlation between sinapic acid and fruit ripeness [45]. It is speculated that the ripening of '*Kiyomi tangor*' citrus may also be related to sinapic acid.

3.3. Natural Variation and Rootstocks Effect Volatile Organic Compounds in the Peel and Pulp of 'Kiyomi tangor' Citrus

Today, consumers increasingly prefer foods with raw, natural and delicious flavors, which are closely related to the content of VOCs in food [46]. Our results revealed that the mutation conferred significantly fewer VOC species and total content in MT pulp. Fruits of MT grafted on Pt rootstock had a significantly higher amount of octanal and contained unique octanoic acid and methyl ester, although both VOC species and total content were significantly lower (Figure 6). Previous studies have found that octanal exhibited sweet, citrus, fatty, pungent, green flavors [47] and octanoic acid and methyl ester exhibited exceptional citrus flavors [48]. These unique substances give the MT/Pt pulp its distinctive flavor, which is presumably related to the Pt rootstock. In contrast, VOCs were significantly higher in MT/Cj peels compared to WT/Cj, especially for D-limonene, β-Myrcene, and valencen. In addition, MT/Pt peel showed a similar pattern to MT/Cj and also contained a unique set of 16 VOCs (Table S2, Figures 7 and 8). These results tentatively suggest that the mutation confers significantly higher VOCs content to MT peel and that Pt rootstock has the same effect. This is similar to previous studies on watermelon [49], avocado [50], and grapes [51], indicating that the effect of rootstocks on VOCs is significant.

4. Materials and Methods

4.1. Plant Materials and Treatment

In this experiment, '*Kiyomi tangor*' (WT) and its mutation (MT) were used as materials. WT grafted on Ziyang xiangcheng (Cj) and Trifoliate orange (Pt) was used as a control (WT/Cj, WT/Pt), and changes in MT fruit grafted on Cj (MT/Cj), and Pt (MT/Pt) were studied. The preceding study conducted by our research group revealed that there was no significant disparity in quality between WT/Cj and WT/Pt fruits at maturity [52]. Therefore, only WT/Cj was selected as the control. Trees were 5 years old and at the same management level in all treatments.

The experiment was conducted with three trees as one treatment, with three replications and a total of 20 healthy fruits collected from each tree from five different positions, with a total of 60 fruits per treatment. Commercially mature fruits were rapidly brought back to the laboratory from Meishan City, Sichuan Province, China. Twelve fruits were randomly selected from each treatment for color determination, and the pulp of the remaining fruits was separated and mixed well, frozen in liquid nitrogen and stored in a −80 °C refrigerator until use.

4.2. Color Measurements

The peel color (L*, a*, b*) was measured using a CR-400 (Konica Minolta Inc., Tokyo, Japan). The color purity index C was calculated by the formula: $C = (a^2 + b^2)^{1/2}$ [53].

4.3. Standard Quality Parameters

Fruit weight was measured by electronic scales and vertical/transverse diameter was measured by vernier calipers. Total soluble solids (TSS) and titratable acids (TA) were measured using a glycolic acid meter (Pocket PAL-BXIACID1; ATAGO, Tokyo, Japan) [54]. V_C was measured according to the method described by previous authors [55]. Take 5 mL of juice, dilute with 1% oxalic acid and fix the volume to 50 mL. 5 mL of the diluted solution was pipetted into a 50 mL triangular flask and titrated with 2,6-dichlorophenol indigo dye solution until a stable light pink color was obtained.

4.4. Sugar and Acid Compounds

The sugar and acid components of the pulp were measured using high performance liquid chromatography (HPLC). The sugar components were determined and analyzed referring to the previous description [56]. A mixed sample of 2 g was accurately weighed and extracted by adding 4 mL of distilled water. After water bath and centrifugation, the sample was transferred to an injection vial through a 0.45 μm aqueous phase filter membrane to analyze the sugar compositions. Acid components were determined according to our previous method [34,57]. The mixed sample was accurately weighed to 0.5 g, ground with 3 mL of pre-chilled 0.2% metaphosphoric acid, and after volume fixation and centrifugation, the supernatant was taken to determine the organic acid compositions.

4.5. Total Flavonoid, Phenol Contents and Phenol Compounds

The total flavonoid contents were determined by a colorimetric method with aluminum chloride [58]. The extract was mixed with 75 μL of 95% ethanol, 10 μL of 10% aluminum chloride, 10 μL of 1.0 M potassium acetate, and 140 μL of distilled water. After incubation, the absorbance was measured at 415 nm. A standard curve was made with rutin, and the data units are μg g^{-1} FW. The Folin–Ciocalteu method was used to quantify the total phenol content, with specific reference to the previous method [59]. The standard curves were made with gallic acid aqueous solution, and the data units are μg g^{-1} FW.

The phenol components were determined by an HPLC method optimized in our laboratory [40]. The mixture of 0.5 g was accurately weighed into a 2 mL centrifuge tube, extracted by adding the extraction solution and protected from light, and the supernatant was used for the analysis of phenol components.

4.6. Pulp and Peel Volatile Organic Compounds (VOCs) Identification and Quantification

Headspace solid-phase microextraction-gas chromatography–mass spectrometry (HS-SPME-GC–MS) was used to extract and detect VOCs in different parts of fruits. Fruit tissues (0.1 g peel and 0.5 g pulp) were ground to a powder using liquid nitrogen and quickly placed in a headspace flask, then 5 mL of saturated sodium chloride solution and 50 μL of 1-hexanol (0.1%, *v*/*v*) were added and VOCs were measured after equilibration and extraction [54]. The inlet temperature, desorption time and column chamber warming procedures were performed as previously described [60]. The data were processed using the Data Analysis Application (www.agilent.com/chem, accessed on 30 September 2022), database NIST 11.0/14.0 (NIST/EPA/NIH, Gaithersburg, MD, USA). Combined with NIST data and retention index (RI) data to characterize the substances [61]. The relative quantification of each substance was performed in combination with internal standards [62].

4.7. Statistical Analysis

Data were organized using Microsoft Office Excel 2016 (Microsoft, Redmond, Washington, DC, USA), bar graphs were plotted using Origin 2021 (OriginLab Corporation, Northampton, MA, USA), and ANOVA procedure using IBM SPSS 23.0 (SPSS, Inc., Chicago, IL, USA) was used for significance of differences analysis. Heat map, Venn diagram and PCA analysis were performed on the website (https://www.omicstudio.cn/tool, accessed 30 October 2022).

5. Conclusions

In conclusion, in terms of fruit appearance quality, the mutation conferred a distinctly protuberant character to the MT peel and a smaller fruit size. When MT was grafted on Pt rootstock, this protuberant character was more pronounced, with significantly larger fruit size and redder peel color. In terms of intrinsic quality, the mutation resulted in significantly lower TSS in MT, but significantly higher V_C content in MT/Pt fruit. In terms of antioxidants, the differences in total phenols and total flavonoid contents between WT/Cj and MT/Cj peels were not significant, but the pulp was significantly lower. MT/Pt peel and pulp total phenols and total flavonoid contents were significantly higher than WT/Cj. In terms of VOCs, the mutation reduced the type and content of MT/Cj pulp VOCs, and the MT/Pt treatment was significantly lower. However, in the peel, the differences in the types of MT/Cj peel VOCs were not significant compared to WT/Cj, but the contents were significantly higher, especially for β Myrcene and valencen. In addition, MT/Pt peel contained a unique set of 16 VOCs. The findings reported here contribute to a better understanding of the differences in fruit quality of *'Kiyomi tangor'* citrus before and after mutation and provide implications for the selection of MT citrus rootstocks.

Supplementary Materials: The following supporting information can be downloaded at: https://www.mdpi.com/article/10.3390/ijms242316810/s1.

Author Contributions: Conceptualization, T.W., Z.Z. and L.D.; software, T.W.; validation, T.W., Z.Z. and L.D.; formal analysis, T.W. and L.D.; investigation, W.L., Y.Y., S.H. and J.W.; data curation, M.Z. and G.S.; writing—original draft preparation, T.W. and Z.Z.; writing—review and editing, T.W. and L.D.; project administration, Z.W. and B.X. All authors have read and agreed to the published version of the manuscript.

Funding: This research was funded by the Sichuan Science and Technology Program [Grant No. 2021YFYZ0023-14] and the Sichuan Provincial Science and Technology Department Science and Technology Plan Project [Grant No. 2021YFN0025].

Institutional Review Board Statement: Not applicable.

Informed Consent Statement: Not applicable.

Data Availability Statement: Data are contained within the article.

Acknowledgments: We are very grateful to Jun Liu for providing and taking care of the plant material.

Conflicts of Interest: The authors declare no conflict of interest.

References

1. Pellati, F.; Benvenuti, S. Fast high-performance liquid chromatography analysis of phenethylamine alkaloids in Citrus natural products on a pentafluorophenylpropyl stationary phase. *J. Chromatogr. A* **2007**, *1165*, 58–66. [CrossRef]
2. Goldenberg, L.; Yaniv, Y.; Porat, R.; Carmi, N. Mandarin fruit quality: A review. *J. Sci. Food Agric.* **2018**, *98*, 18–26. [CrossRef]
3. Pasdaran, A.; Hamedi, A.; Shiehzadeh, S.; Hamedi, A. A review of citrus plants as functional foods and dietary supplements for human health, with an emphasis on meta-analyses, clinical trials, and their chemical composition. *Clin. Nutr. ESPEN* **2023**, *54*, 311–336. [CrossRef]
4. Koltunow, A.M.; Hidaka, T.; Robinson, S.P. Polyembryony in Citrus. Accumulation of seed storage proteins in seeds and in embryos cultured in vitro. *Plant Physiol.* **1996**, *110*, 599–609. [CrossRef]
5. Zeng, J.W.; Gao, C.Y.; Deng, G.M.; Jiang, B.; Yi, G.J.; Peng, X.X.; Zhong, Y.; Zhou, B.R.; Liu, K. Transcriptome analysis of fruit development of a citrus late-ripening mutant by microarray. *Sci. Hortic.* **2012**, *134*, 32–39. [CrossRef]
6. Ban, S.; Jung, J.H. Somatic Mutations in Fruit Trees: Causes, Detection Methods, and Molecular Mechanisms. *Plants* **2023**, *12*, 1316. [CrossRef]
7. Gülşen, O.; Roose, M.L. Lemons: Diversity and relationships with selected Citrus genotypes as measured with nuclear genome markers. *J. Am. Soc. Hortic. Sci.* **2001**, *126*, 309–317. [CrossRef]
8. Liu, Q.; Zhu, A.; Chai, L.J.; Zhou, W.J.; Yu, K.Q.; Ding, J.; Xu, J.; Deng, X.X. Transcriptome analysis of a spontaneous mutant in sweet orange *Citrus sinensis* (L.) Osbeck during fruit development. *J. Exp. Bot.* **2009**, *60*, 801–813. [CrossRef]
9. Luo, T.; Xu, K.Y.; Luo, Y.; Chen, J.J.; Sheng, L.; Wang, J.Q.; Han, J.W.; Zeng, Y.L.; Xu, J.; Chen, J.M.; et al. Distinct Carotenoid and Flavonoid Accumulation in a Spontaneous Mutant of Ponkan (*Citrus reticulata* Blanco) Results in Yellowish Fruit and Enhanced Postharvest Resistance. *J. Agric. Food Chem.* **2015**, *63*, 8601–8614. [CrossRef]
10. Liu, Y.Z.; Liu, Q.; Xiong, J.J.; Deng, X.X. Difference of a citrus late-ripening mutant (*Citrus sinensis*) from its parental line in sugar and acid metabolism at the fruit ripening stage. *Sci. China Ser. C-Life Sci.* **2007**, *50*, 511–517. [CrossRef]
11. Butkeviciute, A.; Abukauskas, V.; Janulis, V.; Kviklys, D. Phenolic Content and Antioxidant Activity in Apples of the 'Galaval' Cultivar Grown on 17 Different Rootstocks. *Antioxidants* **2022**, *11*, 266. [CrossRef] [PubMed]
12. Ikinci, A.; Bolat, I.; Ercisli, S.; Kodad, O. Influence of rootstocks on growth, yield, fruit quality and leaf mineral element contents of pear cv. 'Santa Maria' in semi-arid conditions. *Biol. Res.* **2014**, *47*, 8. [CrossRef] [PubMed]

13. Daza, A.; Garcia-Galavis, P.A.; Grande, M.J.; Santamaria, C. Fruit quality parameters of 'Pioneer' Japanese plums produced on eight different rootstocks. *Sci. Hortic.* **2008**, *118*, 206–211. [CrossRef]
14. Shahkoomahally, S.; Chang, Y.R.; Brecht, J.K.; Chaparro, J.X.; Sarkhosh, A. Influence of rootstocks on fruit physical and chemical properties of peach cv. UFSun. *Food Sci. Nutr.* **2021**, *9*, 401–413. [CrossRef] [PubMed]
15. Rebolledo-Martinez, A.; Peralta-Antonio, N.; Rebolledo-Martinez, L.; Becerril-Roman, E.A.; Rebolledo-Garcia, R.L. Effect of rootstock in tree growth, dry matter, flowering, yield and quality of 'Manila' mango. *Sci. Hortic.* **2019**, *251*, 155–161. [CrossRef]
16. Alfaro, J.M.; Bermejo, A.; Navarro, P.; Quinones, A.; Salvador, A. Effect of Rootstock on Citrus Fruit Quality: A Review. *Food Rev. Int.* **2021**, *39*, 2835–2853. [CrossRef]
17. Benjamin, G.; Tietel, Z.; Porat, R. Effects of Rootstock/Scion Combinations on the Flavor of Citrus Fruit. *J. Agric. Food Chem.* **2013**, *61*, 11286–11294. [CrossRef]
18. Cardenosa, V.; Barros, L.; Barreira, J.C.M.; Arenas, F.; Moreno-Rojas, J.M.; Ferreira, I. Different Citrus rootstocks present high dissimilarities in their antioxidant activity and vitamins content according to the ripening stage. *J. Plant Physiol.* **2015**, *174*, 124–130. [CrossRef]
19. Han, S.; Kim, H.M.; Lee, S. Simultaneous determination of polymethoxyflavones in Citrus species, *Kiyomi tangor* and Satsuma mandarin, by high performance liquid chromatography. *Food Chem.* **2012**, *134*, 1220–1224. [CrossRef]
20. Chang, Y.H.; Seo, J.; Song, E.; Choi, H.J.; Shim, E.; Lee, O.; Hwang, J. Bioconverted Jeju Hallabong tangor (Citrus kiyomi x ponkan) peel extracts by cytolase enhance antioxidant and anti-inflammatory capacity in RAW 264.7 cells. *Nutr. Res. Pract.* **2016**, *10*, 131–138. [CrossRef]
21. Belasco, R.; Edwards, T.; Munoz, A.J.; Rayo, V.; Buono, M.J. The Effect of Hydration on Urine Color Objectively Evaluated in CIE L*a*b* Color Space. *Front. Nutr.* **2020**, *7*, 9. [CrossRef] [PubMed]
22. Bertelli, A.; Biagi, M.; Corsini, M.; Baini, G.; Cappellucci, G.; Miraldi, E. Polyphenols: From Theory to Practice. *Foods* **2021**, *10*, 2595. [CrossRef] [PubMed]
23. Finicelli, M.; Squillaro, T.; Galderisi, U.; Peluso, G. Polyphenols, the Healthy Brand of Olive Oil: Insights and Perspectives. *Nutrients* **2021**, *13*, 3831. [CrossRef] [PubMed]
24. Mazur, S.P.; Nes, A.; Wold, A.B.; Remberg, S.F.; Aaby, K. Quality and chemical composition of ten red raspberry (*Rubus idaeus* L.) genotypes during three harvest seasons. *Food Chem.* **2014**, *160*, 233–240. [CrossRef] [PubMed]
25. Lado, J.; Gambetta, G.; Zacarias, L. Key determinants of citrus fruit quality: Metabolites and main changes during maturation. *Sci. Hortic.* **2018**, *233*, 238–248. [CrossRef]
26. Perez-Roman, E.; Borreda, C.; Usach, A.L.G.; Talon, M. Single-nucleotide mosaicism in citrus: Estimations of somatic mutation rates and total number of variants. *Plant Genome* **2022**, *15*, 13. [CrossRef]
27. Huang, Q.C.; Liu, J.M.; Hu, C.X.; Wang, N.A.; Zhang, L.; Mo, X.F.; Li, G.G.; Liao, H.H.; Huang, H.M.; Ji, S.F.; et al. Integrative analyses of transcriptome and carotenoids profiling revealed molecular insight into variations in fruits color of *Citrus reticulata* Blanco induced by transplantation. *Genomics* **2022**, *114*, 11. [CrossRef]
28. Zheng, X.J.; Zhu, K.J.; Sun, Q.; Zhang, W.Y.; Wang, X.; Cao, H.B.; Tan, M.L.; Xie, Z.Z.; Zeng, Y.L.; Ye, J.L.; et al. Natural Variation in CCD4 Promoter Underpins Species-Specific Evolution of Red Coloration in Citrus Peel. *Mol. Plant.* **2019**, *12*, 1294–1307. [CrossRef]
29. Garcia-Munoz, M.C.; Henao-Rojas, J.C.; Moreno-Rodriguez, J.M.; Botina-Azain, B.L.; Romero-Barrera, Y. Effect of rootstock and environmental factors on fruit quality of Persian lime (*Citrus latifolia* Tanaka) grown in tropical regions. *J. Food Compos. Anal.* **2021**, *103*, 7. [CrossRef]
30. Wang, J.H.; Liu, L.Q.; Dong, J.H.; Zhu, L.F.; Wang, Y.C.; Hu, G.B.; Shi, S.Y. Inheritance of Fruit Weight, Size and Organic Acid Content in a Distant Hybrid Population of Longan 'Huanongzao' and Lychee 'Ziniangxi'. *Horticulturae* **2022**, *8*, 999. [CrossRef]
31. Larrinaga, A.R.; Guitian, P. Intraspecific variation in fruit size and shape in Corema album (Ericaceae) along a latitudinal gradient: From fruits to populations. *Biol. J. Linnean Soc.* **2016**, *118*, 940–950. [CrossRef]
32. Cimen, B.; Yesiloglu, T.; Incesu, M.; Yilmaz, B. Studies on mutation breeding in citrus: Improving seedless types of 'Kozan' common orange by gamma irradiation. *Sci. Hortic.* **2021**, *278*, 10. [CrossRef]
33. Zheng, T.; Zhang, S.H.; Leng, X.P.; Sadeghnezhad, E.; Li, T.; Pervaiz, T.; Liu, F.Q.; Jia, H.F.; Fang, J.G. Profiling Analysis of Volatile and Non-volatile Compounds in *Vitis vinifera* Berries (cv. Chardonnay) and Spontaneous Bud Mutation. *Front. Nutr.* **2021**, *8*, 16. [CrossRef] [PubMed]
34. Liao, L.; Dong, T.T.; Qiu, X.; Rong, Y.; Wang, Z.H.; Zhu, J. Nitrogen nutrition is a key modulator of the sugar and organic acid content in citrus fruit. *PLoS ONE* **2019**, *14*, e0223356. [CrossRef] [PubMed]
35. Zhou, Y.; He, W.Z.; Zheng, W.L.; Tan, Q.L.; Xie, Z.Z.; Zheng, C.S.; Hu, C.X. Fruit sugar and organic acid were significantly related to fruit Mg of six citrus cultivars. *Food Chem.* **2018**, *259*, 278–285. [CrossRef]
36. Li, L.; Liu, X.; Qiu, Z.; Zhao, G. Microbial synthesis of plant polyphenols. *Chin. J. Biotechnol.* **2021**, *37*, 2050–2076. [CrossRef]
37. Wang, Y.; Liu, X.-J.; Chen, J.-B.; Cao, J.-P.; Li, X.; Sun, C.-D. Citrus flavonoids and their antioxidant evaluation. *Crit. Rev. Food Sci. Nutr.* **2022**, *62*, 3833–3854. [CrossRef]
38. Han, X.; Wang, Y.; Lu, H.C.; Yang, H.Y.; Li, H.Q.; Gao, X.T.; Pei, X.X.; He, F.; Duan, C.Q.; Wang, J. The combined influence of rootstock and vintage climate on the grape and wine flavonoids of *Vitis vinifera* L. cv. Cabernet Sauvignon in eastern China. *Front. Plant Sci.* **2022**, *13*, 15. [CrossRef]

39. Parvaneh, T.; Abedi, B.; Davarynejad, G.H.; Moghadam, E.G. Enzyme activity, phenolic and flavonoid compounds in leaves of Iranian red flesh apple cultivars grown on different rootstocks. *Sci. Hortic.* **2019**, *246*, 862–870. [CrossRef]
40. Huang, S.; Ye, S.; Xiong, B.; Liu, X.; Xu, Y.; Dong, Z.; Wang, Z. Study of different content of phenolic compounds in citrus pulps and peels. *Food Ferment. Ind.* **2018**, *44*, 241–247. [CrossRef]
41. Chen, Y.; Pan, H.L.; Hao, S.X.; Pan, D.M.; Wang, G.J.; Yu, W.Q. Evaluation of phenolic composition and antioxidant properties of different varieties of Chinese citrus. *Food Chem.* **2021**, *364*, 10. [CrossRef] [PubMed]
42. Ignat, I.; Volf, I.; Popa, V.I. A critical review of methods for characterisation of polyphenolic compounds in fruits and vegetables. *Food Chem.* **2011**, *126*, 1821–1835. [CrossRef] [PubMed]
43. Lee, H.; Oh, I.N.; Kim, J.; Jung, D.H.; Cuong, N.P.; Kim, Y.; Lee, J.; Kwon, O.; Park, S.U.; Lim, Y.; et al. Phenolic compound profiles and their seasonal variations in new red phenotype head-forming Chinese cabbages. *LWT-Food Sci. Technol.* **2018**, *90*, 433–439. [CrossRef]
44. Hou, J.X.; Liang, L.; Su, M.Y.; Yang, T.M.; Mao, X.J.; Wang, Y.X. Variations in phenolic acids and antioxidant activity of navel orange at different growth stages. *Food Chem.* **2021**, *360*, 7. [CrossRef] [PubMed]
45. Siriamornpun, S.; Kaewseejan, N. Quality, bioactive compounds and antioxidant capacity of selected climacteric fruits with relation to their maturity. *Sci. Hortic.* **2017**, *221*, 33–42. [CrossRef]
46. Gong, X.H.; Huang, J.; Xu, Y.Q.; Li, Z.B.; Li, L.; Li, D.; Belwal, T.; Jeandet, P.; Luo, Z.S.; Xu, Y.Q. Deterioration of plant volatile organic compounds in food: Consequence, mechanism, detection, and control. *Trends Food Sci. Technol.* **2023**, *131*, 61–76. [CrossRef]
47. Li, J.; Huang, Q.; Zheng, X.C.; Ge, Z.K.; Lin, K.; Zhang, D.D.; Chen, Y.; Wang, B.; Shi, X.W. Investigation of the Lactic Acid Bacteria in Kazak Cheese and Their Contributions to Cheese Fermentation. *Front. Microbiol.* **2020**, *11*, 13. [CrossRef]
48. Zhang, Y.; Wang, G.; Dong, J.; Cui, F.; Xu, X.; Han, Z. Ana lysis of A rom a Com ponen ts in Five S traw berry Cu ltivars. *Acta Hortic. Sin.* **2008**, *35*, 433–438. [CrossRef]
49. Guler, Z.; Candir, E.; Yetisir, H.; Karaca, F.; Solmaz, I. Volatile organic compounds in watermelon (*Citrullus lanatus*) grafted onto 21 local and two commercial bottle gourd (*Lagenaria siceraria*) rootstocks. *J. Horticult. Sci. Biotechnol.* **2014**, *89*, 448–452. [CrossRef]
50. Ceballos, R.; Rioja, T. Rootstock affects the blend of biogenic volatile organic compounds emitted by 'Hass' avocado. *Chil. J. Agric. Res.* **2019**, *79*, 330–334. [CrossRef]
51. Palai, G.; Caruso, G.; Gucci, R.; D'Onofrio, C. Deficit irrigation differently affects aroma composition in berries of *Vitis vinifera* L. (cvs Sangiovese and Merlot) grafted on two rootstocks. *Aust. J. Grape Wine Res.* **2022**, *28*, 590–606. [CrossRef]
52. Dong, T.T.; Xiong, B.; Huang, S.J.; Liao, L.; Qiu, X.; Sun, G.C.; He, Y.Z.Z.; Duan, C.W.; Wang, X.J.; Zhang, X.; et al. Investigation of the cause of reduced sugar content in *Kiyomi tangor* fruit of Ziyang xiangcheng (*Citrus junos* Sieb. ex Tanaka) rootstock. *Sci. Rep.* **2019**, *9*, 11. [CrossRef] [PubMed]
53. Voss, D.H. Relating Colorimeter Measurement of Plant Color to the Royal Horticultural Society Colour Chart. *HortScience* **1992**, *27*, 1256–1260. [CrossRef]
54. Wang, T.; Xiong, B.; Zheng, Z.; Qin, Z.; Deng, L.; Zheng, W.; Zhang, M.; Sun, G.; He, S.; Wang, J.; et al. Natural Variation Confers 'Aiyuan 38' Citrus Mutant a New Color and Unique Flavor. *Int. J. Mol. Sci.* **2023**, *24*, 8816. [CrossRef] [PubMed]
55. Kelebek, H.; Selli, S. Identification of phenolic compositions and the antioxidant capacity of mandarin juices and wines. *J. Food Sci. Technol. Mysore* **2014**, *51*, 1094–1101. [CrossRef] [PubMed]
56. Sun, B.; Di, H.; Zhang, J.; Xia, P.; Huang, W.; Jian, Y.; Zhang, C.; Zhang, F. Effect of light on sensory quality, health-promoting phytochemicals and antioxidant capacity in post-harvest baby mustard. *Food Chem.* **2021**, *339*, 128057. [CrossRef]
57. Bi, X.; Liao, L.; Deng, L.; Jin, Z.; Huang, Z.; Sun, G.; Xiong, B.; Wang, Z. Combined Transcriptome and Metabolome Analyses Reveal Candidate Genes Involved in Tangor (*Citrus reticulata* × *Citrus sinensis*) Fruit Development and Quality Formation. *Int. J. Mol. Sci.* **2022**, *23*, 5457. [CrossRef]
58. Sulaiman, S.F.; Sajak, A.A.B.; Ooi, K.L.; Seow, E.M. Effect of solvents in extracting polyphenols and antioxidants of selected raw vegetables. *J. Food Compos. Anal.* **2011**, *24*, 506–515. [CrossRef]
59. Molan, A.L.; Flanagan, J.; Wei, W.; Moughan, P.J. Selenium-containing green tea has higher antioxidant and prebiotic activities than regular green tea. *Food Chem.* **2009**, *114*, 829–835. [CrossRef]
60. Shen, S.L.; Yin, X.R.; Zhang, B.; Xie, X.L.; Jiang, Q.; Grierson, D.; Chen, K.S. CitAP2.10 activation of the terpene synthase CsTPS1 is associated with the synthesis of (+)-valencene in 'Newhall' orange. *J. Exp. Bot.* **2016**, *67*, 4105–4115. [CrossRef]
61. Zhang, X.X.; Dai, Z.; Fan, X.J.; Liu, M.; Ma, J.F.; Shang, W.T.; Liu, J.G.; Strappe, P.; Blanchard, C.; Zhou, Z.K. A study on volatile metabolites screening by HS-SPME-GC-MS and HS-GC-IMS for discrimination and characterization of white and yellowed rice. *Cereal Chem.* **2020**, *97*, 496–504. [CrossRef]
62. Feng, X.Y.; Wang, H.W.; Wang, Z.R.; Huang, P.M.; Kan, J.Q. Discrimination and characterization of the volatile organic compounds in eight kinds of huajiao with geographical indication of China using electronic nose, HS-GC-IMS and HS-SPME-GC-MS. *Food Chem.* **2022**, *375*, 11. [CrossRef] [PubMed]

International Journal of
Molecular Sciences

MDPI

Article

Comparative Analysis of Acetylated Flavonoids' Chemopreventive Effects in Different Cancer Cell Lines

Daigo Urakawa [1], Yuki Shioiridani [2], Shinya Igata [2], De-Xing Hou [1,2] and Kozue Sakao [1,2,*]

1 The United Graduate School of Agriculture Sciences, Kagoshima University, Kagoshima 890-0065, Japan; k6044078@kadai.jp (D.U.); k8469751@kadai.jp (D.-X.H.)
2 Graduate School of Agriculture, Forestry and Fisheries, Kagoshima University, Kagoshima 890-0065, Japan
* Correspondence: k0939002@kadai.jp

Abstract: Flavonoids, a class of natural compounds with anticancer activity, exhibit varying biological activities and potencies based on their structural differences. Acylation, including acetylation of flavonoids, generally increases their structural diversity, which is closely related to the diversity of bioactivity within this group of compounds. However, it remains largely unknown how acetylation affects the bioactivity of many flavonoids. Based on our previous findings that *O*-acetylation enhances quercetin's bioactivity against various cancer cells, we synthesized 12 acetylated flavonoids, including seven novel compounds, to investigate their anticancer activities in the MDA-MB-231, HCT-116, and HepG2 cell lines. Our results showed that acetylation notably enhanced the cell proliferation inhibitory effect of quercetin and kaempferol across all cancer cell lines tested. Interestingly, while the 5,7,4′-*O*-triacetate apigenin (3Ac-A) did not show an enhanced the effect of inhibition of cell proliferation through acetylation, it exhibited significantly strong anti-migration activity in MDA-MB-231 cells. In contrast, the 7,4′-*O*-diacetate apigenin (2Ac-Q), which lacks acetylation at the 5-position hydroxy group, showed enhanced cell proliferation inhibitory effect but had weaker anti-migration effects compared to 3Ac-A. These results indicated that acetylated flavonoids, especially quercetin, kaempferol, and apigenin derivatives, are promising for anticancer applications, with 3Ac-A potentially having unique anti-migration pathways independent of apoptosis induction. This study highlights the potential application of flavonoids in novel chemopreventive strategies for their anti-cancer activity.

Keywords: flavonoid; acetylation; modification; anti-cancer; anti-migration; chemoprevention

1. Introduction

Flavonoids are compounds found in many vegetables and fruits, such as onions, apples, and broccoli [1,2]. Flavonoids are known for their many bioactive compounds, including anticancer, antioxidant, anti-inflammatory, and anti-obesity effects [3–6]. Also, flavonoids are categorized into the following six groups based on the Flavan skeleton, and according to their structural differences: flavanol, flavone, flavanone, flavonol, isoflavone, and anthocyanidin. It has been reported that the B ring's catechol structure, the carbonyl group at position 4′, the double bond at C2–C3, and the hydroxyl group at position 3 play an important role in these bioactivities [7]. Furthermore, the diversity of secondary metabolites, including flavonoid subclasses, results from structural modifications such as hydroxylation, methylation, acylation, and glycosylation of the basic skeleton. These modifications, in turn, affect the physicochemical properties, stability, solubility, and bioavailability of the compounds [8–10]. The bioactivity of flavonoids is, thus, strongly related to their chemical structure.

One of the biological activities of flavonoids that has been extensively studied is their anticancer effect. Many types of flavonoids exhibit anticancer effects by inhibiting cancer cell proliferation and metastasis, inhibiting angiogenesis, or inducing apoptosis [11–14].

It has also been shown that a diet rich in flavonoids, which are epigenetic modifiers, is effective in preventing the development and progression of cancer cells [15,16]. Additionally, a report indicates an inverse correlation between flavonoid intake and mortality in patients diagnosed with invasive breast cancer [17]. However, the effective utilization of these flavonoids is hindered by major obstacles such as low membrane permeability, limited systemic bioavailability, rapid metabolic conversion, and oxidative degradation [18]. To overcome these challenges, scientific modifications of flavonoids have been investigated [19]. The aforementioned methylation, glycosylation, and acylation of flavonoids have been well studied as means of scientific modification, not only in plants.

Flavonoid glycosylation modulates the physiological activity of modified forms by altering the solubility and stability of the molecules. Glycosylated compounds can increase drug solubility by more than twofold, enhancing uptake in vitro [20]. However, while glycosylated genistein exhibits higher solubility than its parent compound, it shows no significant improvement in anticancer activity [21], indicating that glycosylation alone may be insufficient to enhance biological activity. Additionally, the direct glycosylation processes of some flavonoids can result in lower yields due to decomposition during synthesis [22], presenting challenges for scientific modification.

Flavonoid methylation is widely observed in nature. In the case of quercetin, for example, methylated derivatives ranging from mono-*O*-methylated to penta-*O*-methylated quercetin have been identified, and their biological activities have been evaluated [23]. Methylated flavonoids generally exhibit higher metabolic stability, oral bioavailability, and biological activity compared to their unmethylated counterparts [10,24]. This suggests that methylation is a beneficial modification for enhancing the efficacy of flavonoids. Research on methylated flavonoids extends beyond major compounds like quercetin [25] and apigenin to include less common derivatives such as tricetin [26]. Many studies have investigated their structure–activity relationships, providing valuable insights into how the position and number of methyl groups can enhance or diminish their biological activity [27,28].

In plants, acylation predominantly occurs as either aromatic or aliphatic acylation, often modifying sugar residues [29,30]. This modification is important for enhancing molecular stability, improving water solubility, and increasing resistance to enzymatic degradation [31,32]. Acetylation has also been identified as a modification of sugar residues, contributing to the cytoprotective effects [33] and anti-inflammatory effects in vivo [34]. On the other hand, while flavonoids with direct modification of the hydroxy group, such as *O*-glycosylation or *O*-methylation, are common in plants, *O*-acylated flavonoids are less frequently observed. Therefore, it is necessary and valuable to obtain *O*-acylated flavonoids through synthesis [35]. The *O*-acylation of polyphenols via enzyme-catalyzed and chemical methods can improve their stability, reduce degradation, increase membrane permeability, and enhance their bioactivity, such as inducing apoptosis in cancer cells [36–39]. These methods allow the use of a wide range of acyl donors and offer numerous benefits that are not achievable through natural means. The acetyl group, the smallest acyl donor, can easily and effectively modify hydroxy groups in flavonoids like methyl groups, often enhancing their biological activity. For example, acetylating just a single hydroxy group on the B ring of myricetin reduced its oxidation rate by approximately 3.1-fold [40]. Our previous studies also have shown that acetylating quercetin improves its uptake and metabolic stability in HepG2 cells, enhancing apoptosis induction compared to unmodified quercetin [41]. Furthermore, acetylated quercetin exhibited stronger apoptotic effects in human leukemia HL-60 cells and MCF-7 and MDA-MB-231 breast cancer cells by inhibiting proliferation, activating caspase-3 and PARP, and inducing DNA fragmentation. In contrast, methylated quercetin showed a reduced apoptosis-inducing effect in these experiments, suggesting that acetylation may, in certain contexts, offer comparative advantages over methylation for enhancing apoptotic activity [23,42]. Similar anticancer-promoting effects have been reported for acetylated chrysin and luteolin [43,44]. However, despite the potential of acetylated flavonoids to offer benefits comparable to methylated flavonoids, research on acetylated derivatives of other flavonoids remains limited. This has created a notable gap in understanding the structure-activity relationships of acetylated flavonoids, especially when compared to the extensive studies conducted on methylated flavonoids. As research in acylation advances, exploring the potential benefits of acetylated flavonoids may reveal advantages comparable to or even surpassing those of methylated counterparts. In this study, based on previous studies on acetyl derivatives of quercetin, eight flavonoids were selected from flavonols, flavones, and flavanones, focusing on the structure of quercetin. We synthesized acetylated derivatives, comprising seven novel compounds, and conducted a comparative analysis of their chemopreventive effects in different cancer cell lines.

In 2022, the second and third leading causes of cancer-related deaths in men were colorectal cancer (9.3% of total cancer deaths) and liver cancer (7.8% of total cancer deaths), respectively. For women, the leading cause was breast cancer (6.9% of total cancer deaths), followed by colorectal cancer (7.8% of total cancer deaths) [45]. Quercetin has been extensively studied for its chemopreventive properties against these three cancers [46]. Furthermore, our findings have already shown that *O*-acetylation of quercetin enhances its anticancer activity against liver cancer (HepG2) and breast cancer (MDA-MB-231 and MCF-7). Therefore, to extensively investigate the effects of acetyl modification, we tested its ability to inhibit cell proliferation in MDA-MB-231 (breast cancer), HCT-116 (colon cancer), and HepG2 (liver cancer) cells, and examined the structure–activity relationships that enhance physiological activity. Furthermore, we have identified the effect of flavonoid acetylation on the inhibition of cancer cell migration. MDA-MB-231 cells are well-known for their high metastatic potential in breast cancer. Our previous studies showed that acetylated quercetin induced higher apoptotic cell death in MDA-MB-231 cells compared to unmodified quercetin. However, its effect on metastasis inhibition remains unclear. Several reports have investigated the anti-metastatic potential of flavonoids [47]. The eight flavonoids used in this study—kaempferol [48], quercetin [49], myricetin [50], luteolin [51], chrysin [52], apigenin [53], Naringenin [54], and Taxifolin [55]—have demonstrated metastasis inhibitory effects in breast cancer cells. Therefore, elucidating the impact of acetylation on these metastasis inhibitory effects will provide new insights into the chemopreventive potential of flavonoids.

To the best of our knowledge, this is the first study to investigate the impact of acetylation on the anticancer and antimigration activities of multiple flavonoids, with a focus on structure–activity relationships. Additionally, our evaluation of apigenin revealed that its acetylation can confer unique anti-migratory properties distinct from other flavonoids. This finding highlights the potential of acetylated flavonoids in developing new strategies for chemical prevention, particularly in the context of anticancer effects.

2. Results

2.1. Synthesis and Structural Characterization of Acetyl Flavonoids

Acetyl-flavonols (compounds **2**, **4**, and **6**), acetyl-flavones (compounds **8**, **10**, **12**, and **20**), and acetyl-flavanones (compounds **14** and **16**) were synthesized by reacting the parent flavonoids with acetic anhydride in the presence of pyridine as a catalyst (Figure 1). The products were purified by recrystallization. As described below, the chemical structures were determined by ^{1}H-NMR and FT-IR analyses.

Successful acetylation was confirmed by the disappearance of the hydroxyl group signals in the ^{1}H-NMR spectra of the parent flavonoids (Table S1) and/or the appearance of new signals around 2.3 ppm corresponding to the acetyl groups in the spectra of the acetylated derivatives. Additionally, the integral ratios for each compound were consistent with the expected structures. In the FT-IR spectra, acetylation was evidenced by the disappearance of the hydroxyl group absorption band (3000–3500 cm^{-1}) present in the parent flavonoids and the appearance of a characteristic carbonyl absorption band around 1700 cm^{-1} from the acetyl groups. Additionally, the C-O stretching vibrations of the acetyl groups were also observed in the region of 1100–1300 cm^{-1}. These spectral changes clearly indicate that the hydroxyl groups of the flavonoids were acetylated. Detailed spectral data for each acetyl-flavonoid are provided, including the compound name, physical state, yield, melting point (m.p.), ^{1}H-NMR spectrum, and FT-IR spectrum, as shown below.

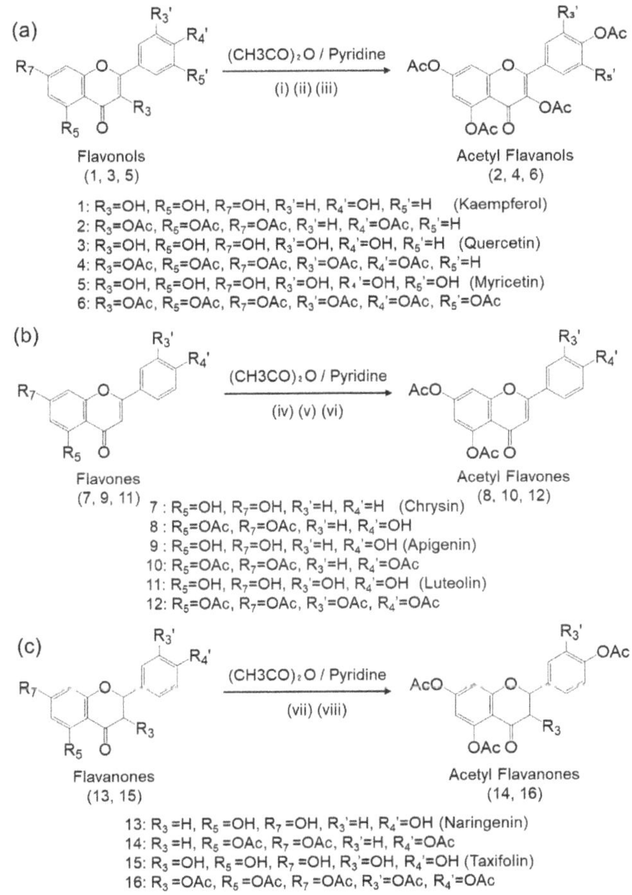

Figure 1. Scheme for the acetylation of flavonols, flavones, and flavanones. (**a**) The conditions for the acetylation of flavonols were as follows: for compound **2**, (i) $(CH_3CO)_2O$ (5 eq.), 25 °C, 12 h; for compound **4**, (ii) $(CH_3CO)_2O$ (6 eq.), 25 °C, 2 h; and for compound **6**, (iii) $(CH_3CO)_2O$ (7 eq.), 25 °C, 12 h. (**b**) The conditions for the acetylation of flavones were as follows: for compound **8**, (iv) $(CH_3CO)_2O$ (3 eq.), reflux, 7 h; for compound **10**, (v) $(CH_3CO)_2O$ (4 eq.), reflux, 7 h; and for compound **12**, (vi) $(CH_3CO)_2O$ (5 eq.), reflux, 12 h. (**c**) The conditions for the acetylation of flavanones were as follows: for compound **14**, (vii) $(CH_3CO)_2O$ (100 eq.), 25 °C, 5 h; for compound **16**, (viii) $(CH_3CO)_2O$ (8 eq.), 37 °C, 1 h.

Compound **2**: 3,5,7,4′-*O*-tetraacetate kaempferol (4Ac-K), White powder; Yield: 71.0 ± 2.12 (%); m.p. 126 °C; ^{1}H-NMR (600 MHz, DMSO-D$_6$) δ: 7.96 (2H, d, *J* = 8.7 Hz), 7.65 (1H, d, *J* = 2.1 Hz), 7.39 (2H, d, *J* = 8.7 Hz), 7.18 (1H, d, *J* = 2.1 Hz), 2.33 (12H, dd, *J* = 10.6, 1.3 Hz). FT-IR (cm^{-1}): 1761 (C=O), 1154 (C-O). A ^{1}H-NMR signal derived from an acetyl group was observed at 2.33 ppm, with an integration ratio of 12H, suggesting the presence of four *O*-acetyl.

Compound **4**: 3,5,7,3′,4-*O*-pentaacetate quercetin (5Ac-Q), white powder; yield: 70.8 ± 11.7 (%); m.p. 226 °C; ^{1}H-NMR (600 MHz, $CDCl_3$) δ: 7.71 (2H, dt, *J* = 16.9, 6.1 Hz), 7.35 (1H, dd, *J* = 10.3, 5.3 Hz), 6.88 (1H, d, *J* = 2.1 Hz), 2.35 (15H, dd, *J* = 3.3, 1.2 Hz). FT-IR (cm^{-1}): 1760 (C=O), 1175 (C-O). A ^{1}H-NMR signal derived from an acetyl group was observed at 2.35 ppm, with an integration ratio of 15H, suggesting the presence of five *O*-acetyl.

Compound **6**: 3,5,7,3′,4′,5′-*O*-hexaacetate myricetin (6Ac-M), white powder; yield: 67.2 ± 10.7 (%); m.p. 231 °C; ^{1}H-NMR (600 MHz, 600 MHz, DMSO-D$_6$) δ: 7.80 (2H, d, *J* = 0.7 Hz), 7.66 (1H, dd, *J* = 2.2, 0.7 Hz), 7.18 (1H, dd, *J* = 2.2, 0.7 Hz), 2.34 (18H, s). FT-IR (cm^{-1}): 1765 (C=O), 1131 (C-O). A ^{1}H-NMR signal derived from an acetyl group was observed at 2.34 ppm, with an integration ratio of 18H, suggesting the presence of six *O*-acetyl.

Compound **8**: 5,7-*O*-diacetate chrysin (2Ac-C), pale-white powder; yield: 57.7 ± 2.38 (%); m.p. 190 °C; ^{1}H-NMR (600 MHz, DMSO-D$_6$) δ: 8.09 (2H, t, *J* = 4.2 Hz), 7.64 (1H, dd, *J* = 2.2, 0.7 Hz), 7.60 (2H, d, *J* = 7.9 Hz), 7.17–7.09 (2H, m), 6.94 (1H, s), 2.34 (6H, d, *J* = 4.2 Hz). FT-IR (cm^{-1}): 1766 (C=O), 1182 (C-O). A ^{1}H-NMR signal derived from an acetyl group was observed at 2.34 ppm, with an integration ratio of 6H, suggesting the presence of two *O*-acetyl.

Compound **10**: 5,7,4′-*O*-triacetate apigenin (3Ac-A), brown powder; yield: 55.2 ± 1.97 (%); m.p. 197 °C; ^{1}H-NMR (600 MHz, CDCl$_3$) δ: 7.47 (2H, d, *J* = 8.6 Hz), 7.16 (2H, t, *J* = 4.3 Hz), 6.79–6.78 (1H, m), 6.55 (1H, dd, *J* = 2.5, 1.4 Hz), 5.50 (1H, dd, *J* = 13.7, 2.5 Hz), 3.04 (1H, dd, *J* = 16.7, 13.7 Hz), 2.78 (1H, dd, *J* = 16.7, 2.7 Hz), 2.35 (9H, dd, *J* = 44.9, 6.3 Hz). FT-IR (cm^{-1}): 1751 (C=O), 1188 (C-O). A ^{1}H-NMR signal derived from an acetyl group was observed at 2.35 ppm, with an integration ratio of 9H, suggesting the presence of three *O*-acetyl.

Compound **12**: 5,7,3′,4′-*O*-tetraacetate luteolin (4Ac-L), white powder; yield: 71 ± 4.24 (%) %; m.p. 214 °C; ^{1}H-NMR (600 MHz, CDCl$_3$) δ: 7.78 (2H, dd, *J* = 8.5, 2.1 Hz), 7.74 (2H, d, *J* = 2.2 Hz), 7.39 (1H, d, *J* = 8.5 Hz), 6.86 (1H, d, *J* = 2.0 Hz), 2.36–2.34 (12H, m). FT-IR (cm^{-1}): 1764 (C=O), 1185 (C-O). A ^{1}H-NMR signal derived from an acetyl group was observed at 2.34–2.36 ppm, with an integration ratio of 12H, suggesting the presence of four *O*-acetyl.

Compound **14**: 5,7,4′-*O*-triacetate naringenin (3Ac-N), white powder; yield: 25.8 ± 6.22 (%); m.p. 127 °C; ^{1}H-NMR (600 MHz, CDCl$_3$) δ: 7.47 (2H, d, *J* = 8.6 Hz), 7.16 (2H, d, *J* = 6.6 Hz), 6.79–6.78 (1H, m), 6.54 (1H, t, *J* = 1.1 Hz), 5.50 (1H, dd, *J* = 13.7, 2.5 Hz), 3.06 (1H, s), 3.04 (1H, d, *J* = 3.1 Hz), 3.01 (1H, s), 2.39 (9H, s). FT-IR (cm^{-1}): 1748 (C=O), 1169 (C-O). A ^{1}H-NMR signal derived from an acetyl group was observed at 2.39 ppm, with an integration ratio of 9H, suggesting the presence of three *O*-acetyl.

Compound **16**: 3,5,7,3′,4′-*O*-pentaacetate taxifolin (5Ac-T), white powder; yield: 72.0 ± 11.3 (%); m.p. 138 °C; ^{1}H-NMR (600 MHz, CDCl$_3$) δ: 7.39 (1H, t, *J* = 4.4 Hz), 7.29 (2H, d, *J* = 2.1 Hz), 6.79 (1H, t, *J* = 1.1 Hz), 6.61 (1H, t, *J* = 1.1 Hz), 5.66 (1H, d, *J* = 12.3 Hz), 5.43 (1H, d, *J* = 12.3 Hz), 2.34 (15H, d, *J* = 26.7 Hz). FT-IR (cm^{-1}): 1760 (C=O), 1177 (C-O). A ^{1}H-NMR signal derived from an acetyl group was observed at 2.34 ppm, with an integration ratio of 15H, suggesting the presence of five *O*-acetyl.

2.2. Acetylated Flavonoids Inhibit the Proliferation of Three Types of Cancer

Many previously reported acetylated flavonoids, including those in our study, have demonstrated increased inhibition of cell proliferation compared to their parent compounds. In this study, we assessed the cell proliferation inhibitory effects of acetylated forms of various flavonoids on the MDA-MB-231, HCT-116, and HepG2 cell lines. We determined the 50% Inhibition Concentration (IC$_{50}$) after 48 h of treatment, with acetylated quercetin showing the most pronounced inhibitory effect on cell proliferation in our previous investigation [23,41]. The IC$_{50}$ values of each parent compound and its acetylated derivatives for these three cell types were summarized in Table 1.

Table 1. Antiproliferative activity of flavonoids and acetyl flavonoids against cancer cells.

Compound	Cell Line MDA-MB-231	HCT-116	HepG2
	Flavonols (IC_{50} μM)		
K	46.7 ± 6.8	34.9 ± 1.0	33.4 ± 2.1
4Ac-K	33.6 ± 0.6	28.5 ± 0.5	23.2 ± 0.1
Q	24.3 ± 9.6	23.5 ± 2.8	28.2 ± 5.0
5Ac-Q	17.4 ± 10.8	15.7 ± 2.6	15.5 ± 0.2
M	27.2 ± 0.3	Over 160	Over 160
6Ac-M	50.9 ± 5.5	81.7 ± 36	76.6 ± 20.4
	Flavone (IC_{50} μM)		
C	37.5 ± 5.0	27.5 ± 2.9	25.3 ± 2.8
2Ac-C	38.4 ± 5.4	32.2 ± 1.3	23.8 ± 9.0
A	27.1 ± 5.2	19.0 ± 4.2	25.8 ± 1.9
3Ac-A	31.1 ± 7.2	21.9 ± 2.6	6.5 ± 1.0
L	12.9 ± 3.6	9.3 ± 2.2	10.2 ± 2.0
4Ac-L	20.2 ± 6.3	12.2 ± 3.7	12.4 ± 3.5
	Flavanone (IC_{50} μM)		
N	Over 160	120.4 ± 39.8	118.5 ± 30.8
3Ac-N	156.4 ± 23.1	94.3 ± 23.1	97.9 ± 10.6
T	Over 160	Over 160	Over 160
5Ac-T	128.0 ± 23.1	125.8 ± 23.1	152.2 ± 23.1

IC_{50} ± SD values in μM after 48 h incubation. IC_{50} values were calculated from dose-response curves.

First, the cell proliferation inhibition effect of each compound on MDA-MB-231 cells was measured using the trypan blue exclusion assay. As Table 1 shows, the inhibitory concentrations (IC_{50}) of the original compounds, flavonols such as kaempferol, quercetin, and myricetin, were 46.7 μM, 24.3 μM, and 27.2 μM, respectively. The IC_{50} values of their acetylated derivatives, 4Ac-K, 5Ac-Q, and 6Ac-M, were 33.6 μM, 17.4 μM, and 50.9 μM, respectively. These results indicate that acetylation enhanced the tumor cell growth inhibitory potential of kaempferol and quercetin by approximately 1.39-fold and 1.40-fold, respectively. In contrast, the acetylation of myricetin resulted in a decreased inhibitory effect, with the IC_{50} value increasing by 1.87-fold. For flavones such as chrysin, apigenin, and luteolin, the IC_{50} values were 37.5 μM, 27.1 μM, and 12.9 μM, respectively. Their acetylated derivatives, 2Ac-C, 3Ac-A, and 4Ac-L had IC_{50} values of 38.4 μM, 31.1 μM, and 20.2 μM, respectively. These data suggest that acetylation did not significantly affect the ability of chrysin and apigenin to inhibit tumor cell proliferation, with changes in IC_{50} values being minimal (1.02-fold and 1.15-fold increase, respectively). However, the acetylation of luteolin decreased its inhibitory effect, increasing the IC_{50} value by 1.57-fold. Flavanones such as naringenin and taxifolin had IC_{50} values above 160 μM. The IC_{50} values of their acetylated derivatives, 3Ac-N and 5Ac-T, were 156.4 μM and 128.0 μM, respectively. The IC_{50} values for both naringenin and taxifolin exceeded 160 μM, and acetylation slightly improved their inhibitory effect, reducing the IC_{50} values to 156.4 μM (1.02-fold) and 128.0 μM (1.25-fold), respectively. This suggests that acetylation may have enhanced the inhibitory effect on cell proliferation for these flavanones, although the changes were not substantial. Overall, these results demonstrate that acetylation can either enhance or reduce the anticancer activity of flavonoids depending on the specific compound. Kaempferol and quercetin derivatives showed significant improvements, while myricetin and luteolin derivatives exhibited a decreased effectiveness. The impact of acetylation on flavones and flavanones was less pronounced but still indicated the potential modulation of bioactivity.

The cell growth inhibitory effects of each flavonoid and its acetylated derivatives on HCT-116 cells were subsequently determined in a similar manner. The IC_{50} values of kaempferol and quercetin were 34.85 μM and 23.45 μM, respectively, while that of myricetin exceeded 160 μM. The IC_{50} values of their acetylated derivatives, 4Ac-K, 5Ac-Q, and 6Ac-M, were 28.53 μM, 15.66 μM, and 81.66 μM, respectively. These results suggest that acetylation enhanced the tumor cell proliferation inhibitory ability of kaempferol and quercetin by approximately 1.22-fold and 1.50-fold, respectively. In contrast, acetylation of myricetin reduced the IC_{50} value to 81.66 μM, which was about twice as strong as the parent compound. However, it should be noted that the IC_{50} value of the parent compound is above 160 μM, indicating it was not highly potent to begin with. The IC_{50} values of the chrysin, apigenin, and luteolin were 27.5 μM, 19.0 μM, and 9.3 μM, respectively. Upon acetylation, their inhibitory potency was moderately affected, with IC_{50} values of 32.2 μM for 2Ac-C, 21.9 μM for 3Ac-A, and 12.2 μM for 4Ac-L. These results suggest that acetylation led to slight increases in IC_{50} values by factors of approximately 1.17, 1.15, and 1.31, respectively, indicating a modest reduction in inhibitory effect due to acetylation. The IC_{50} of the naringenin was 120.4 μM, and that of taxifolin exceeded 160 μM. The IC_{50} values of 3Ac-N and 5Ac-T, were 94.3 μM and 125.8 μM, respectively. These results suggest that acetylation of taxifolin reduces its IC_{50} value by at least about 1.27-fold and increases its ability to inhibit tumor cell proliferation. The ability of naringenin to inhibit HCT-116 cell proliferation was also increased by acetylation, with an approximately 1.28-fold decrease in IC_{50} value. Similar to the evaluation using MDA-MB-231 cells, acetylation demonstrated improvements in cell proliferation inhibition for kaempferol and quercetin, with a slight decrease observed for flavones and an enhancement for flavanones. Myricetin showed at least a 2-fold increase in activity.

Furthermore, the inhibitory effect on cell proliferation in HepG2 cells was also measured using the same method. Kaempferol and quercetin exhibited IC_{50} values of 33.38 μM and 28.16 μM, respectively, with myricetin exceeding 160 μM. Acetylated derivatives 4Ac-K, 5Ac-Q, and 6Ac-M showed IC_{50} values of 23.2 μM, 15.5 μM, and 76.6 μM, respectively, indicating an approximately 1.4-fold, 1.8-fold, and 2.1-fold enhancement in inhibitory activity compared to their respective parent compounds. For flavones, chrysin, apigenin, and luteolin demonstrated IC_{50} values of 25.3 μM, 25.8 μM, and 10.2 μM, respectively. The IC_{50} values of 2Ac-C, 3Ac-A, and 4-Ac-L were 23.8 μM, 6.5 μM, and 12.4 μM, respectively. These results indicate that the acetylation of apigenin enhances its tumor cell proliferation inhibitory activity by approximately 4-fold, while the acetylation of luteolin decreases its inhibitory activity by about 1.2-fold. Chrysin showed a slight improvement in its inhibitory effect with an IC_{50} reduction of about 1.1-fold. The flavone naringenin showed an IC_{50} of 120.4 μM and taxifolin exceeded 160 μM. The acetylated derivatives, 3Ac-N and 5Ac-T, showed IC_{50} values of 97.9 μM and 152.2 μM, respectively, suggesting an approximately 1.2-fold and at least 1.1-fold increase in inhibitory activity over the parent compound, respectively.

2.3. *Effects of Flavonoid Acetylation on Cell Cycle Progression-Related Proteins*

Based on the demonstrated enhancement of cell proliferation inhibitory ability through acetylation in all three cell types by kaempferol and quercetin, experiments were conducted to investigate the relationship between cell proliferation inhibitory ability and the cell cycle-related proteins. In cellular experiments aimed at identifying cell cycle stages and the movement of cell cycle-related proteins, accurately discerning the behavior of cells in different stages when mixed poses a challenge. Therefore, Western blotting was conducted using cells synchronized in the S phase with hydroxyurea (HU) in the present experiment. Synchronization via HU treatment induces cell arrest in the late G1/early S phase, and subsequent release from HU-mediated arrest yields a cell population that uniformly progresses through S and G2/M phases.

Each compound's cytoplasmic signaling proteins associated with cell cycle progression relative to MDA-MB-231 cells were detected by Western blotting. As shown in Figure 2a, kaempferol and its acetylated derivative 4Ac-K showed the upregulation of CyclinA (S phase) and CyclinB1 (G2/M phase) at 12 h compared to the control group at the same time point. Compared to the 0 h baseline, the fold changes of CyclinA and Cyclin B1 in the control group decreased to 0.07 and 0.42, respectively, at 12 h. In contrast, treatment with kaempferol for 12 h showed fold changes of 0.39 and 1.06, respectively. Similarly, treatment with 4Ac-K resulted in fold changes of 0.52 for cyclin A and 0.82 for cyclin B1.

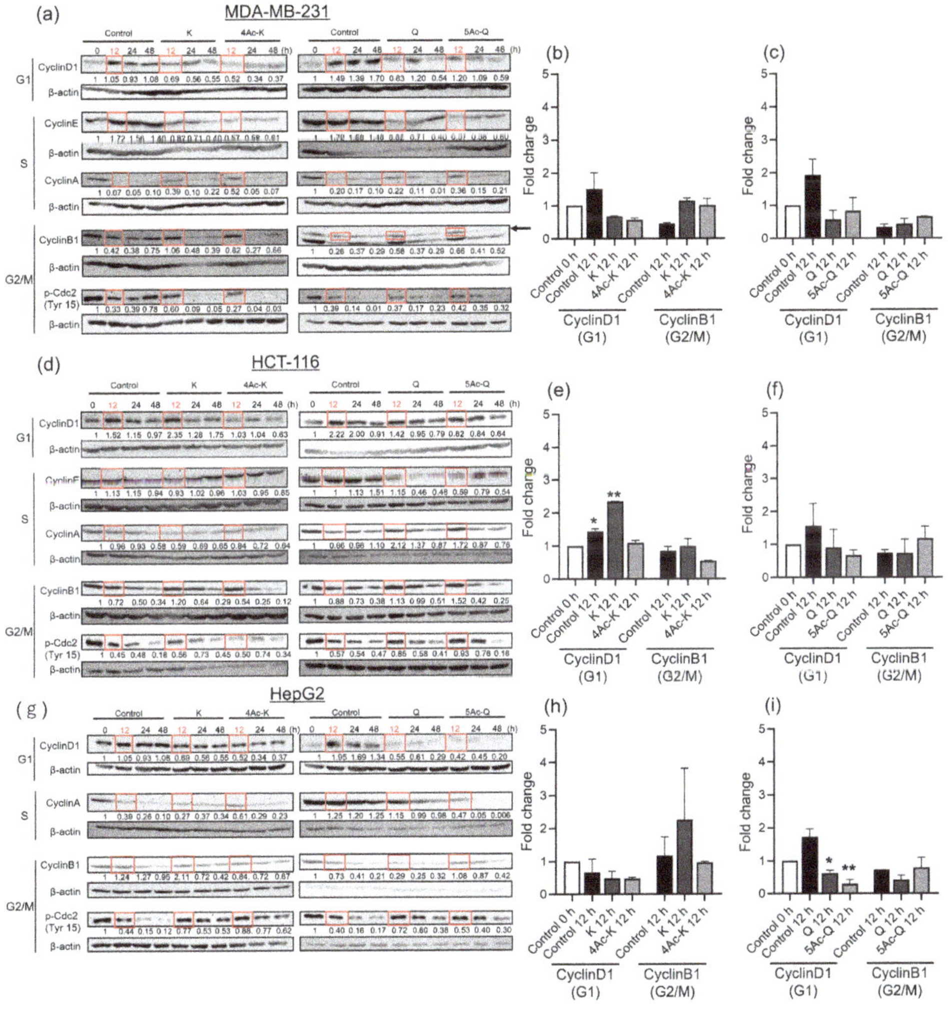

Figure 2. Western blot of cytoplasmic signaling proteins related to cell cycle progression. Results are shown for checkpoint promoters for G1-phase progression (CyclinD1) and regulators for S-phase progression (CyclinA/E); regulators for G2/M transition (CyclinB1 and p-Cdc2) in MDA-MB-231 (**a**), HCT-116 (**d**), and HepG2 (**g**) cell lines. Blots were stripped and re-probed with anti-actin antibody to correct for differences in protein loading. Numbers on bottom of bands are fold changes in expression level relative to corresponding DMSO-treated control. Red square indicate the 12 h blots in which the cycle was arrested. Quantitative data for protein expression of kaempferol and 4Ac-K (**b**,**e**,**h**) and quercetin and 5Ac-Q (**c**,**f**,**i**) in each cell, normalized by β-actin expression. Each value in the graph represents the mean ± SD (n = 2–4) of repeated experiments. * $p < 0.05$, ** $p < 0.01$, indicating a significant difference between the control and each sample. For Western blot analysis, 1 mM hydroxyurea was added 24 h prior to sample addition and all cell cycles were synchronized in S phase. Cells were treated with 40 μM of kaempferol (K) or 4Ac-K, and quercetin (Q) or 5Ac-Q for indicated time, then the harvested cell lysate was used. Western blotting was performed two to four times using independently prepared lysates.

Setting the 12 h control value as 1, the fold change in CyclinA with kaempferol treatment was 5.57, and with 4Ac-K treatment it was 7.43. For Cyclin B1, the fold change with kaempferol treatment was 2.52, and with 4Ac-K treatment it was 1.95. The checkpoint control factor for G1 transition, CyclinD1, was downregulated to approximately half of the control level at the 12 h point with both kaempferol and 4Ac-K (Figure 2b). Additionally, although it was not statistically significant between the samples, both showed an increase in p-Cdc2 compared to the control. Quercetin reduced the expression of CyclinD1and CyclinE, and slightly increased CyclinB1 at 12 h, while 5Ac-Q decreased CyclinE to more than one third of the control and slightly increased CyclinA, CyclinB1, and p-Cdc2 (Figure 2a,c). These findings suggested that Kaempferol, 4Ac-K, quercetin, and 5Ac-Q may regulate cell cycle-related proteins associated with G1 or G2/M phases.

In HCT-116 cells, as shown in Figure 2d, treatment with kaempferol for 12 h reduced CyclinE and CyclinA expression, while increasing CyclinD1 levels by 1.5-fold and CyclinB1 levels by 1.7-fold (Figure 2e). Additionally, phosphorylated-Cdc2 (p-Cdc2) levels increased by 1.2-fold. Subsequently, CyclinD1 and CyclinB1 decreased over time in a time-dependent manner. Conversely, 4Ac-K did not significantly alter CyclinE and CyclinA expression compared to the controls, but it exhibited lower fold changes for CyclinD1 (0.67-fold) and CyclinB1 (0.75-fold) compared to the controls. The CyclinD1 and CyclinB1 levels continued to decrease over time, suggesting a G2/M phase arrest induced by 4Ac-K. These results suggest that kaempferol and 4Ac-K may influence different cell cycle-related proteins.

After 12 h of treatment, quercetin and 5Ac-Q exhibited similar effects on checkpoint-related proteins (Figure 2d,f). Both compounds significantly decreased CyclinD1 levels (0.63-fold for quercetin and 0.37-fold for 5Ac-Q) and increased CyclinA (3.2-fold for quercetin and 2.6-fold for 5Ac-Q), CyclinB1 (1.3-fold for quercetin and 1.7-fold for 5Ac-Q), and p-Cdc2 (1.5-fold for quercetin and 1.6-fold for 5Ac-Q). Subsequently, CyclinD1, CyclinB1, and p-Cdc2 levels decreased over time, suggesting that Cdc2 inactivation was accompanied by an increased accumulation of CyclinB1 in the cytosol. These findings indicate that proteins associated with the G2/M phase were similarly affected by both quercetin and 5Ac-Q treatments.

In HepG2 cells (Figure 2g), both kaempferol and 4Ac-K showed decreased CyclinD1 levels (0.65-fold and 0.49-fold, respectively) compared to the controls after 12 h. Kaempferol increased CyclinB1 levels by 1.7-fold and p-Cdc2 levels by 1.8-fold, while 4Ac-K exhibited fold changes of 0.7-fold for CyclinB1 and 2.0-fold for p-Cdc2. The fold changes in CyclinD1 and p-Cdc2 proteins were greater for 4Ac-K compared to kaempferol. In contrast, both quercetin and 5Ac-Q decreased CyclinD1 levels to nearly a quarter of the control. Quercetin showed an increase in p-Cdc2 levels by 1.8-fold, with no significant change in CyclinB1 levels over 12–24 h. On the other hand, 5Ac-Q exhibited an increase in CyclinB1 levels by 1.5-fold and p-Cdc2 levels by 1.3-fold at 12 h, followed by a decrease. These findings suggest that in HepG2 cells, acetyl derivatives affect cell cycle-associated proteins in the G2/M phase similar to the parent and child products.

2.4. Screening of Acetylated Flavonoids for Anti-Migration Properties

We evaluated whether acetylation of flavonoids also affects the anti-migration potential of MDA-MB-231 cells. In the wound healing assay, we set the sample treatment time to 6 h since wounds heal within 9 h. Additionally, to observe a significant difference, we treated the cells at twice the IC_{50} concentration obtained from the 48 h cell proliferation inhibition assay, as lower concentrations were insufficient to produce measurable effects. The flavonols, all acetylated compounds, showed anti-migration activity. 4Ac-K exhibited the highest anti-migration activity, reducing cell migration to 167.7% compared to the control. Kaempferol and 5Ac-Q reduced cell migration to 131.0% and 122.0%, respectively, while 6Ac-M reduced migration to 121.0% (Figure 3b). For the flavones, 3Ac-A, synthesized from apigenin, demonstrated significantly higher anti-migration activity than the other flavonols, reducing cell migration to 153.1%. Luteolin and 4Ac-L also showed higher anti-migration activity than the control, reducing cell migration to 135.1% and

129.0%, respectively (Figure 3c). Among the flavanones, neither the parent compounds nor their acetylated derivatives exhibited significant migration inhibition. 3Ac-N showed lower anti-migration activity than the original compound, with cell migration at 96.1%. 5Ac-T showed a slight increase in anti-migration activity, reducing migration to 112.0%, though this difference was not statistically significant (Figure 3d). Statistical analysis was performed using a one-way ANOVA followed by Dunnett's multiple comparison test.

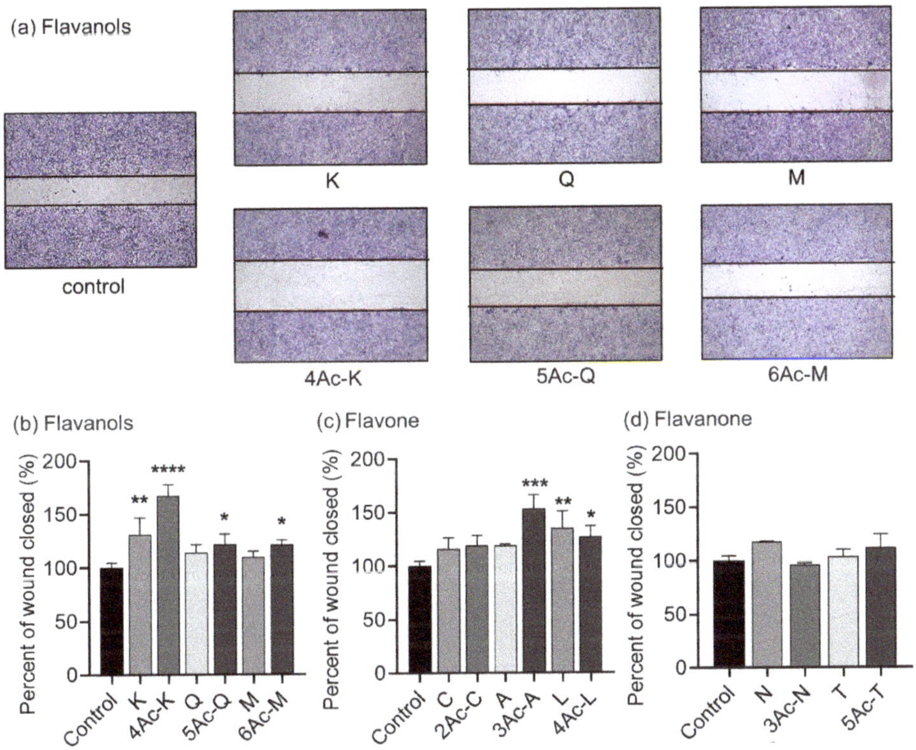

Figure 3. Assessment of anti-cellular migration using wound healing assay. (**a**) Representative images of scratch assay showing the effect of flavonols (K: kaempferol, Q: quercetin, M: myricetin) and acetyl flavonols treatments (6 h treatment) on wound healing in MDA-MB-231 cells. (**b**–**d**), the wound healing quantification of flavonols, flavones (C: chrysin, A: apigenin, L: luteolin), flavanones (N: naringenin, T: taxifolin) and their acetyl derivatives. The results shown (mean ± SD) are representative of two independent experiments (n = 9). * mark denoted significant differences (* $p < 0.05$, ** $p < 0.01$, *** $p < 0.001$, **** $p < 0.0001$) between the control and each compound using one-way ANOVA followed by Dunnett's multiple comparison test.

2.5. *Synthesis of Acetylated Compounds with Retained 5-Hydroxy Group*

Our previous study reported that among the acetyl derivatives of quercetin, 4Ac-Q, which retained the hydroxy group at position 5 of the A ring, enhanced the ability to induce apoptosis [23]. Therefore, we newly synthesized compounds **17**, **18**, **19**, and **20** (Figure 4), which also retain the hydroxy group at position 5 of the A ring, expecting to further enhance cell proliferation inhibition and antimigration potential.

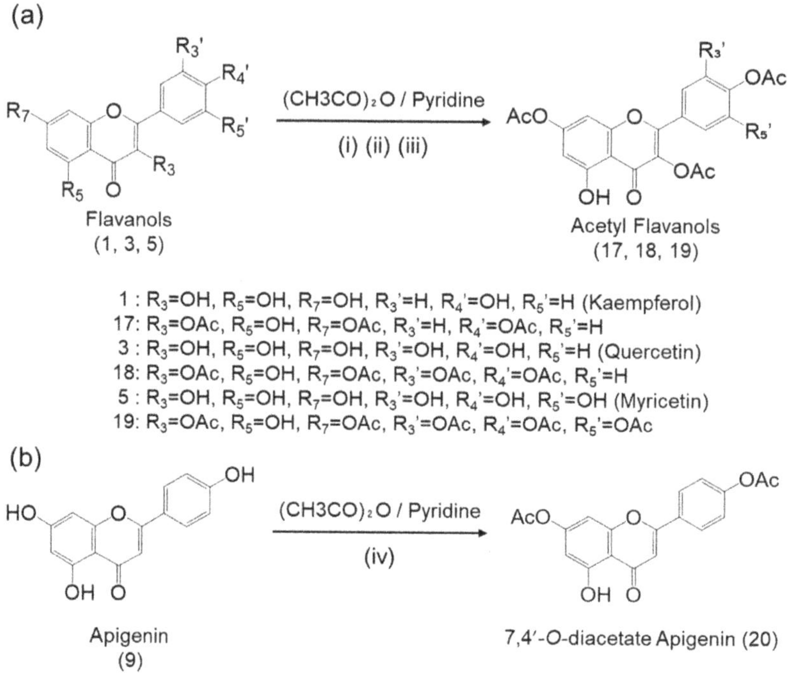

Figure 4. Selective acetylation of hydroxyl groups other than the 5th position hydroxyl group. (**a**) Reagents and condition for compound **17**, (i): $(CH_3CO)_2O$ (3.2 eq), 27 °C, 5 min; for compound **18**, (ii): $(CH_3CO)_2O$ (4.2 eq), 27 °C, 7 min; for compound **19**, (iii): $(CH_3CO)_2O$ (5.2 eq), 27 °C, 12 min; and (**b**) for compound **20**, (iv): $(CH_3CO)_2O$ (2 eq), 0 °C, 5 min.

A hydroxy group at position 5 of the A ring is not easily acetylated due to hydrogen bonding with the carbonyl group at position 4. By shortening the reaction time, controlled acetylation at this position was achieved. Consequently, compounds **17**, **18**, and **19** were successfully synthesized while retaining the hydroxy group at position 5. The ^{1}H-NMR results confirmed that the signal around 12 ppm, corresponding to the hydroxy group at position 5 of the A-ring in the parent compound, remained unchanged, while new signals derived from acetyl groups appeared around 2.3 ppm. Furthermore, the integral ratio corresponding to three protons (3H: equivalent to one $-COCH_3$ group) was reduced compared to acetylating all hydroxy groups of each parent compound, confirming successful acetylation without modification at the 5 position. Similarly, compound **20** retained the hydroxy group at position 5 of the A ring successfully by adjusting both the temperature and reaction time. The ^{1}H-NMR results for compound **20** indicated a decrease in the integral ratio of the protons (3H) corresponding to one acetyl group signal around 2.3 ppm, consistent with the results for compounds **17**, **18**, and **19**. The signal around 13 ppm corresponding to the hydroxy group at position 5 of the A ring of apigenin was outside the measurement range and could not be confirmed. Furthermore, FT-IR analysis of compounds **17**, **18**, **19**, and **20** revealed an additional absorption peak specific to the C=O stretch around 1700 cm^{-1}.

Compound **17**: 3,7,4′-*O*-triacetate kaempferol (3Ac-K), pale-yellow powder; yield: 53.0 ± 2.82 (%); m.p. 206°C; ^{1}H-NMR (600 MHz, $CDCl_3$) δ: 12.15 (1H, s), 7.88 (2H, d, *J* = 8.8 Hz), 7.28 (2H, d, *J* = 8.8 Hz), 6.85 (1H, d, *J* = 2.0 Hz), 6.60 (1H, d, *J* = 2.0 Hz), 2.35 (9H, t, *J* = 5.5 Hz). FT-IR (cm^{-1}): 1770 (C=O), 1147 (C-O). A ^{1}H-NMR signal derived from an acetyl group was observed at 2.35 ppm, with an integration ratio of 9H, suggesting the presence of three *O*-acetyl groups. Additionally, a peak derived from the hydroxy group at position 5 was observed at 12.15 ppm with an integration ratio of 1H.

Compound **18**: 3,7,3′,4′-*O*-tetraacetate quercetin (4A-Q), pale-yellow powder; yield: 72.1 ± 3.25 (%); m.p. 199 °C; ^{1}H-NMR (600 MHz, $CDCl_3$) δ: 12.10 (1H, s), 7.75 (2H, dd, *J* = 10.7, 1.8 Hz), 7.37 (1H, d, *J* = 8.3 Hz), 6.86 (1H, d, *J* = 1.6 Hz), 6.61 (1H, d, *J* = 1.6 Hz), 2.36 (12H, dd, *J* = 12.0, 10.2 Hz). FT-IR (cm^{-1}): 1764 (C=O), 1204 (C-O). A ^{1}H-NMR signal derived from an acetyl group was observed at 2.36 ppm, with an integration ratio of 12H, suggesting the presence of three *O*-acetyl groups. Additionally, a peak derived from the hydroxy group at position 5 was observed at 12.1 ppm with an integration ratio of 1H.

Compound **19**: 3,7,3′,4′,5′-*O*-pentaacetate myricetin (5Ac-M), pale-yellow powder; Yield: 61.0 ± 5.64 (%); m.p.211°C; ^{1}H-NMR ($CDCl_3$) δ: 12.07 (1H, s), 7.65 (2H, s), 6.86 (1H, d, *J* = 2.0 Hz), 6.61 (1H, d, *J* = 2.0 Hz), 2.37 (15H, dd, *J* = 21.5, 17.6 Hz). 1775 (C=O), 1179 (C-O). A ^{1}H-NMR signal derived from an acetyl group was observed at 2.37 ppm, with an integration ratio of 15H, suggesting the presence of three *O*-acetyl groups. Additionally, a peak derived from the hydroxy group at position 5 was observed at 12.07 ppm with an integration ratio of 1H.

Compound **20**: 7,4′-*O*-diacetate apigenin (2Ac-A) brown powder; yield: 67.5 ± 6.57 (%); m.p. 217 °C; ^{1}H-NMR (ACETONE-D_6) δ: 8.20–8.17 (2H, m), 7.52–7.51 (1H, m), 7.39–7.35 (2H, m), 7.03–7.02 (1H, m), 6.98–6.95 (1H, m), 6.62–6.61 (1H, m), 2.84 (5H, tt, *J* = 12.4, 2.3 Hz), 2.33–2.29 (6H, m). FT-IR (cm^{-1}): 1753 (C=O), 1189 (C-O). A ^{1}H-NMR signal derived from an acetyl group was observed at 2.29–2.33 ppm, with an integration ratio of 6H, suggesting the presence of three *O*-acetyl groups.

2.6. Effect of Position 5 Hydroxyl Group on Flavonoids' Anti-Proliferative Activity

The cell viability of each sample against MDA-MB-231 cells was measured using a trypan blue assay. The IC_{50} values of kaempferol and its acetyl derivatives 3Ac-K and 4Ac-K for inhibition of cell growth were 46.7 μM, 42.2 μM, and 33.6 μM, respectively (Figure 5a). This indicated that 4Ac-K was 1.39-fold more effective in inhibiting cell proliferation than kaempferol. Conversely, 3Ac-K showed only a slight increase in potency compared to the parent compound. The IC_{50} values for quercetin and its acetyl derivatives 4Ac-Q and 5Ac-Q were 24.3 μM, 16.6 μM, and 17.4 μM, respectively (Figure 5b). Quercetin's derivative 4Ac-Q exhibited a 1.46-fold improvement in cell growth inhibition compared to quercetin, whereas 5Ac-Q showed a 1.40-fold increase. Among the derivatives, 4Ac-Q consistently demonstrated higher cell growth inhibition within the range of 20–120 μM. Similarly, the IC_{50} values for myricetin and its acetyl compounds 5Ac-M and 6Ac-M were 27.2 μM, 47.3 μM, and 50.9 μM (Figure 5c). Acetylation decreased the inhibitory potency of myricetin compared to the parent compound, regardless of the presence or absence of the hydroxy group at the 5-position. The fold decreases were 1.74-fold for 5Ac-M and 1.87-fold for 6Ac-M. Figure 5d showed that the IC_{50} values for apigenin and its acetyl compounds 2Ac-A and 3Ac-A were 27.1 μM, 16.7 μM, and 31.1 μM. The 2Ac-A derivative exhibited a 1.62-fold enhancement in inhibitory effect compared to apigenin, while 3Ac-A showed a decrease in efficacy by 1.15-fold.

The IC_{50} values indicated that acetylation at the 5-position hydroxy group generally decreased the inhibitory potency of the flavonoids. For kaempferol, 4Ac-K (5-acetylated) was more effective than the parent compound, suggesting an exception. In contrast, quercetin (4Ac-Q vs. 5Ac-Q), myricetin (5Ac-M vs. 6Ac-M), and apigenin (2Ac-A vs. 3Ac-A) all showed that retaining the 5-position hydroxy group enhanced cell growth inhibition compared to their 5-acetylated derivatives. Thus, the presence of the 5-position hydroxy group generally enhances the inhibitory activity against cell growth.

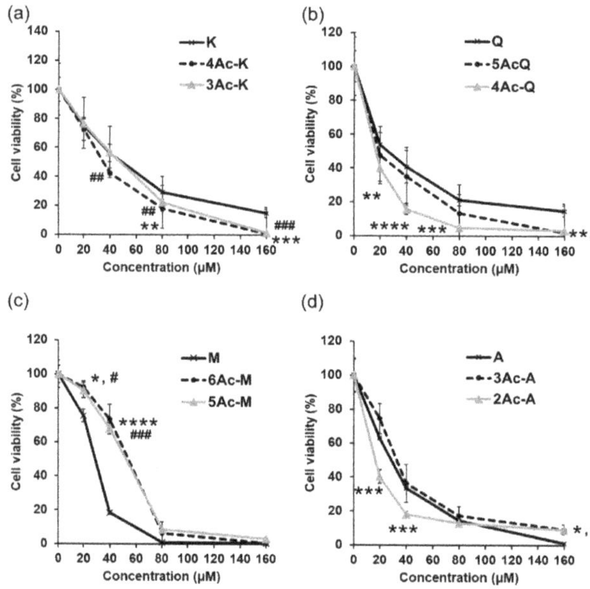

(e)

	IC_{50} value (μM)
K	46.7 ± 6.8
3Ac-K	42.2 ± 0.2
4Ac-K	33.6 ± 0.6
Q	24.3 ± 9.6
4Ac-Q	16.6 ± 0.1
5Ac-Q	17.4 ± 10.8
M	27.2 ± 0.3
5Ac-M	47.3 ± 6.2
6Ac-M	50.9 ± 5.5
A	27.1 ± 5.2
2Ac-A	13.0 ± 0.2
3Ac-A	31.1 ± 7.2

Figure 5. Effect of the existence of the 5-position hydroxyl group of acetylated flavonoids on the proliferation of MDA-MB-231 cells. Cells were treated for 48 h with compounds at concentrations ranging from 0–160 μM or 0.1% DMSO (control). The treated compounds are (**a**) kaempferol (K), (**b**) quercetin (Q), (**c**) myricetin (M), and (**d**) apigenin (A) as parent compounds and their acetyl derivatives, respectively. (**e**) IC_{50} values in μM. IC_{50} values were calculated from dose- – response curves. The data shown represent the means ± SD of three or more independent experiments. *, # $p < 0.05$, **, ## $p < 0.01$, ***, ### $p < 0.001$, **** $p < 0.0001$: significant differences between control and each compound (*: 3Ac-K, 4Ac-Q, 5Ac-Q or 2Ac-A, #: 4Ac-K, 5Ac-Q, 6Ac-Q or 3Ac-A) by Tukey's multiple range test.

2.7. Effects of 5-Position Hydroxyl Group Acetylation on Anti-Migration Activity

Next, the effects of acetylation of the hydroxyl group at 5-position on the antimigration activity of each flavonoid were evaluated using a wound healing assay. As depicted in Figure A1, kaempferol and apigenin exhibited statistically significant anti-migration activity in their parent forms, which was further enhanced upon acetylation. In contrast, quercetin and myricetin did not show significant anti-migration activity at double their IC_{50} concentrations, but both demonstrated significant enhancement in anti-migration effects upon acetylation. Notably, 4Ac-K and 3Ac-A, which have the 5-position acetylated, and 4Ac-Q and 5Ac-M, which retain the 5-position hydroxy group, exhibited the highest anti-migration effects within their respective groups. Thus, the results suggest that there is no clear correlation between the presence or absence of acetylation at the hydroxy group in position 5 and increased anti-migration activity.

Each acetyl derivative that demonstrated excellent anti-invasive capacity in the wound healing assay was further evaluated in a transwell migration assay, measuring the effect of 24 h treatment with concentrations as low as half the IC_{50} concentration. This was done to determine whether these compounds could exhibit anti-migratory effects even at lower concentrations and shorter treatment durations (Figure 6). The results indicated that 3Ac-A exhibited the highest anti-migratory capacity, with a migration rate of 15.8% compared to the control. Conversely, 5Ac-M showed the lowest anti-migratory ability, with a migration rate of 86%. Comparing the acetyl derivatives of kaempferol, quercetin, and myricetin, which have flavonol as their basic backbone, 4Ac-K, 4Ac-Q, and 5Ac-M showed the highest anti-migration ability, in that order, respectively. This was consistent with the results of the wound healing assay.

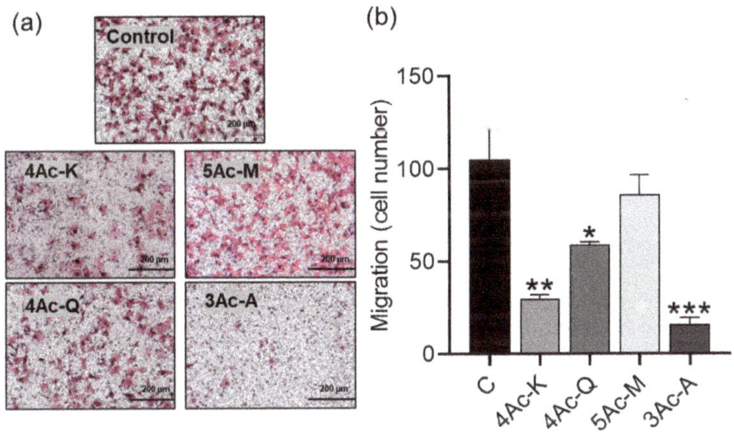

Figure 6. Assessment of anti-cellular migration using transwell assay. (**a**): Representative images from transwell chamber assay depicting migration by MDA-MB-231 cells treated for 24 h with DMSO (control) or the indicating samples. (**b**): Quantitation of migrated MDA-MB-231 cells from experiment shown in panel (**a**). Ability to inhibit cell migration was evaluated at half the IC_{50} concentration of each compound. The microscope magnification was 20× and the scale bar indicates 200 μm. Experiment was repeated at least three times in triplicate and combined results were shown as mean ± SD (n = 9). Significantly different ($p < 0.05$) compared with corresponding DMSO-treated control and each sample. * $p < 0.05$, ** $p < 0.01$, *** $p < 0.001$: significant differences between control and each compound by Dunnett's multiple comparison test.

2.8. Acetylated Flavonoids Induce Apoptosis

The apoptosis-inducing ability of 4Ac-K, 4Ac-Q, 5Ac-M, and 3Ac-A on MDA-MB-231 cells was evaluated by flow cytometry. The results of the anti-cell migration assay using a transwell assay, in which cells were treated with half the IC_{50} concentration for 24 h, showed statistically significant anti-migration effects for 3Ac-A, 4Ac-K, and 4Ac-Q, in that order. Consequently, cells were treated with half the IC_{50} value of each compound for 24 h to assess apoptosis induction, followed by Annexin/PI staining.

At half the IC_{50} dose, 4Ac-K exhibited the highest ability to induce apoptosis, at 2.2-fold that of the control (Figure 7b). In contrast, 3Ac-A, 4Ac-Q, and 5Ac-M showed no significant difference compared to the control. Since 3Ac-A and 4Ac-Q, which demonstrated significant effects in the anti-migration assay, did not induce significant apoptosis at half the IC_{50} dose, cells were also treated with the IC_{50} concentration for 24 h to determine if higher concentrations would induce apoptosis. At this concentrations, 4Ac-K maintained the strongest apoptosis-inducing ability, followed by 3Ac-A and 4Ac-Q, all of which showed significant apoptosis induction relative to the control (Figure 7d). Although 5Ac-M induced apoptosis, it did not significantly differ from the control.

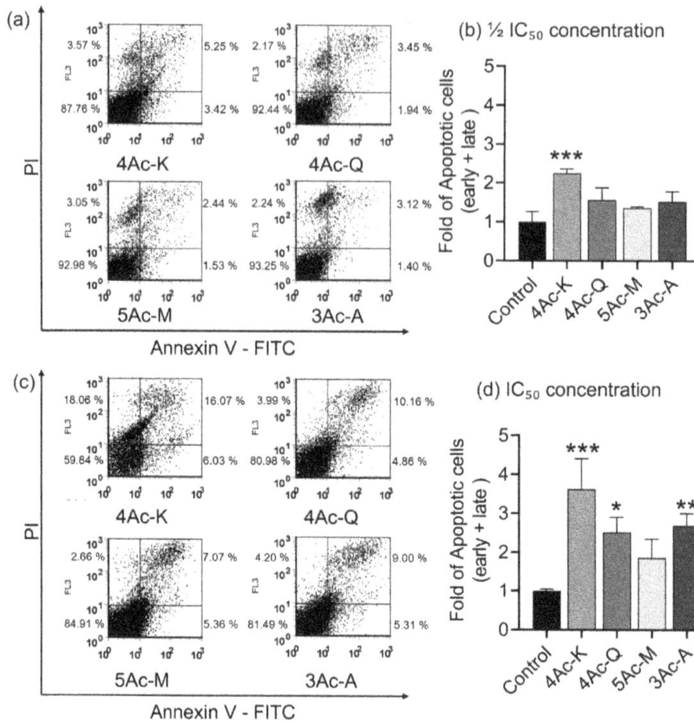

Figure 7. Effect of acetylated kaempferol, quercetin, myricetin, and apigenin on the apoptosis induction in MDA-MB-231 cells. (**a**,**c**) Representative flow histograms depicting the apoptotic fraction. Quantitation of % apoptotic fraction (early + late apoptotic cells) in MDA-MB-231 cells. Measurements in (**b**,**d**) are for a 24 h treatment at half the IC_{50} concentration and IC_{50}, respectively. Data represented as the means ± SD; * mark denotes significant differences (* $p < 0.005$, ** $p < 0.01$, *** $p < 0.001$) between the control and each sample. Statistically analyzed by Dunnett's multiple comparison test.

While Figure 6b showed that 3Ac-A had stronger anti-migration ability against MDA-MB-231 cells compared to other acetyl derivatives, it did not show significant apoptosis induction at the same concentration (Figure 7b). This suggested that 3Ac-A possesses excellent antimigration activity independent of its apoptosis-inducing ability. Consequently, a comparison of anti-migration and apoptosis-inducing activities was conducted using 2Ac-A and 3Ac-A (Figure S1). The results indicated that 2Ac-A, treated at the IC_{50} concentration for 48 h, had superior apoptosis-inducing ability compared to 3Ac-A, while it exhibited inferior anti-migration ability. This strengthens the hypothesis that 3Ac-A, unlike other acetyl derivatives, has antimigration ability independent of apoptosis induction.

3. Discussion

3.1. Cell Proliferation Inhibitory Ability and Structure–Activity Relationships

As shown in Figure S1, all cell proliferation assays demonstrated that the flavonoids, excluding myricetin, inhibited proliferation in a dose-dependent manner. Their acetylated forms also exhibited similar dose-dependent inhibition patterns, albeit with varying degrees of potency. Of particular interest were the distinct growth inhibition patterns observed for myricetin and acetylated myricetin in the MDA-MB-231 cell line upon treatment with the acetylated flavonoids. Myricetin inhibited proliferation rapidly at 40 μM, and at 80 μM, the cells became round, aggregated, and floating, while acetylated myricetin showed a slow sigmoidal pattern of inhibition down to 80 μM. One of the major differences between myricetin and the other flavonoids used in this study is the presence of three hydroxyl groups on the B ring. As a result, acetylated myricetin has a bulkier steric structure compared to other flavonoids and undergoes a significant change in polarity. However, considering the consistent dose-dependent growth inhibition patterns observed in the HCT-116 and HepG2 cells, it is unlikely that the increased steric hindrance due to acetylation is the direct cause of the change in the growth inhibition pattern. The enhanced antitumor effects of quercetin-3-*O*-glucoside acylated with long-chain fatty acids suggest that the anticancer activity is influenced not only by the drug's lipophilicity but also by its affinity for cellular receptors [56]. Therefore, it is plausible that the number of acetyl groups on the B ring of acetylated flavonoids may have led to chemical and physical interactions specific to the cell type. Cho et al. reported the differing metabolism of quercetin derivatives (with *O*-acylated pivaloxymethyl groups) in HCT-116 and MCF-7 breast cancer cells; it is important to consider the differences in the types and expression levels of metabolic enzymes and metabolic pathways in individual cells. To elucidate these hypotheses, it is necessary to (1) vary the number of acetyl modifications on the hydroxyl groups of the B ring of myricetin, (2) measure the cellular uptake of acetylated myricetin, and (3) track the metabolites of acetylated myricetin in each cell type.

Numerous reports suggest that the acetylation of flavonoids enhances cellular uptake, leading to an inhibitory effect on cancer cell growth. For instance, our study demonstrated an increased uptake of quercetin into HepG2 cells upon quercetin acetylation [41], and Lambert et al. observed an increased cellular uptake and an enhanced growth inhibition in HCT-116 cells upon EGCG acetylation [57]. However, our present study revealed exceptions to this trend. In certain cell types and with specific flavonoids, the cell proliferation inhibitory effect decreased after acetylation, depending on the cell type. These findings are summarized in Table 2.

Table 2. Increase/decrease in cell proliferation inhibition ability with acetylation.

Sample	MDA-MB-231	HCT-116	HepG2
	Flavonols		
3Ac-K	↑	-	-
4Ac-K	↑	↑	↑
4Ac-Q	↑	-	-
5Ac-Q	↑	↑	↑
5Ac-M	↓	-	-
6Ac-M	↓	↑	↑
	Flavone		
2Ac-C	↓	↓	↑
2Ac-A	↑	-	-
3Ac-A	↓	↓	↑
4Ac-L	↓	↓	↓
	Flavanone		
3Ac-N	↑	→	→
5Ac-T	↑	↑	↑

Comparison results of IC_{50} values are indicated by the direction of the arrows; gray upward arrows indicate a slight increase. Horizontal arrows indicated no statistically significant change.

The strength of cell proliferation inhibitory effects was ranked as follows: apigenin > kaempferol > naringenin. Similarly, luteolin showed stronger cell proliferation inhibitory effects compared to quercetin and taxifolin. Importantly, the order of inhibitory activity did not change after acetylation for any of these flavonoids.

Interestingly, both apigenin and luteolin derivatives containing chrysin exhibited decreased inhibitory activity upon acetylation in the MDA-MB-231 and HCT-116 cells. Conversely, acetylation of flavonols, except for myricetin in the MDA-MB-231 cells, contributed to enhanced inhibitory activity. All flavonols showed enhanced activity upon acetylation, particularly the cell proliferation inhibitory activity of taxifolin, which was reduced to a minimum of 125.8 μM by acetylation, even though the half-life inhibitory concentration (IC_{50}) was above 160 μM in all cell types. Furthermore, comparison of the IC_{50} values in Figure 5 suggested that for many flavonoids, avoiding acetylation of the 5-OH group in the A ring may help maintain their activity. The structural difference between flavonols and flavones lies in the presence or absence of a hydroxyl group at the C-3 position of the C ring. It has been reported that the absence of a hydroxyl group at this position enhances cell proliferation inhibitory effects [58,59]. This supports our findings that luteolin, compared to quercetin, and apigenin, compared to kaempferol, inhibit the proliferation of various cancer cell types at lower IC_{50} values. Furthermore, acetylation of the hydroxyl group at C-3 in the C ring resulted in enhanced cell proliferation inhibitory activity. Consequently, the compounds acetylated at the C-3 hydroxyl group, such as 4Ac-K and 5Ac-Q, showed no significant difference in IC_{50} values compared to the compounds lacking the C-3 hydroxyl group, such as 3Ac-A and 4Ac-L. This suggests that acetylating the hydroxyl group at C-3 in the C ring to a non-hydroxyl group enhances growth inhibitory activity. Consistent with this observation, studies by Zhang et al. [60] and Karancsi et al. [61] have demonstrated that methylation at the C3 position of the C ring plays a crucial role in enhancing the growth inhibitory activity of Rhamnetin (3′,7-dimethylquercetin) and 3-*O*-methylquercetin. On the other hand, apigenin and naringenin, as well as quercetin and taxifolin, share differences in the presence or absence of a C2–C3 double bond. The presence of the C2–C3 double bond is crucial for the inhibitory activity of flavonoids. Studies have demonstrated that quercetin and taxifolin, when evaluated in human leukemia HL-60 cells, exhibit higher proliferation inhibitory activity when the C ring contains a double bond C-ring [62]. Additionally, it has been reported that the presence of a C2–C3 double bond in citrus-derived flavonoids is essential for exerting anticancer effects against cancer cells [63]. Our findings are consistent with these observations. Taxifolin, a flavanone, lacks a C2–C3 double bond, making it susceptible to inactivation through strong hydrogen bonding with polymers [64]. Our study suggests that this inactivation may be mitigated by acetylation. It is conceivable that protection of hydroxyl groups by acetylation disrupts interactions via hydrogen bonding. However, considering the possibility of deacetylation within cells followed by inactivation, the cell proliferation inhibitory effects might result from interactions as acetylated derivatives with receptors on cell membranes. Future studies should evaluate interactions between acetylated flavonoids and receptors before deacetylation induction.

From these results, it is suggested that to enhance the cell proliferation inhibitory activity through acetylation, attention should be paid to the acetylation of the C-3 hydroxyl group and the absence of acetyl groups at the 5-OH position in the A ring. However, as shown in Table 2, there are numerous exceptions to these observations, including cell specificity. To comprehensively understand the delicate effects of acetylation on flavonoid activity, further studies are needed to (1) expand the types of flavonoids targeted for acetylation, (2) increase the diversity of cell types evaluated, (3) measure the impact on cellular uptake and metabolism, and (4) elucidate the interactions of acetylated derivatives with cell membrane receptors. Such research will strengthen our understanding of the nuanced effects of acetylation on flavonoid activity.

3.2. Effects of Acetylation of Quercetin and Kaempferol on the Cell Cycle

Flavonoids are known to possess cell proliferation inhibitory properties, and this mechanism has been reported in multiple studies [65–67].

In MDA-MB-231 cells, kaempferol induced an increase in CyclinB1 and p-Cdc2 levels, along with a decrease in CyclinD1. This suggests that kaempferol caused a G2/M phase arrest via inhibition of Cdc2, leading to the reduced CyclinD1 levels associated with the G1 phase. Consistent with this, a study by Zhu et al. reported a decrease in G1 phase cells and an increase in G2/M phase cells over time upon the addition of kaempferol to MDA-MB-231 cells [68]. Comparison of 4Ac-K with the controls revealed a decrease in CyclinD1 levels and an increase in CyclinA and CyclinB1 at 12 h, suggesting a potential influence on cell cycle progression. Previous studies have indicated that the inhibition of the nuclear translocation of the CyclinB1/Cdc2 complex prolongs G2 phase duration [69]. A regulator of these processes is 14-3-3σ; thus, detecting 14-3-3σ would be essential to ascertain whether the observed increase in CyclinB1 following 4Ac-K treatment correlates with cell cycle delay.

Quercetin treatment for 12 h led to decreased CyclinD1 and CyclinE levels, while CyclinB and p-Cdc2 levels showed no significant change. Chekuri et al. treated MDA-MB-231 cells with 10 μg/mL of quercetin extracted from A. indica plant leaves for 24 h, reporting an increase in the G1 phase [70]. Our findings diverge, suggesting quercetin's 24 h treatment affects the G2/M phase in MDA-MB-231 cells. The disparity might stem from using pure quercetin in our study versus plant extracts in Chekuri et al., potentially explaining different effects on the cell cycle. Leong et al. reported no significant cell cycle impact with 24 h treatment of 200 μM quercetin compared to controls [71]. This disparity in results could be attributed to their use of a higher concentration of quercetin than ours and their utilization of flow cytometry for cell cycle quantification.

The 12 h treatment of 5Ac-Q mirrored the findings seen with 4Ac-K, implying a potential G2/M phase arrest mediated by CyclinB1 and CyclinA accumulation through p-Cdc2 inactivation. Acetylation altered the dynamics of proteins such as CyclinD1 and CyclinB1, suggesting their involvement in cell cycle arrest and regulation, independent or dependent of their behavior in parent compounds.

In HCT-116 cells, the addition of kaempferol for 12 h resulted in an increased expression of CyclinB1, which is involved in the progression from G2 to M phase, while p-Cdc2 expression levels were comparable to those of controls. The increased expression of CyclinD1 also suggests that progression to the G1 phase is taking place, as Budisan et al. reported similar results, in which the addition of kaempferol to HCT-116 cells for 24 h reduced the percentage of cells in the G1 phase and increased the number of cells in the G2 phase [72]. This was consistent with our results for 24 h treatment with kaempferol. The addition of 4Ac-K to HCT-116 cells resulted in decreased CyclinD1 and CyclinB1 levels and increased phosphorylated inactive Cdc2, unlike the parent compound, indicating a potential arrest of the cell cycle in the G2/M phase. Chiou et al. reported that treatment with acetylated EGCG (AcEGCG) activates ERK, induces Cdc 25A degradation, and inhibits cyclin B1/CDK1 complex formation [73]. Similarly, at 12 h and 24 h, quercetin and 5Ac-Q increased CyclinB1 and the inactive form, p-Cdc2, suggesting that Cdc2 inactivation is accompanied by increased accumulation of CyclinB1 in the cytosol. These results indicate that all samples in HCT-116 affect the G2/M phase, with the acetyl compound possibly having a stronger effect on the G2/M phase than the parent compound.

For HepG2 cells, 4Ac-K exhibited increased phosphorylated Cdc2 (p-Cdc2) levels, along with decreased CyclinD1 levels after 12 h compared to kaempferol, indicating a pronounced effect on the G2/M phase.

Quercetin and 5Ac-Q treatments for 12 and 24 h resulted in decreased CyclinD1 and CyclinB1 levels, as well as increased p-Cdc2 levels, which also led to a slight decrease in CyclinA levels associated with the S phase. This suggests a potential impact on the G2/M phase, consistent with findings from Xu et al., who reported that 24 h treatment with 10 μM quercetin increased the proportion of cells in the G2/M phase [74]. Similar studies have shown that acylated quercetin-3-*O*-glucoside increased the number of cells in the G2/M phase in HepG2 cells [56]. Additionally, our results indicate that 5Ac-Q may affect the G2/M phase more strongly than quercetin. Furthermore, our previous study demonstrated a delayed metabolite effect of quercetin and 4Ac-Q in HepG2 cells after 6 h, indicating an enhanced uptake [41]. Together, these findings suggest distinct effects of quercetin and 5Ac-Q treatments on the cell cycle.

Treatment of MDA-MB-231 cells with quercetin and 5Ac-Q and of HCT-116 cells with kaempferol and 4Ac-K for 12 h suggested varying effects on the cell cycle. Treatment of HCT-116 and HepG2 cells with 5Ac-Q indicated a stronger impact on the G2/M phase compared to treatment with quercetin. These findings underscore one outcome supporting the enhanced cell proliferation inhibition by acetylation of quercetin and kaempferol. While the influence of acetylation on the cell cycle of each flavonol was suggested, further studies are needed to accurately correlate the differences in cell proliferation inhibition between parent compounds and acetylated derivatives with cell cycle arrest. The quantification of cell numbers in each phase using flow cytometry, along with the comparison of the metabolites of the acetylated compounds in cells compared to those of their parent compounds, will be essential to achieve a more definitive understanding.

3.3. *Anti-Migration and Apoptosis-Inducing Activities of Acetylated Flavonoids*

In the assessment of anti-migration activity using a wound healing assay and a migration assay, the acetyl derivatives containing apigenin exhibited notable anti-migration effects. Apigenin has shown the inhibition of IL-6 in MDA-MB-231 cells and the suppression of metastasis by hindering HGF-promoted invasive growth [75–77]. Similarly, in this study, an acetyl derivative of kaempferol and quercetin also displayed superior migration inhibitory activity compared to its parent compound. Quercetin and kaempferol, parent compounds, have been reported to impede breast cancer cell invasion notably by inhibiting MMP-3 [78]. As there are currently no reports on the mechanisms of action for the anti-migration effects of acetylated flavonoids, it would be most effective to base future comparative studies on the known mechanisms of their parent compounds.

Notably, 4Ac-K exhibited potent apoptosis induction at low doses, suggesting a possible correlation between its anti-migration activity and apoptosis induction. In this study, both 2Ac-A and 4Ac-K showed markedly superior apoptosis induction and anti-migration ability. The mechanism by which apigenin induces apoptosis may also be related to anti-angiogenesis and anti-migration in human hepatocellular carcinoma cells [79]. However, the findings from Figures 6, 7 and S1 suggest that 3Ac-A may have anti-migration properties independent of its apoptosis-inducing ability. Further studies with apoptosis inhibitors and related agents are needed to clarify whether 3Ac-A operates through pathways independent of apoptosis induction. The reported anti-metastatic effects of natural compounds such as isothiocyanates, achieved by downregulation of the IAP family, are noteworthy [80]. Our previous studies have shown that 4Ac-Q strongly downregulates the IAP family [23,41], suggesting a possible link between IAP family regulation and anti-migration effects. Given this possibility, it is plausible that acetylated apigenin derivatives may also regulate the IAP family. Investigating this potential relationship could help clarify the connection between IAP family regulation and the anti-migration effects of these compounds. However, it should also be noted that apigenin induces apoptosis through various pathways other than caspase-3 activation [58]. By elucidating these additional mechanisms, a more comprehensive understanding of the apoptosis-inducing and anti-migratory pathways of apigenin, 2Ac-A, and 3Ac-A may be achieved.

3.4. *Anti-Migration Activity and Structure–Activity Relationships*

The compounds with enhanced anti-migratory activity upon acetylation included 4Ac-K, 5Ac-Q, 6Ac-M, and 5Ac-T. Similar to the results of the inhibitory effect on cell proliferation, the acetylation modification at the C-ring 3 position may also contribute to the enhancement of anti-migratory activity. Yamauchi et al. reported that methylation of the hydroxyl group at position 3 of the C ring of quercetin inhibited the migration ability of B16 melanoma cells [25]. One of the primary contributions of the acetyl group is likely the protection of the hydroxyl group at position 3. Additionally, the fact that acetylated derivatives of apigenin, which lacks a hydroxyl group at position 3, exhibited significantly increased activity suggests that there may be further contributions beyond the protection of the 3-position hydroxyl group. Elucidating these additional contributions will be one of the most important research tasks in the future, as 3Ac-Q demonstrated anti-migratory effects even at concentrations that did not induce apoptosis.

Comparing 4Ac-K with 5Ac-Q and 6Ac-M, 4Ac-K showed the highest anti-migration effect. Similarly, among 2Ac-C, 3Ac-A, and 4Ac-L, 3Ac-A demonstrated the highest migration inhibition. Thus, the presence of an acetyl group at the 4′-position of the B ring contributed to the enhanced anti-migration activity. However, the slightly reduced inhibitory effect of 3Ac-N, whose parent compound is naringenin with a 4′-OH group, suggests that a C2–C3 double bond is required as a basic backbone. This experiment alone could not determine the effect of the presence or absence of a hydroxyl group at position 5 of the A ring on anti-migration ability. It is suggested that further insights may be obtained by synthesizing and comparing new compounds, such as acetylated derivatives (3Ac-L and 2Ac-N), in which the hydroxyl group at position 5 of luteolin and naringenin remains.

In terms of the contribution of acetylation to the enhanced induction of apoptosis, the comparison among 4Ac-K, 4Ac-Q, and 5Ac-M suggests that acetyl modification at position 4′ of the B ring is likely responsible for the increased activity. Specifically, compounds like 2Ac-A and 4Ac-K, which exhibited significant apoptosis induction, share the structural feature of having an acetyl group only at position 4′ of the B ring, presumed to enhance apoptosis induction, at least in the context of MDA-MB-231 cells. Furthermore, comparing the apoptosis-inducing abilities of 2Ac-A and 3Ac-A (Figure S2) revealed that the presence of a hydroxyl group at position 5 of the A ring is necessary, consistent with the results observed in the evaluation of cell proliferation inhibition.

4. Materials and Methods

4.1. Chemicals, Antibodies, and Reagents

Quercetin for the synthesis of 4Ac-Q (purity $\geq$ 90%) was purchased from Sigma-Aldrich (St. Louis, MO, USA), and quercetin for the bioactivity assay (purity > 99%) was purchased from LKT Laboratories (St. Paul, MN, USA). Kaempferol, Myricetin, Chrysin, and Naringenin were purchased from TCI Co., Ltd. (Tokyo, Japan). Apigenin and luteolin were obtained from Wako Co., Ltd. (Osaka, Japan) and Funakoshi (Tokyo, Japan), respectively. Trypan blue was purchased from Nacalai Tesque (Tokyo, Japan). Hematoxylin-Eosin Stain kit was from Cosmo Bio Co., Ltd. (Tokyo, Japan). The antibodies against Cyclin A1 (sc-271682), Cyclin B1 (sc-245), Cyclin D1 (sc-20044), Cyclin E (sc-377100), P21 (sc-756), P27 (sc-528), and p-Cdc2 p34 (sc-136014) were obtained from Santa Cruz Biotechnology (Dallas, TX, USA). Chk1 (#2360), p-chk1 (#2348), and anti-rabbit (#7074) secondary antibodies were from Cell Signaling Technology (Beverly, MA, USA). The antibody against β-actin (A5441) was from Sigma-Aldrich (St. Louis, MO, USA). Transwell Permeable Support (8 mm polycarbonate membrane) chambers used for cell migration assay were purchased from Corning (Corning, NY, USA).

4.2. Synthesis of Acetylated Derivatives of Flavonoids

4.2.1. 3,5,7,4′-. O-tetraacetate kaempferol (4Ac-K) (Compound **2**)

Kaempferol (**1**) (0.480 g, 1.67 mmol) was dissolved in pyridine (4 mL), acetic anhydride (5 eq, 0.8 mL) was added, and the solution was stirred at room temperature for 12 h. After the reaction was completed, 30 mL of cold water was added to the stirred solution to quench the reaction. The resulting precipitate was collected by filtration and dried overnight. For purification, each filtered precipitate was dissolved in 50 mL of methanol with heating. The solution was then filtered, cooled, and recrystallized. After recrystallization, HPLC was performed to confirm the purity of the obtained compounds. Compounds with a purity of more than 97% were used.

4.2.2. 3,5,7,3′,4-. O-pentaacetate quercetin (5Ac-Q) (Compound **4**)

Quercetin (**3**) (0.638 g, 2.11 mmol) was dissolved in pyridine (4 mL), acetic anhydride (6 eq, 1.2 mL) was added, and the solution was stirred at room temperature for 1 h. After the reaction was completed, cold water (30 mL) was added to the stirred solution to quench the reaction. The resulting precipitate was collected by filtration and dried overnight. For purification, each filtered precipitate was dissolved in methanol (75 mL) with heating. The solution was then filtered, cooled, and recrystallized. After recrystallization, HPLC was performed to confirm the purity of the obtained compounds. Compounds with a purity of more than 97% were used.

4.2.3. 3,5,7,3′,4′,5′-. O-hexaacetate myricetin (6Ac-M) (Compound **6**)

Myricetin (**5**) (0.498 g, 1.56 mmol) was dissolved in pyridine (4 mL), acetic anhydride (7 eq, 1.1 mL) was added, and the solution was stirred at room temperature for 12 h. After the reaction was completed, cold water (30 mL) was added to the stirred solution to quench the reaction. The resulting precipitate was collected by filtration and dried overnight. For purification, each filtered precipitate was dissolved in methanol (150 mL) with heating. The solution was then filtered, cooled, and recrystallized. The recrystallized product was collected and used as a sample.

4.2.4. 5,7-. O-diacetate chrysin (2Ac-C) (Compound **8**)

Chrysin (**7**) (0.502 g, 1.97 mmol) was dissolved in pyridine (8 mL), acetic anhydride (3 eq, 0.555 mL) was added, and the solution was refluxed for 7 h. After the reaction was completed, cold water (50 mL) was added to the stirred solution to quench the reaction. The resulting precipitate was collected by filtration and dried overnight. For purification, each filtered precipitate was dissolved in methanol (100 mL) with heating. The solution was then filtered, cooled, and recrystallized. The recrystallized product was collected and used as a sample.

4.2.5. 5,7,4′-. O-triacetate apigenin (3Ac-A) (Compound **10**)

Apigenin (**9**) (0.4868 g, 1.80 mmol) was dissolved in pyridine (12 mL), acetic anhydride (4 eq, 0.680 mL) was added, and the solution was refluxed for 7 h. After the reaction was completed, cold water (50 mL) was added to the stirred solution to quench the reaction. The resulting precipitate was collected by filtration and dried overnight. For purification, each filtered precipitate was dissolved in methanol (150 mL) with heating. The solution was then filtered, cooled, and recrystallized. The recrystallized product was collected and used as a sample.

4.2.6. 5,7,3′,4′-. O-tetraacetate luteolin (4Ac-L) (Compound **12**)

Luteolin (**11**) (0.496 g, 1.73 mmol) was dissolved in pyridine (20 mL), acetic anhydride (5 eq, 0.810 mL) was added, and the solution was refluxed for 7 h. After the reaction was completed, cold water (150 mL) was added to the stirred solution to quench the reaction. The resulting precipitate was collected by filtration and dried overnight. For purification, each filtered precipitate was dissolved in methanol (150 mL) with heating. The solution was then filtered, cooled, and recrystallized. The recrystallized product was collected and used as a sample.

4.2.7. 5,7,4′-. O-triacetate naringenin (3Ac-N) (Compound **14**)

Naringenin (**13**) (0.511 g, 1.87 mmol) was dissolved in pyridine (27 mL), acetic anhydride (100 eq, 17.5 mL) was added, and the solution was stirred at room temperature for 7 h. After the reaction was completed, cold water (150 mL) was added to the stirred solution to quench the reaction. The resulting precipitate was collected by filtration and dried overnight. For purification, each filtered precipitate was dissolved in methanol (50 mL) with heating. The solution was then filtered, cooled, and recrystallized. The recrystallized product was collected and used as a sample.

4.2.8. 3,5,7,3,′4′-. O-pentaacetate taxifolin (5Ac-T) (Compound **16**)

Taxifolin (**15**) (0.187 g, 0.614 mmol) was dissolved in pyridine (27 mL), acetic anhydride (8 eq, 0.46 mL) was added, and the solution was stirred at 37 °C for 1 h. After the reaction was completed, cold water (20 mL) was added to the stirred solution to quench the reaction. The resulting precipitate was collected by filtration and dried overnight. For purification, each filtered precipitate was dissolved in methanol (30 mL) with heating. The solution was then filtered, cooled, and recrystallized. The recrystallized product was collected and used as a sample.

4.2.9. 3,7,4′-. O-triacetate kaempferol (3Ac-K) (Compound **17**)

Kaempferol (**1**) (0.106 g, 1.67 mmol) was dissolved in pyridine (4 mL), acetic anhydride (3 eq, 0.104 mL) was added, and the solution was stirred at room temperature for 7 min. After the reaction was completed, cold water (30 mL) was added to the stirred solution to quench the reaction. The resulting precipitate was collected by filtration and dried overnight. For purification, each filtered precipitate was dissolved in methanol (30 mL) with heating. The solution was then filtered, cooled, and recrystallized. The recrystallized product was collected and used as a sample.

4.2.10. 3,7,3′,4′-. O-tetraacetate quercetin (4A-Q) (Compound **18**)

Quercetin (**3**) (0.600 g, 1.98 mmol) was dissolved in pyridine (4 mL), acetic anhydride (4.2 eq, 0.79 mL) was added, and the solution was stirred at room temperature for 12 min. After the reaction was completed, cold water (30 mL) was added to the stirred solution to quench the reaction. The resulting precipitate was collected by filtration and dried overnight. For purification, each filtered precipitate was dissolved in methanol (50 mL) with heating. The solution was then filtered, cooled, and recrystallized. The recrystallized product was collected and used as a sample.

4.2.11. 3,7,3′,4′,5′-. O-pentaacetate myricetin (5Ac-M) (Compound **19**)

Myricetin (**5**) (0.346 g, 1.09 mmol) was dissolved in pyridine (4 mL), acetic anhydride (5 eq, 0.51 mL) was added, and the solution was stirred at room temperature for 7 min. After the reaction was completed, cold water (30 mL) was added to the stirred solution to quench the reaction. The resulting precipitate was collected by filtration and dried overnight. For purification, each filtered precipitate was dissolved in methanol (150 mL) with heating. The solution was then filtered, cooled, and recrystallized. The recrystallized product was collected and used as a sample.

4.2.12. 7,4′-. O-diacetate apigenin (2Ac-A) (Compound **20**)

Apigenin (**9**) (0.0314 g, 0.116 mmol) was dissolved in pyridine (4 mL), acetic anhydride (2 eq, 0.022 mL) was added, and the solution was stirred on ice for 5 min. After the reaction was completed, cold water (50 mL) was added to the stirred solution to quench the reaction. The resulting precipitate was collected by filtration and dried overnight. For the purification of acetylated apigenin, it was dissolved in acetone. The solution was then concentrated using an evaporator. Hexane was added, and the mixture was left overnight to crystallize.

4.3. Structure Identification of Acetylated Flavonoids

All acetylated flavonoids were determined by ^{1}H-NMR recorded on a spectrometer (JEOL ECA-600, 600 MHz for 1H, JEOL, Tokyo, Japan) using tetramethylsilane (TMS) as an internal standard. All acetylated flavonoids except 2Ac-A were determined using Chloroform-D1, and 2Ac-A was measured using Acetone-D6 as the solvent. The infrared absorption wavelengths of each compound were measured using an FT-IR/IRT-3000 ATR-30-Z spectrophotometer (Japan Spectroscopic Corporation, Tokyo, Japan). Attenuated total reflection (ATR) was performed by placing a powder sample on the entire surface of the ATR crystal and pressing the sample hard against the prism while compressing it. The measuring range was 400–4000 cm^{-1}. The purity of each compound was confirmed by HPLC to be at least 97%.

4.4. Cell Culture

The human breast cancer cell line MDA-MB-231 was obtained from the American Type Culture Collection (Manassas, VA, USA) and growth conditions were optimized at 37 °C in a 5% CO_2 atmosphere. Cells were cultured in Roswell Park Memorial Institute medium (RPMI 1640) supplemented with 10% fetal bovine serum, 1% penicillin-streptomycin-neomycin (PSN), and 1.5% glutamine. The human hepatocellular carcinoma cell line HepG2 was obtained from the RIKEN Cell Bank (RCB1648) (Tsukuba, Ibaraki, Japan), and the human colon carcinoma cell line HCT-116 was obtained from the American Type Culture Collection. These cells were cultured in DMEM medium supplemented with 10% fetal bovine serum and 1% penicillin-streptomycin-glutamine (PSG) at 37 °C in a 5% CO_2 atmosphere. All cells were treated with the indicated time and dose samples after 24 h of culture.

4.5. Cell Viability Assay

Cell viability was determined by trypan blue dye exclusion assay: MDA-MB-231 (5.0×10^4 cells/well), HCT-116 (1.96×10^5 cells/well) or HepG2 (2.4×10^5/well) cells were plated in each well of 6-well plates. Cells were treated with each flavonoid and acetylated flavonoids for 48 h. After cell collection by trypsin, each cell solution was stained with 100 μL of 0.4% trypan blue solution for 1 min. Cells were counted on a hemocytometer for viable cells, and cell viability was expressed as the optical density ratio of treatment to control.

4.6. Western Blot Analysis

The Western blot analysis procedure followed our previously established protocols [81]. Briefly, MDA-MB-231 (8.1×10^4 cells/well), HCT-116 (2.27×10^5 cells/well) or HepG2 (2.78×10^5/well) cells were each plated in 6-well plates. Then, 1 mM hydroxyurea was added 24 h prior to sample addition and all cell cycles were synchronized in S phase. After 24 h, kaempferol, 4Ac-K, quercetin, and 5Ac-Q were treated with 40 μM. After 12, 24, and 48 h, cells were collected and lysed using RIPA buffer. RIPA buffer consisted of 50 mM Tris-HCl (pH 8.0), 150 mM NaCl, 1 mM EDTA, 1% Nonidet P-40, 0.25% Na-deoxycholate, 1 mM sodium fluoride, 1 mM sodium orthovanadate, 1 mM phenylmethylsulfhonyl fluoride, and a proteinase inhibitor cocktail (Nacalai Tesque, Kyoto, Japan). Equal concentrations of lysed proteins were subjected to SDS-PAGE and transferred electrophoretically to PVDF membranes (GE Healthcare UK, Amersham, Buckinghamshire, UK) for 2 h. After blocking, the membranes were incubated overnight at 4 °C with specific primary antibodies, then exposed to the corresponding HRP-conjugated secondary antibodies for 1 h at room temperature. Each membrane was then detected using chemiluminescence and was scanned and evaluated using LumiVision software (version 140.ex) (TAITEC, Saitama, Japan).

4.7. Detection of Apoptosis Using Flow Cytometry

Quantitation of apoptotic cell death was assessed by eBioscience™ AnnexinV-FITC Apop Kit, purchased from Invitrogen-Life Technologies (Carlsbad, CA, USA). In brief, after 24 or 48 h of treatment with each sample, MDA-MB-231 cells were collected and suspended in binding buffer and FITC-Annexin V/PI stain; after 15 min of staining, cells were analyzed using flow cytometry at FL1 (530 nm) and FL3 (630 nm) (CyFlow@, Sysmex Partec GmbH, Gorlitz, Germany).

4.8. Wound Healing Assay

MDA-MB-231 cells were plated into each well of 6-well plate and the cells were cultured approximately confluent monolayer. Each well was scratched by 200 μL pipette tip and treated with desired concentration of samples. Cells were allowed to migrate for 6 h, then, the cells were removed from the medium, fixed by methanol, and stained by Giemsa stain solution (Sigma-Aldrich, St. Louis, MO, USA). Migration of cells was quantified by measuring distances between the borders of cells; image analysis was performed using Image J software (version 1.50i) (Bethesda, MD, USA). At least three nonoverlapping areas per well were examined for wound healing.

4.9. Cell Migration Assay

The migration ability of MDA-MB-231 was performed by using polycarbonate membrane transwell inserts (Corning, NY, USA). Membrane pore size was 8.0 μm. MDA-MB-231 (3.0×10^5/well) cells were suspended in serum-free medium containing DMSO or the test agents and placed in the upper compartment of the 6-well type transwell chamber. After incubation for 24 h, the cells on the upper surface of the transwell were removed by cotton swab. The motile cells from the bottom face of the filter were fixed with methanol and stained with hematoxylin and eosin. The evaluation of migration was analyzed by taking a picture with KEYENCE microscope (BZ-X810, KEYENCE, Osaka Oosaka, Japan) at 20 objective magnifications and counting migrated cells.

4.10. Statistical Analysis

Statistical analysis of the data was determined by a one-way analysis of variance (ANOVA) test followed by Tukey's multiple comparison test and Dunnett's test as described in each figure legend, employing GraphPad Prism 9.5.1 software (San Diego, CA, USA). The data are expressed as the means ± standard deviation (SD). All experiments were performed in biological triplicates ($n = 3$) with at least three replicates. The significance level was set as p-values of less than 0.05.

5. Conclusions

Among the 12 acetylated flavonoids tested, the acetyl derivatives of kaempferol and quercetin showed enhanced cell proliferation inhibition in all three cancer cell types. The evaluation of anti-migratory and apoptosis-inducing activities against MDA-MB-231 cells suggested that apigenin may acquire a unique anti-migratory activity different from that of other flavonoids through acetylation.

The presence of a hydroxyl group at the 5 position of flavonoid structures has been shown to potentially enhance cell proliferation inhibition and antimigration activity through acetylation. However, further verification of this effect is needed across compounds. Acetylated flavonoids may need to be manufactured with the appropriate number and position of acetyl groups to maintain or enhance their biological activity. Furthermore, future research should focus on synthesizing a wider variety of acetylated flavonoids and accumulating more data on their cellular uptake patterns, metabolic rates, and metabolites.

Supplementary Materials: The following supporting information can be downloaded at: https://www.mdpi.com/article/10.3390/ijms25147689/s1.

Author Contributions: Conceptualization, D.-X.H. and K.S.; methodology, K.S., Y.S., S.I. and D.U.; validation, K.S. and D.-X.H.; formal analysis, K.S., D.U. and S.I.; investigation, K.S., Y.S., S.I. and D.U.; data curation, K.S. and D.-X.H.; resources, writing—original draft preparation, K.S. and D.U.; writing—review and editing, D.-X.H.; supervision, D.-X.H. and K.S.; project administration, K.S. and D.-X.H.; funding acquisition, K.S. All authors have read and agreed to the published version of the manuscript.

Funding: This work was supported by JSPS KAKENHI Grant Number JP19K05931 and JP16K1870.

Institutional Review Board Statement: Not applicable.

Informed Consent Statement: Not applicable.

Data Availability Statement: The data used to support the findings of this study are available from the corresponding author upon request.

Conflicts of Interest: The authors declare no conflicts of interest.

Appendix A

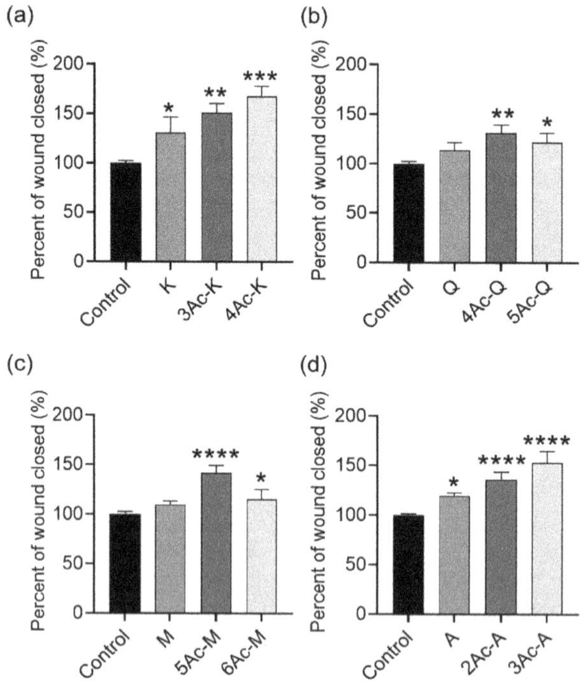

Figure A1. Assessment of anti-cellular migration using wound healing assay. Wound healing was quantified after 6 h of treatment with concentrations of (**a**) kaempferol, (**b**) quercetin, (**c**) myricetin, (**d**) apigenin and their acetyl derivatives at 2× IC_{50} values. The results shown (mean ± SD) are representative of two independent experiments (n = 9). * mark denoted significant differences: * $p < 0.005$, ** $p < 0.01$, *** $p < 0.001$, **** $p < 0.0001$ between control and each compound. Statistically analyzed by Dunnett's multiple comparison test.

References

1. Aherne, S.A.; O'Brien, N.M. Dietary flavonols: Chemistry, food content, and metabolism. *Nutrition* **2002**, *18*, 75–81. [CrossRef] [PubMed]
2. Bei, R.; Masuelli, L.; Turriziani, M.; Volti, L.G.; Malaguarnera, M.; Galvano, F. Impaired Expression and Function of Signaling Pathway Enzymes by Anthocyanins: Role on Cancer Prevention and Progression. *Curr. Enzym. Inhib.* **2009**, *5*, 184–197. [CrossRef]
3. Dong, J.; Zhang, X.; Zhang, L.; Bian, H.X.; Xu, N.; Bao, B.; Liu, J. Quercetin reduces obesity-associated ATM infiltration and inflammation in mice: A mechanism including AMPKα1/SIRT1. *J. Lipid Res.* **2014**, *55*, 363–374. [CrossRef] [PubMed]
4. Zhao, X.; Chen, R.; Shi, Y.; Zhang, X.; Tian, C.; Xia, D. Antioxidant and Anti-Inflammatory Activities of Six Flavonoids from Smilax glabra Roxb. *Molecules* **2020**, *25*, 5295. [CrossRef] [PubMed]
5. Maleki, S.J.; Crespo, J.F.; Cabanillas, B. Anti-inflammatory effects of flavonoids. *Food Chem.* **2019**, *299*, 125124. [CrossRef] [PubMed]
6. Ontiveros, M.; Rinaldi, D.; Marder, M.; Espelt, M.V.; Mangialavori, I.; Vigil, M.; Rossi, J.P.; Ferreira-Gomes, M. Natural flavonoids inhibit the plasma membrane Ca^{2+}-ATPase. *Biochem. Pharmacol.* **2019**, *166*, 1–11. [CrossRef] [PubMed]
7. Jiang, W.-J.; Daikonya, A.; Ohkawara, M.; Nemoto, T.; Noritake, R.; Takamiya, T.; Kitanaka, S.; Iijima, H. Structure-activity relationship of the inhibitory effects of flavonoids on nitric oxide production in RAW264.7 cells. *Bioorgan. Med. Chem.* **2017**, *25*, 779–788. [CrossRef] [PubMed]
8. Desmet, S.; Morreel, K.; Dauwe, R. Origin and Function of Structural Diversity in the Plant Specialized Metabolome. *Plants* **2021**, *10*, 2393. [CrossRef] [PubMed]
9. Alseekh, S.; Perez de Souza, L.; Benina, M.; Fernie, A.R. The style and substance of plant flavonoid decoration; towards defining both structure and function. *Phytochemistry* **2020**, *174*, 112347. [CrossRef]

10. Wang, S.; Alseekh, S.; Fernie, A.R.; Luo, J. The Structure and Function of Major Plant Metabolite Modifications. *Mol. Plant* **2019**, *12*, 899–919. [CrossRef]
11. Lin, Y.-C.; Tsai, P.-H.; Lin, C.-Y.; Cheng, C.-H.; Lin, T.-H.; Lee, K.P.H.; Huang, K.-Y.; Chen, S.-H.; Hwang, J.-J.; Kandaswami, C.C.; et al. Impact of Flavonoids on Matrix Metalloproteinase Secretion and Invadopodia Formation in Highly Invasive A431-III Cancer Cells. *PLoS ONE* **2013**, *8*, e71903. [CrossRef] [PubMed]
12. Tu, S.-H.; Chen, L.-C.; Ho, Y.-S. An apple a day to prevent cancer formation: Reducing cancer risk with flavonoids. *J. Food Drug Anal.* **2017**, *25*, 119–124. [CrossRef] [PubMed]
13. Kim, D.H.; Khan, H.; Ullah, H.; Hassan, S.T.S.; Šmejkal, K.; Efferth, T.; Mahomoodally, M.F.; Xu, S.; Habtemariam, S.; Filosa, R.; et al. MicroRNA targeting by quercetin in cancer treatment and chemoprotection. *Pharmacol. Res.* **2019**, *147*, 104346. [CrossRef] [PubMed]
14. Tian, S.-S.; Jiang, F.-S.; Zhang, K.; Zhu, X.-X.; Jin, B.; Lu, J.-J.; Ding, Z.-S. Flavonoids from the leaves of Carya cathayensis Sarg. inhibit vascular endothelial growth factor-induced angiogenesis. *Fitoterapia* **2014**, *92*, 34–40. [CrossRef]
15. Fatima, N.; Baqri, S.S.R.; Bhattacharya, A.; Koney, N.K.; Husain, K.; Abbas, A.; Ansari, R.A. Role of Flavonoids as Epigenetic Modulators in Cancer Prevention and Therapy. *Front. Genet.* **2021**, *12*, 758733. [CrossRef] [PubMed]
16. Busch, C.; Burkard, M.; Leischner, C.; Lauer, U.M.; Frank, J.; Venturelli, S. Epigenetic activities of flavonoids in the prevention and treatment of cancer. *Clin. Epigenetics* **2015**, *7*, 64. [CrossRef]
17. Fink, B.N.; Steck, S.E.; Wolff, M.S.; Britton, J.A.; Kabat, G.C.; Gaudet, M.M.; Abrahamson, P.E.; Bell, P.; Schroeder, J.C.; Teitelbaum, S.L.; et al. Dietary Flavonoid Intake and Breast Cancer Survival among Women on Long Island. *Cancer Epidemiol. Biomark. Prev.* **2007**, *16*, 2285–2292. [CrossRef]
18. Amawi, H.; Ashby, C.R., Jr.; Tiwari, A.K. Cancer chemoprevention through dietary flavonoids: What's limiting? *Chin. J. Cancer* **2017**, *36*, 50. [CrossRef]
19. Lewandowska, U.; Fichna, J.; Gorlach, S. Enhancement of anticancer potential of polyphenols by covalent modifications. *Biochem. Pharmacol.* **2016**, *109*, 1–13. [CrossRef]
20. Fernández, C.; Nieto, O.; Rivas, E.; Montenegro, G.; Fontenla, J.A.; Fernández-Mayoralas, A. Synthesis and biological studies of glycosyl dopamine derivatives as potential antiparkinsonian agents. *Carbohydr. Res.* **2000**, *327*, 353–365. [CrossRef]
21. Koirala, N.; Pandey, R.P.; Van Thang, D.; Jung, H.J.; Sohng, J.K. Glycosylation and subsequent malonylation of isoflavonoids in *E. coli*: Strain development, production and insights into future metabolic perspectives. *J. Ind. Microbiol. Biotechnol.* **2014**, *41*, 1647–1658. [CrossRef] [PubMed]
22. Needs, P.W.; Williamson, G. Syntheses of daidzein-7-yl β-d-glucopyranosiduronic acid and daidzein-4′,7-yl di-β-d-glucopyranosiduronic acid. *Carbohydr. Res.* **2001**, *330*, 511–515. [CrossRef] [PubMed]
23. Sakao, K.; Hamamoto, S.; Urakawa, D.; He, Z.; Hou, D.X. Anticancer Activity and Molecular Mechanisms of Acetylated and Methylated Quercetin in Human Breast Cancer Cells. *Molecules* **2024**, *29*, 2408. [CrossRef] [PubMed]
24. Wen, L.; Jiang, Y.; Yang, J.; Zhao, Y.; Tian, M.; Yang, B. Structure, bioactivity, and synthesis of methylated flavonoids. *Ann. N. Y. Acad. Sci.* **2017**, *1398*, 120–129. [CrossRef] [PubMed]
25. Yamauchi, K.; Mitsunaga, T.; Afroze, S.H.; Uddin, M.N. Structure—Activity Relationships of Methylquercetin on Anti-migration and Anti-proliferation Activity in B16 Melanoma Cells. *Anticancer. Res.* **2017**, *37*, 1575–1579. [CrossRef]
26. Ninomiya, M.; Nishida, K.; Tanaka, K.; Watanabe, K.; Koketsu, M. Structure-activity relationship studies of 5,7-dihydroxyflavones as naturally occurring inhibitors of cell proliferation in human leukemia HL-60 cells. *J. Nat. Med.* **2013**, *67*, 460–467. [CrossRef] [PubMed]
27. Koirala, N.; Thuan, N.H.; Ghimire, G.P.; Thang, D.V.; Sohng, J.K. Methylation of flavonoids: Chemical structures, bioactivities, progress and perspectives for biotechnological production. *Enzym. Microb. Technol.* **2016**, *86*, 103–116. [CrossRef]
28. Chen, L.; Teng, H.; Xie, Z.; Cao, H.; Cheang, W.S.; Skalicka-Woniak, K.; Georgiev, M.I.; Xiao, J. Modifications of dietary flavonoids towards improved bioactivity: An update on structure–activity relationship. *Crit. Rev. Food Sci. Nutr.* **2018**, *58*, 513–527. [CrossRef] [PubMed]
29. Lu, Y.; He, Y.; Zhu, S.; Zhong, X.; Chen, D.; Liu, Z. New Acylglycosides Flavones from Fuzhuan Brick Tea and Simulation Analysis of Their Bioactive Effects. *Int. J. Mol. Sci.* **2019**, *20*, 494. [CrossRef]
30. El-Kersh, D.M.; Abou El-Ezz, R.F.; Fouad, M.; Farag, M.A. Unveiling Natural and Semisynthetic Acylated Flavonoids: Chemistry and Biological Actions in the Context of Molecular Docking. *Molecules* **2022**, *27*, 5501. [CrossRef]
31. Suzuki, H.; Nakayama, T.; Yonekura-Sakakibara, K.; Fukui, Y.; Nakamura, N.; Yamaguchi, M.-a.; Tanaka, Y.; Kusumi, T.; Nishino, T. cDNA Cloning, Heterologous Expressions, and Functional Characterization of Malonyl-Coenzyme A:Anthocyanidin 3-O-Glucoside-6″-O-Malonyltransferase from Dahlia Flowers. *Plant Physiol.* **2002**, *130*, 2142–2151. [CrossRef] [PubMed]
32. Sendri, N.; Bhandari, P. Anthocyanins: A comprehensive review on biosynthesis, structural diversity, and industrial applications. *Phytochem. Rev.* **2024**, *23*, 1–62. [CrossRef]
33. Liu, X.; Shi, Y.; Hu, H.; Li, J.; Li, C.; Yang, L.; Wang, Z. Three new C-glycosyflavones with acetyl substitutions from Swertia mileensis. *J. Nat. Med.* **2018**, *72*, 922–928. [CrossRef] [PubMed]
34. Le, J.; Lu, W.; Xiong, X.; Wu, Z.; Chen, W. Anti-Inflammatory Constituents from Bidens frondosa. *Molecules* **2015**, *20*, 18496–18510. [CrossRef] [PubMed]
35. Sakao, K.; Hou, D.-X. Effects and Mechanisms of Acylated Flavonoid on Cancer Chemopreventive Activity. *Curr. Pharmacol. Rep.* **2020**, *6*, 286–305. [CrossRef]

36. Massi, A.; Bortolini, O.; Ragno, D.; Bernardi, T.; Sacchetti, G.; Tacchini, M.; De Risi, C. Research Progress in the Modification of Quercetin Leading to Anticancer Agents. *Molecules* **2017**, *22*, 1270. [CrossRef] [PubMed]
37. Sadilova, E.; Stintzing, F.C.; Carle, R. Thermal Degradation of Acylated and Nonacylated Anthocyanins. *J. Food Sci.* **2006**, *71*, C504–C512. [CrossRef]
38. Jokioja, J.; Yang, B.; Linderborg, K.M. Acylated anthocyanins: A review on their bioavailability and effects on postprandial carbohydrate metabolism and inflammation. *Compr. Rev. Food Sci. Food Saf.* **2021**, *20*, 5570–5615. [CrossRef] [PubMed]
39. Tu, F.; Xie, C.; Li, H.; Lei, S.; Li, J.; Huang, X.; Yang, F. Effect of in vitro digestion on chestnut outer-skin and inner-skin bioaccessibility: The relationship between biotransformation and antioxidant activity of polyphenols by metabolomics. *Food Chem.* **2021**, *363*, 130277. [CrossRef]
40. Cho, S.; Kong, B.; Jung, Y.; Shin, J.; Park, M.; Chung, W.J.; Ban, C.; Kweon, D.H. Synthesis and physicochemical characterization of acyl myricetins as potential anti-neuroexocytotic agents. *Sci. Rep.* **2023**, *13*, 5136. [CrossRef]
41. Sakao, K.; Saruwatari, H.; Minami, S.; Hou, D.X. Hydroxyl Group Acetylation of Quercetin Enhances Intracellular Absorption and Persistence to Upregulate Anticancer Activity in HepG2 Cells. *Int. J. Mol. Sci.* **2023**, *24*, 16652. [CrossRef] [PubMed]
42. Sakao, K.; Fujii, M.; Hou, D.X. Acetyl derivate of quercetin increases the sensitivity of human leukemia cells toward apoptosis. *Biofactors* **2009**, *35*, 399–405. [CrossRef] [PubMed]
43. Zheng, X.; Meng, W.-D.; Xu, Y.-Y.; Cao, J.-G.; Qing, F.-L. Synthesis and anticancer effect of chrysin derivatives. *Bioorgan. Med. Chem. Lett.* **2003**, *13*, 881–884. [CrossRef] [PubMed]
44. Hoang, T.K.-D.; Huynh, T.K.-C.; Nguyen, T.-D. Synthesis, characterization, anti-inflammatory and anti-proliferative activity against MCF-7 cells of O-alkyl and O-acyl flavonoid derivatives. *Bioorgan. Chem.* **2015**, *63*, 45–52. [CrossRef] [PubMed]
45. Bray, F.; Laversanne, M.; Sung, H.; Ferlay, J.; Siegel, R.L.; Soerjomataram, I.; Jemal, A. Global cancer statistics 2022: GLOBOCAN estimates of incidence and mortality worldwide for 36 cancers in 185 countries. *CA Cancer J Clin* **2024**, *74*, 229–263. [CrossRef] [PubMed]
46. Maugeri, A.; Calderaro, A.; Patanè, G.T.; Navarra, M.; Barreca, D.; Cirmi, S.; Felice, M.R. Targets Involved in the Anti-Cancer Activity of Quercetin in Breast, Colorectal and Liver Neoplasms. *Int. J. Mol. Sci.* **2023**, *24*, 2952. [CrossRef] [PubMed]
47. Adinew, G.M.; Taka, E.; Mendonca, P.; Messeha, S.S.; Soliman, K.F.A. The Anticancer Effects of Flavonoids through miRNAs Modulations in Triple-Negative Breast Cancer. *Nutrients* **2021**, *13*, 1212. [CrossRef] [PubMed]
48. Li, S.; Yan, T.; Deng, R.; Jiang, X.; Xiong, H.; Wang, Y.; Yu, Q.; Wang, X.; Chen, C.; Zhu, Y. Low dose of kaempferol suppresses the migration and invasion of triple-negative breast cancer cells by downregulating the activities of RhoA and Rac1. *OncoTargets Ther.* **2017**, *10*, 4809–4819. [CrossRef] [PubMed]
49. Shahi Thakuri, P.; Gupta, M.; Singh, S.; Joshi, R.; Glasgow, E.; Lekan, A.; Agarwal, S.; Luker, G.D.; Tavana, H. Phytochemicals inhibit migration of triple negative breast cancer cells by targeting kinase signaling. *BMC Cancer* **2020**, *20*, 4. [CrossRef]
50. Sharma, P.; Khan, M.A.; Najmi, A.K.; Chaturvedi, S.; Akhtar, M. Myricetin-induced apoptosis in triple-negative breast cancer cells through inhibition of the PI3K/Akt/mTOR pathway. *Med. Oncol.* **2022**, *39*, 248. [CrossRef]
51. Cao, D.; Zhu, G.-Y.; Lu, Y.; Yang, A.; Chen, D.; Huang, H.-J.; Peng, S.-X.; Chen, L.-W.; Li, Y.-W. Luteolin suppresses epithelial-mesenchymal transition and migration of triple-negative breast cancer cells by inhibiting YAP/TAZ activity. *Biomed. Pharmacother.* **2020**, *129*, 110462. [CrossRef] [PubMed]
52. Huang, C.; Wei, Y.X.; Shen, M.C.; Tu, Y.H.; Wang, C.C.; Huang, H.C. Chrysin, Abundant in Morinda citrifolia Fruit Water-EtOAc Extracts, Combined with Apigenin Synergistically Induced Apoptosis and Inhibited Migration in Human Breast and Liver Cancer Cells. *J. Agric. Food Chem.* **2016**, *64*, 4235–4245. [CrossRef] [PubMed]
53. Rahmani, A.H.; Alsahli, M.A.; Almatroudi, A.; Almogbel, M.A.; Khan, A.A.; Anwar, S.; Almatroodi, S.A. The Potential Role of Apigenin in Cancer Prevention and Treatment. *Molecules* **2022**, *27*, 6051. [CrossRef] [PubMed]
54. Qi, Z.; Kong, S.; Zhao, S.; Tang, Q. Naringenin inhibits human breast cancer cells (MDA-MB-231) by inducing programmed cell death, caspase stimulation, G2/M phase cell cycle arrest and suppresses cancer metastasis. *Cell. Mol. Biol.* **2021**, *67*, 8–13. [CrossRef] [PubMed]
55. Li, J.; Hu, L.; Zhou, T.; Gong, X.; Jiang, R.; Li, H.; Kuang, G.; Wan, J.; Li, H. Taxifolin inhibits breast cancer cells proliferation, migration and invasion by promoting mesenchymal to epithelial transition via β-catenin signaling. *Life Sci.* **2019**, *232*, 116617. [CrossRef]
56. Sudan, S.; Rupasinghe, H.V. Antiproliferative activity of long chain acylated esters of quercetin-3-O-glucoside in hepatocellular carcinoma HepG2 cells. *Exp. Biol. Med.* **2015**, *240*, 1452–1464. [CrossRef] [PubMed]
57. Lambert, J.D.; Sang, S.; Hong, J.; Kwon, S.-J.; Lee, M.-J.; Ho, C.-T.; Yang, C.S. Peracetylation as a Means of Enhancing in Vitro Bioactivity and Bioavailability of Epigallocatechin-3-Gallate. *Drug Metab. Dispos.* **2006**, *34*, 2111. [CrossRef] [PubMed]
58. Tavsan, Z.; Kayali, H.A. Flavonoids showed anticancer effects on the ovarian cancer cells: Involvement of reactive oxygen species, apoptosis, cell cycle and invasion. *Biomed. Pharmacother.* **2019**, *116*, 109004. [CrossRef] [PubMed]
59. Plochmann, K.; Korte, G.; Koutsilieri, E.; Richling, E.; Riederer, P.; Rethwilm, A.; Schreier, P.; Scheller, C. Structure–activity relationships of flavonoid-induced cytotoxicity on human leukemia cells. *Arch. Biochem. Biophys.* **2007**, *460*, 1–9. [CrossRef]
60. Zhang, Q.; Yang, W.; Liu, J.; Liu, H.; Lv, Z.; Zhang, C.; Chen, D.; Jiao, Z. Identification of Six Flavonoids as Novel Cellular Antioxidants and Their Structure-Activity Relationship. *Oxidative Med. Cell. Longev.* **2020**, *2020*, 4150897. [CrossRef]

61. Karancsi, Z.; Kovács, D.; Palkovicsné Pézsa, N.; Gálfi, P.; Jerzsele, Á.; Farkas, O. The Impact of Quercetin and Its Methylated Derivatives 3-o-Methylquercetin and Rhamnazin in Lipopolysaccharide-Induced Inflammation in Porcine Intestinal Cells. *Antioxidants* **2022**, *11*, 1265. [CrossRef] [PubMed]
62. Rusak, G.; Gutzeit, H.O.; Müller, J.L. Structurally related flavonoids with antioxidative properties differentially affect cell cycle progression and apoptosis of human acute leukemia cells. *Nutr. Res.* **2005**, *25*, 143–155. [CrossRef]
63. Kawaii, S.; Tomono, Y.; Katase, E.; Ogawa, K.; Yano, M. Antiproliferative activity of flavonoids on several cancer cell lines. *Biosci. Biotechnol. Biochem.* **1999**, *63*, 896–899. [CrossRef] [PubMed]
64. Das, A.; Baidya, R.; Chakraborty, T.; Samanta, A.K.; Roy, S. Pharmacological basis and new insights of taxifolin: A comprehensive review. *Biomed. Pharmacother.* **2021**, *142*, 112004. [CrossRef]
65. Shi, M.D.; Shiao, C.K.; Lee, Y.C.; Shih, Y.W. Apigenin, a dietary flavonoid, inhibits proliferation of human bladder cancer T-24 cells via blocking cell cycle progression and inducing apoptosis. *Cancer Cell Int.* **2015**, *15*, 33. [CrossRef] [PubMed]
66. Wu, P.; Meng, X.; Zheng, H.; Zeng, Q.; Chen, T.; Wang, W.; Zhang, X.; Su, J. Kaempferol Attenuates ROS-Induced Hemolysis and the Molecular Mechanism of Its Induction of Apoptosis on Bladder Cancer. *Molecules* **2018**, *23*, 2592. [CrossRef] [PubMed]
67. Tseng, T.H.; Chien, M.H.; Lin, W.L.; Wen, Y.C.; Chow, J.M.; Chen, C.K.; Kuo, T.C.; Lee, W.J. Inhibition of MDA-MB-231 breast cancer cell proliferation and tumor growth by apigenin through induction of G2/M arrest and histone H3 acetylation-mediated p21(WAF1/CIP1) expression. *Environ. Toxicol.* **2017**, *32*, 434–444. [CrossRef] [PubMed]
68. Zhu, L.; Xue, L. Kaempferol Suppresses Proliferation and Induces Cell Cycle Arrest, Apoptosis, and DNA Damage in Breast Cancer Cells. *Oncol. Res.* **2019**, *27*, 629–634. [CrossRef] [PubMed]
69. Nandi, S.K.; Roychowdhury, T.; Chattopadhyay, S.; Basu, S.; Chatterjee, K.; Choudhury, P.; Banerjee, N.; Saha, P.; Mukhopadhyay, S.; Mukhopadhyay, A.; et al. Deregulation of the CD44-NANOG-MDR1 associated chemoresistance pathways of breast cancer stem cells potentiates the anti cancer effect of Kaempferol in synergism with Verapamil. *Toxicol. Appl. Pharmacol.* **2022**, *437*, 115887. [CrossRef]
70. Chekuri, S.; Vyshnava, S.S.; Somisetti, S.L.; Cheniya, S.B.K.; Gandu, C.; Anupalli, R.R. Isolation and anticancer activity of quercetin from Acalypha indica L. against breast cancer cell lines MCF-7 and MDA-MB-231. *3 Biotech* **2023**, *13*, 289. [CrossRef]
71. Leong, K.X.; Chao, S.P.; Siah, P.C.; Lim, S.K.; Khoo, B.Y. Comparative Studies on a Standardized Subfraction of Red Onion Peel Ethanolic Extract (Plant Substance), Quercetin (Pure Compound), and Their Cell Mechanism and Metabolism on MDA-MB-231. *Evid.-Based Complement. Altern. Med.* **2022**, *2022*, 9284063. [CrossRef] [PubMed]
72. Budisan, L.; Gulei, D.; Jurj, A.; Braicu, C.; Zanoaga, O.; Cojocneanu, R.; Pop, L.; Raduly, L.; Barbat, A.; Moldovan, A.; et al. Inhibitory Effect of CAPE and Kaempferol in Colon Cancer Cell Lines-Possible Implications in New Therapeutic Strategies. *Int. J. Mol. Sci.* **2019**, *20*, 1199. [CrossRef] [PubMed]
73. Chiou, Y.-S.; Sang, S.; Cheng, K.-H.; Ho, C.-T.; Wang, Y.-J.; Pan, M.-H. Peracetylated (−)-epigallocatechin-3-gallate (AcEGCG) potently prevents skin carcinogenesis by suppressing the PKD1-dependent signaling pathway in CD34 + skin stem cells and skin tumors. *Carcinogenesis* **2013**, *34*, 1315–1322. [CrossRef] [PubMed]
74. Xu, Z.; Jia, Y.; Liu, J.; Ren, X.; Yang, X.; Xia, X.; Pan, X. Naringenin and Quercetin Exert Contradictory Cytoprotective and Cytotoxic Effects on Tamoxifen-Induced Apoptosis in HepG2 Cells. *Nutrients* **2022**, *14*, 5394. [CrossRef] [PubMed]
75. Lee, W.-J.; Chen, W.-K.; Wang, C.-J.; Lin, W.-L.; Tseng, T.-H. Apigenin inhibits HGF-promoted invasive growth and metastasis involving blocking PI3K/Akt pathway and β4 integrin function in MDA-MB-231 breast cancer cells. *Toxicol. Appl. Pharmacol.* **2008**, *226*, 178–191. [CrossRef] [PubMed]
76. Lee, H.H.; Jung, J.; Moon, A.; Kang, H.; Cho, H. Antitumor and Anti-Invasive Effect of Apigenin on Human Breast Carcinoma through Suppression of IL-6 Expression. *Int. J. Mol. Sci.* **2019**, *20*, 3143. [CrossRef] [PubMed]
77. Naponelli, V.; Rocchetti, M.T.; Mangieri, D. Apigenin: Molecular Mechanisms and Therapeutic Potential against Cancer Spreading. *Int. J. Mol. Sci.* **2024**, *25*, 5569. [CrossRef] [PubMed]
78. Phromnoi, K.; Yodkeeree, S.; Anuchapreeda, S.; Limtrakul, P. Inhibition of MMP-3 activity and invasion of the MDA-MB-231 human invasive breast carcinoma cell line by bioflavonoids. *Acta Pharmacol. Sin.* **2009**, *30*, 1169–1176. [CrossRef] [PubMed]
79. Kim, B.R.; Jeon, Y.K.; Nam, M.J. A mechanism of apigenin-induced apoptosis is potentially related to anti-angiogenesis and anti-migration in human hepatocellular carcinoma cells. *Food Chem. Toxicol.* **2011**, *49*, 1626–1632. [CrossRef]
80. Sakao, K.; Desineni, S.; Hahm, E.R.; Singh, S.V. Phenethyl isothiocyanate suppresses inhibitor of apoptosis family protein expression in prostate cancer cells in culture and in vivo. *Prostate* **2012**, *72*, 1104–1116. [CrossRef]
81. Tanigawa, S.; Fujii, M.; Hou, D.X. Stabilization of p53 is involved in quercetin-induced cell cycle arrest and apoptosis in HepG2 cells. *Biosci. Biotechnol. Biochem.* **2008**, *72*, 797–804. [CrossRef] [PubMed]

International Journal of *Molecular Sciences*

MDPI

Article

Blackberry Juice Fermented with Two Consortia of Lactic Acid Bacteria and Isolated Whey: Physicochemical and Antioxidant Properties during Storage

Liliana Lugo-Zarate [1,2], Luis Delgado-Olivares [2], Nelly del Socorro Cruz-Cansino [2], Luis Guillermo González-Olivares [3], Nayeli Shantal Castrejón-Jiménez [4], Diego Estrada-Luna [1] and Angélica Saraí Jiménez-Osorio [1,*]

1 Área Académica de Enfermería, Instituto de Ciencias de la Salud, Universidad Autónoma del Estado de Hidalgo, Circuito Actopan Tilcuautla s/n. Ex Hacienda La Concepción, San Agustín Tlaxiaca 42160, Hidalgo, Mexico; lu279379@uaeh.edu.mx (L.L.-Z.); destrada_luna@uaeh.edu.mx (D.E.-L.)

2 Área Académica de Nutrición, Instituto de Ciencias de la Salud, Universidad Autónoma del Estado de Hidalgo, Circuito Actopan Tilcuautla s/n. Ex Hacienda La Concepción, San Agustín Tlaxiaca 42160, Hidalgo, Mexico; ldelgado@uaeh.edu.mx (L.D.-O.); ncruz@uaeh.edu.mx (N.d.S.C.-C.)

3 Área Académica de Química, Instituto de Ciencias Básicas e Ingeniería, Universidad Autónoma del Estado de Hidalgo, Mineral de la Reforma 42184, Hidalgo, Mexico; lgonzales@uaeh.edu.mx

4 Área Académica de Medicina Veterinaria y Zootecnia, Instituto de Ciencias Agropecuarias, Universidad Autónoma del Estado de Hidalgo, Av. Universidad km 1 Ex Hacienda de Aquetzalpa A.P. 32, Tulancingo 43600, Hidalgo, Mexico; nayeli_castrejon@uaeh.edu.mx

* Correspondence: angelica_jimenez@uaeh.edu.mx; Tel.: +52-7717172000 (ext. 4323)

Abstract: Fermenting fruit juices with lactic acid bacteria (LAB) is a sustainable method to enhance fruit harvests and extend shelf life. This study focused on blackberries, rich in antioxidants with proven health benefits. In this research, we examined the effects of fermentation (48 h at 37 °C) at 28 days on whey-supplemented (WH, 1:1) blackberry juice (BJ) inoculated with two LAB mixtures. Consortium 1 (BJWH/C1) included *Levilactobacillus brevis*, *Lactiplantibacillus plantarum*, and *Pediococcus acidilactici*, while consortium 2 (BJWH/C2) comprised *Lacticaseibacillus casei* and *Lacticaseibacillus rhamnosus*. All of the strains were previously isolated from aguamiel, pulque, and fermented milk. Throughout fermentation and storage, several parameters were evaluated, including pH, lactic acid production, viscosity, stability, reducing sugars, color, total phenolic content, anthocyanins, and antioxidant capacity. Both consortia showed a significant increase in LAB count (29–38%) after 16 h. Sample BJWH/C2 demonstrated the best kinetic characteristics, with high regression coefficients ($R^2 = 0.97$), indicating a strong relationship between lactic acid, pH, and fermentation/storage time. Despite some fluctuations during storage, the minimum LAB count remained at 9.8 log CFU/mL, and lactic acid content increased by 95%, with good storage stability. Notably, sample BJWH/C2 increased the total phenolic content during storage. These findings suggest that adding whey enhances biomass and preserves physicochemical properties during storage.

Keywords: lactic acid bacteria; fermentation; blackberry juice; whey; antioxidants

1. Introduction

In recent decades, there has been a growing focus on researching and developing alternative foods and production processes that are viable for human consumption and beneficial to health [1,2]. Fermentation, an ancient process involving the biotransformation of fermentable substrates, such as carbohydrates primarily or proteins, through interactions with bacteria or fungi, has enhanced food's sensory, storage, and nutritional qualities in both industrial and home settings. Additionally, fermentation has proven highly effective in developing functional foods that contain a wide range of metabolites, including organic acids, free amino acids, short-chain fatty acids, bioactive peptides [2,3], volatile compounds, and other substances, that can alter the intestinal microbiota [4]. As a result, many foods with probiotic, antimicrobial, and antioxidant properties have been developed [5,6].

In fruits, natural and spontaneous fermentation is driven by lactic acid bacteria (LAB), such as *Lactobacillus* spp., *Leuconostoc* spp., *Pediococcus* spp., *Weisella* spp., and other autochthonous strains, due to the presence of carbohydrates [7]. On an industrial scale, the use of LAB, like *Lactiplantibacillus plantarum*, *Lacticaseibacillus rhamnosus*, *Lacticaseibacillus casei*, *Lactobacillus gasseri*, and *Lactobacillus acidophilus*, enhances the antioxidative and antibacterial properties of fermented products [8–12] and regulates the levels of secondary metabolites and final products. Functional foods, particularly red fruits, are also explored in fermentation due to their low sugar content and high polyphenolic compound concentration [13]. Fermentation with red fruits has shown significant growth of LAB, including *L. plantarum*, *Pediococcus acidilactici*, *Pediococcus pentosaceus*, and *Leuconostoc mesenteroides* subsp. *mesenteroides*, alongside lactic acid production. However, certain fruits, such as blackberry, possess notable antioxidant, antimicrobial, anti-inflammatory, and antidiabetic properties due to their organic acid content and low pH levels, which can inhibit LAB growth and lactic acid production [14].

On the other hand, whey, a byproduct of milk derivatives, is often considered an organic residue in the industry. However, it has been shown to enhance nutritional quality and promote sustainable food production [15]. This is primarily due to its high protein content, including α-lactalbumin, β-lactoglobulin, immunoglobulins, and lactoferrins [16]. Whey fermentation produces bioactive peptides, exopolysaccharides, organic acids, and antioxidants [17]. Furthermore, whey has been found to enhance antimicrobial activities when combined with certain fruits [18]. Research has also indicated that whey protein can improve glucose levels [19], which may be linked to the stability of phenols and anthocyanins through molecular interactions between antioxidants and whey proteins [20]. Consequently, incorporating whey protein into a matrix could influence LAB substrates and the stability of antioxidant and antibacterial properties during storage.

In fermented foods like yogurt, a decline in LAB concentration during storage at 4 °C for up to 28 days has been observed [21]. However, in fermented juices, such as blackberry juice fermented with *Lactobacillus*, the phenolic content and antioxidant properties have shown improvement under similar conditions [22]. Thus, this study aimed to assess the LAB growth, physicochemical characteristics, and antioxidant capacity of blackberry (*Rubus fruticosus*) juice fermented with whey isolated from milk and inoculated with two different LAB consortia.

2. Results

2.1. Fermentation for 24 and 48 Hours

The lactic acid fermentation experiment involved the controlled inoculation of five individual LAB strains, previously isolated from aguamiel (*L. plantarum* and *Pediococcus acidilactici*), pulque (*Levilactobacillus brevis*), and commercial fermented milk (*Lacticaseibacillus casei* and *Lacticaseibacillus rhamnosus*). These strains have demonstrated inhibitory effects against *Escherichia coli*, *Staphylococcus aureus*, and *Helicobacter pylori* [23]. Each strain was inoculated at a concentration of 9 log CFU/mL into three different fermentation systems: milk whey (WH), blackberry juice (BJ), and a 1:1 mixture of WH and BJ over a 48-h period at 37 °C (Figure 1A–C).

The results showed that *L. brevis* exhibited the highest growth in the WH system, reaching 11.68 ± 0.15 Log 10 CFU/mL at 48 h, followed by *L. plantarum*, which reached 10.90 ± 0.09 Log 10 CFU/mL at 24 h (Figure 1A). In the BJ system, the strains maintained their initial LAB counts throughout the fermentation process (Figure 1B). However, all strains showed significantly increased growth in the mixture of BJ and WH (1:1), averaging 11.94 Log 10 CFU/mL at 24 h (Figure 1C).

Two fermentation systems were developed from the BJWH mixture to compare the effects of LAB strains isolated from pulque and aguamiel (*L. brevis*, *L. plantarum*, and *P. acidilactici*) with those isolated from commercial products (*L. rhamnosus* and *L. casei*). This led to the creation of two consortia: consortium 1 (C1) consisted of *L. brevis*, *L. plantarum*, and *P. acidilactici*, while consortium 2 (C2) included *L. casei* and *L. brevis* (C2). C1 and C2 were inoculated into the three fermentation systems (WH, BJ, and BJWH). The highest growth was observed in both consortia BJWH/C1 and BJWH/C2 at 24 h, with 12.54 ± 0.10 Log 10 CFU/mL and 12.94 ± 0.04 Log 10 CFU/mL, respectively, followed by the WH/C1 system (Figure 1D).

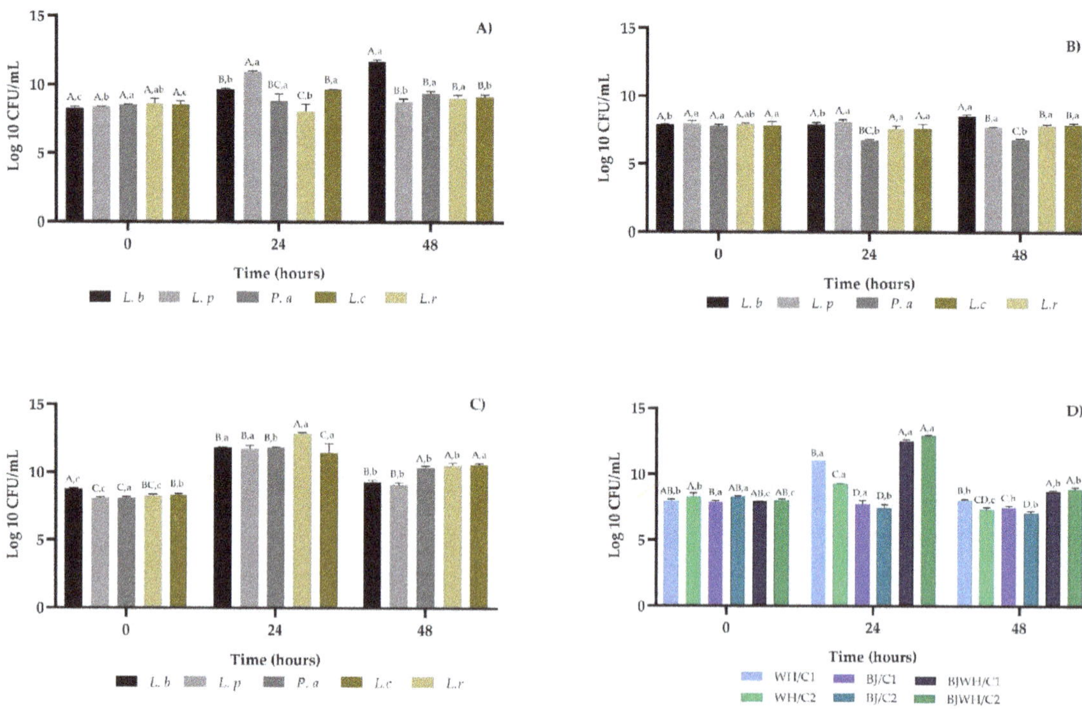

Figure 1. Growth of LAB during fermentation at 0, 24 and 48 h. Individual strains: *Levilactobacillus brevis* (L. b), *Lactiplantibacillus plantarum* (L. p), *Pediococcus acidilactici* (P. a), *Lacticaseibacillus rhamnosus* (L. r), and *Lacticaseibacillus casei* (L. c) were inoculated and fermented in WH (whey protein) (**A**), BJ (blackberry juice) (**B**), and BJWH (blackberry juice + whey protein) (**C**). WH/C1 (whey protein inoculated with consortium 1: L. b, L. p and P. a), WH/C2 (whey protein inoculated with consortium 2: L. r and L. c), BJ/C1 (blackberry juice with consortium 1), BJ/C2 (blackberry juice with consortium 2), BJWH/C1 and BJWH/C2 (blackberry juice + whey protein with consortium 1 and 2, respectively) (**D**). Values represent three experiments' mean ± standard deviation (n = 3). Different capital letters indicate a significant difference ($p < 0.05$) between the different strains or consortium for each fermentation time. Different lowercase letters indicate a significant difference ($p < 0.05$) of each strain or consortium between their different fermentation times.

2.2. Physicochemical Properties of Fermented Samples

After fermentation, pH, lactic acid content, total soluble solids (°Brix), and reducing sugars were measured to assess bacterial growth and the final products of LAB metabolism. The results of these physicochemical properties during the fermentation process (at 0, 24, and 48 h) are presented in Table 1.

Table 1. Physicochemical properties after fermentation in the two consortia of LAB were inoculated in the experimental samples (WH, BJ, and its combination).

Sample	pH			% Lactic Acid			Total Soluble Solids (°Brix)			Reducing Sugars (g/100 mL)		
	0 h	24 h	48 h	0 h	24 h	48 h	0 h	24 h	48 h	0 h	24 h	48 h
WH/												
C1	$5.7 \pm 0.00^{D,a}$	$4.1 \pm 0.00^{BC,b}$	$4.0 \pm 0.01^{E,c}$	$0.02 \pm 0.00^{F,b}$	$0.07 \pm 0.01^{G,a}$	$0.07 \pm 0.00^{FG,a}$	$0.9 \pm 0.06^{D,a}$	$0.6 \pm 0^{F,b}$	$0.4 \pm 0.00^{G,c}$	Nd	Nd	Nd
C2	$5.9 \pm 0.01^{BC,a}$	$3.8 \pm 0.03^{C,b}$	$3.8 \pm 0.04^{F,b}$	$0.02 \pm 0.00^{F,b}$	$0.05 \pm 0.00^{GH,a}$	$0.06 \pm 0.00^{G,a}$	$0.8 \pm 0.06^{D,a}$	$0.4 \pm 0^{G,b}$	$0.0 \pm 0.00^{H,c}$	Nd	Nd	Nd
C-	$5.7 \pm 0.01^{D,a}$	$5.7 \pm 0.00^{A,b}$	$5.7 \pm 0.03^{B,b}$	$0.02 \pm 0.00^{F,b}$	$0.03 \pm 0.00^{H,a}$	$0.03 \pm 0.00^{G,ab}$	1.0 ± 0.00^{D}	1.0 ± 0.00^{E}	1.0 ± 0.00^{F}	Nd	Nd	Nd
BJ/												
C1	$5.9 \pm 0.02^{C,a}$	$3.6 \pm 0.20^{C,b}$	$3.6 \pm 0.04^{H,b}$	$0.4 \pm 0.00^{A,c}$	$1.5 \pm 0.02^{A,a}$	$2.3 \pm 0.02^{A,b}$	$10 \pm 0.06^{A,a}$	$9.07 \pm 0.1^{B,b}$	$8.9 \pm 0.06^{C,b}$	$9.5 \pm 0.00^{A,a}$	$7.9 \pm 0.10^{B,c}$	$8.4 \pm 0.08^{B,b}$
C2	$6.1 \pm 0.04^{A,a}$	$4.7 \pm 0.02^{B,b}$	$4.05 \pm 0.00^{D,c}$	$0.3 \pm 0.00^{B,c}$	$0.7 \pm 0.01^{D,b}$	$1.4 \pm 0.01^{B,a}$	$9.6 \pm 0.30^{AB,a}$	$9.9 \pm 0.1^{A,a}$	$9.5 \pm 0.00^{B,a}$	$8.3 \pm 0.08^{C,a}$	$6.1 \pm 0.10^{C,c}$	$6.5 \pm 0.03^{C,b}$
C-	$5.9 \pm 0.01^{B,a}$	$5.8 \pm 0.03^{A,b}$	$5.6 \pm 0.00^{C,c}$	$0.3 \pm 0.01^{B,b}$	$0.3 \pm 0.00^{E,b}$	$0.4 \pm 0.01^{E,a}$	$9.3 \pm 0.20^{B,b}$	$10 \pm 0.06^{A,a}$	$10 \pm 0.00^{A,a}$	$8.9 \pm 0.07^{B,a}$	$8.6 \pm 0.10^{A,b}$	$9.03 \pm 0.17^{A,a}$
BJWH/												
C1	$5.4 \pm 0.00^{F,a}$	$3.7 \pm 0.01^{C,b}$	$3.5 \pm 0.01^{I,c}$	$0.2 \pm 0.00^{C,c}$	$0.9 \pm 0.05^{B,b}$	$1.2 \pm 0.01^{C,a}$	6.0 ± 0.00^{C}	5.0 ± 0.00^{D}	5.0 ± 0.00^{E}	$3.5 \pm 0.01^{E,a}$	$2.6 \pm 0.00^{D,b}$	$1.8 \pm 0.00^{F,c}$
C2	$5.6 \pm 0.01^{E,a}$	$3.8 \pm 0.01^{C,b}$	$3.7 \pm 0.00^{G,c}$	$0.2 \pm 0.00^{D,c}$	$0.8 \pm 0.01^{C,b}$	$0.9 \pm 0.05^{D,a}$	6.0 ± 0.00^{C}	5.0 ± 0.00^{D}	5.0 ± 0.00^{E}	$3.5 \pm 0.02^{E,a}$	$2.5 \pm 0.03^{E,c}$	$2.7 \pm 0.00^{E,b}$
C-	$5.6 \pm 0.01^{E,a}$	$5.8 \pm 0.60^{A,a}$	$5.8 \pm 0.00^{A,a}$	$0.1 \pm 0.00^{E,a}$	$0.1 \pm 0.00^{F,a}$	$0.13 \pm 0.01^{F,a}$	6.0 ± 0.00^{C}	6.0 ± 0.00^{C}	6.0 ± 0.00^{D}	$3.6 \pm 0.03^{D,a}$	$2.6 \pm 0.03^{E,c}$	$3.4 \pm 0.04^{D,b}$

BJ: blackberry juice; BJWH: blackberry juice + whey protein; WH: whey protein; C1: Consortium 1 of *Levilactobacillis brevis*, *Lactiplantibacillus plantarum*, and *Pediococcus acidilactici*,

C2: Consortium 2 of *Lacticaseibacillus casei* and *Lacticaseibacillus rhamnosus*; C-: Consortium without LAB inoculum; Nd: not detectable. Values represent the mean ± standard deviation of three experiments (n = 3). Different capital letters indicate a significant difference ($p < 0.05$) between the different strains or consortium for each fermentation time. Different lowercase letters indicate a significant difference ($p < 0.05$) of each strain or consortium between their different fermentation times.

Initially, the BJWH system had the lowest pH values, ranging from 5.4 to 5.6. After 24 h, the WH/C2, BJ/C1, BJWH/C1, and BJWH/C2 systems exhibited significantly lower pH values. By 48 h, the pH values remained lower in the BJ/C1, BJWH/C1, and BJWH/C2 systems. The lactic acid percentage increased notably in the BJ/C1 system, followed by BJWH/C1 and BJWH/C2, while the WH/C1 and WH/C2 systems showed no significant changes at 24 and 48 h. This increase in lactic acid was associated with a decrease in total soluble solids by one unit at 24 and 48 h for the BJWH/C1 and BJWH/C2 systems. Additionally, reducing sugars were initially lower in the BJWH systems and decreased most significantly in the BJWH/C1 system by 48 h (Table 1).

2.3. *Physicochemical Properties during Storage*

The endpoint of fermentation was assessed by measuring CFU at intervals 0, 4, 8, 16, 20, 24, 28, 32, and 48 h of incubation at 37 °C for BJWH/C1 and BJWH/C2. Figure 2A shows LAB growth, with BJWH/C1 exhibiting a notable increase from 16 h (13.34 ± 0.14 Log 10 CFU/mL) to 28 h (13.85 ± 0.21 Log 10 CFU/mL). However, a decrease in the LAB count was observed at 20 h for both BJWH/C1 and BJWH/C2, followed by variable behavior in subsequent hours, finally resulting in a lower average count (10.27 Log 10 CFU/mL) at 48 h. Consequently, storage analysis was initiated after 16 h of fermentation of the BJWH/C1 and BJWH/C2 samples.

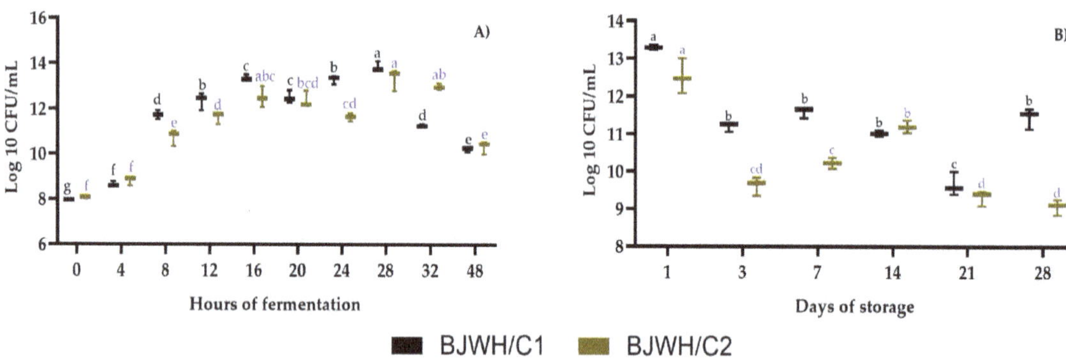

Figure 2. LAB growth during (**A**) fermentation of 48 h and (**B**) storage days after 16 h of fermentation of BJWH/C1 and BJWH/C2 (blackberry juice + whey protein with consortium 1 and 2, respectively). The sample contains C1: *Levilactobacillus brevis*, *Lactiplantibacillus plantarum*, and *Pediococcus acidilactici* and C2: *Lacticaseibacillus casei* and *Lacticaseibacillus rhamnosus*. Values in each point represent three experiments' mean ± standard deviation (n = 3). Different black (BJWH/C1) and blue (BJWH/C2) lowercase letters indicate a significant difference ($p < 0.05$) in different fermentation and storage times.

Fermented samples were stored at 4 °C, and LAB counts were assessed on storage days 3, 7, 14, 21, and 28. While growth patterns varied among samples, BJWH/C1 consistently exhibited higher LAB counts, averaging 11.29 Log 10 CFU/mL on days 3, 7, 14, and 28. Notably, BJWH/C2 showed an increase in LAB count on day 14, followed by a significant decline by day 28 (see Figure 2B).

Kinetic parameters were calculated to determine significant differences between fermentations (see Table 2). BJWH/C2 exhibited more favorable parameters, including a longer generation time than BJWH/C1 (4.20 ± 0.60). However, the fermentation rate (μ) was faster in BJWH/C1, while BJWH/C2 demonstrated a better affinity, as indicated by its constant (k). These observations suggest a higher cellular density in BJWH/C2, particularly in systems containing *L. casei* and *L. rhamnosus*.

Table 2. Kinetic parameters at the time of fermentation (16 h) of blackberry juice supplemented with whey.

Fermentation System	Parameter Value			
	A (Log 10 CFU/mL)	**μ (h^{-1})**	**g**	**k (g/h)**
BJWH/C1	5.39 ± 0.13	0.33 ± 0.00	2.05 ± 0.05	0.48 ± 0.01
BJWH/C2	2.67 ± 0.36 *	0.16 ± 0.02 *	4.20 ± 0.60 *	0.24 ± 0.03 *

Blackberry juice + whey protein inoculated with consortium 1 (*Levilactobacillus brevis*, *Lactiplantibacillus plantarum*, and *Pediococcus acidilactici*, BJWH/C1), consortium 2 (*Lacticaseibacillus casei* and *Lacticaseibacillus rhamnosus*, BJWH/C2) and the negative control without LAB inoculum (BJWH/C-). A: increase in cell density; μ: Grow rate; g: Generation time; k: Growth rate constant. Values represent three experiments' mean ± standard deviation (n = 3). * Indicates significant differences between the samples.

The percentages of lactic acid and pH values were measured from the start of fermentation to the end of the storage (day 28, Figure 3). Lactic acid production rose after 4 h of fermentation for BJWH/C1 (0.12%) and BJWH/C2 (0.16%). The highest levels of lactic acid were observed in the BJWH/C2 group during the final 14 days of storage (Figure 3A). Additionally, while the initial pH averaged 5.6 throughout fermentation, it decreased significantly in the BJWH/C2 system during fermentation and storage. BJWH/C1 showed a substantial reduction in pH within the first eight hours of fermentation but remained relatively stable from 12 h onwards through the storage period (Figure 3B).

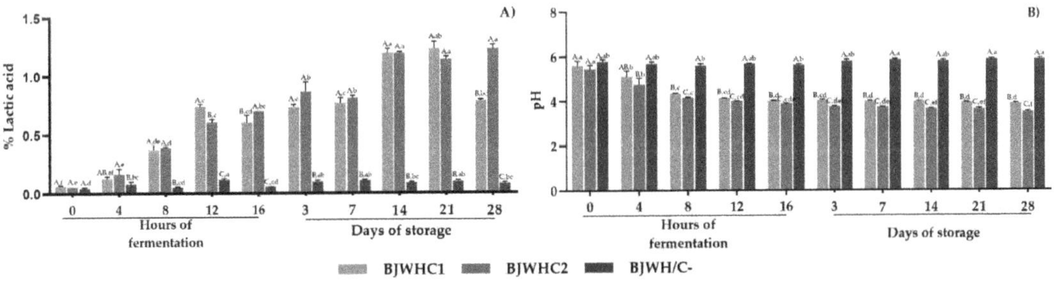

Figure 3. Lactic acid (**A**) and pH (**B**) values across the time of fermentation (16 h) and the time of storage (28 days) in Blackberry juice + whey protein inoculated with consortium 1 of *Levilactobacillus brevis*, *Lactiplantibacillus plantarum*, and *Pediococcus acidilactici* (BJWH/C1), consortium 2 of *Lacticaseibacillus casei* and *Lacticaseibacillus rhamnosus* (BJWH/C2) and the negative control without LAB inoculum (BJWH/C-). Values represent three experiments, mean ± standard deviation (n = 3). Different capital letters indicate significant differences ($p < 0.05$) between the various strains or consortiums for each fermentation and storage time. Different lowercase letters indicate a significant difference ($p < 0.05$) of each strain or consortium between their different fermentation and storage times.

Regression analysis was used to determine the correlation coefficients for lactic acid (Figure 4) and pH (Figure 5) with fermentation and storage time. Data for each factor were modeled with a third-degree polynomial curve. Lactic acid levels increased continuously up to 12 h of fermentation for BJWH/C1 (Figure 4A) and up to 16 h for BJWH/C2 (Figure 4C), with similar trends extending up to 21 days of storage for BJWH/C1 (Figure 4B). Most curves exhibited an R^2 value greater than 0.97, except for lactic acid production in BJWH/C2 during storage (Figure 4D), which had an R^2 of 0.80. pH levels decreased with extended fermentation or storage time, reflecting their relationship with lactic acid production (Figure 5). This analysis highlights a strong connection between lactic acid and pH, significantly influenced by time.

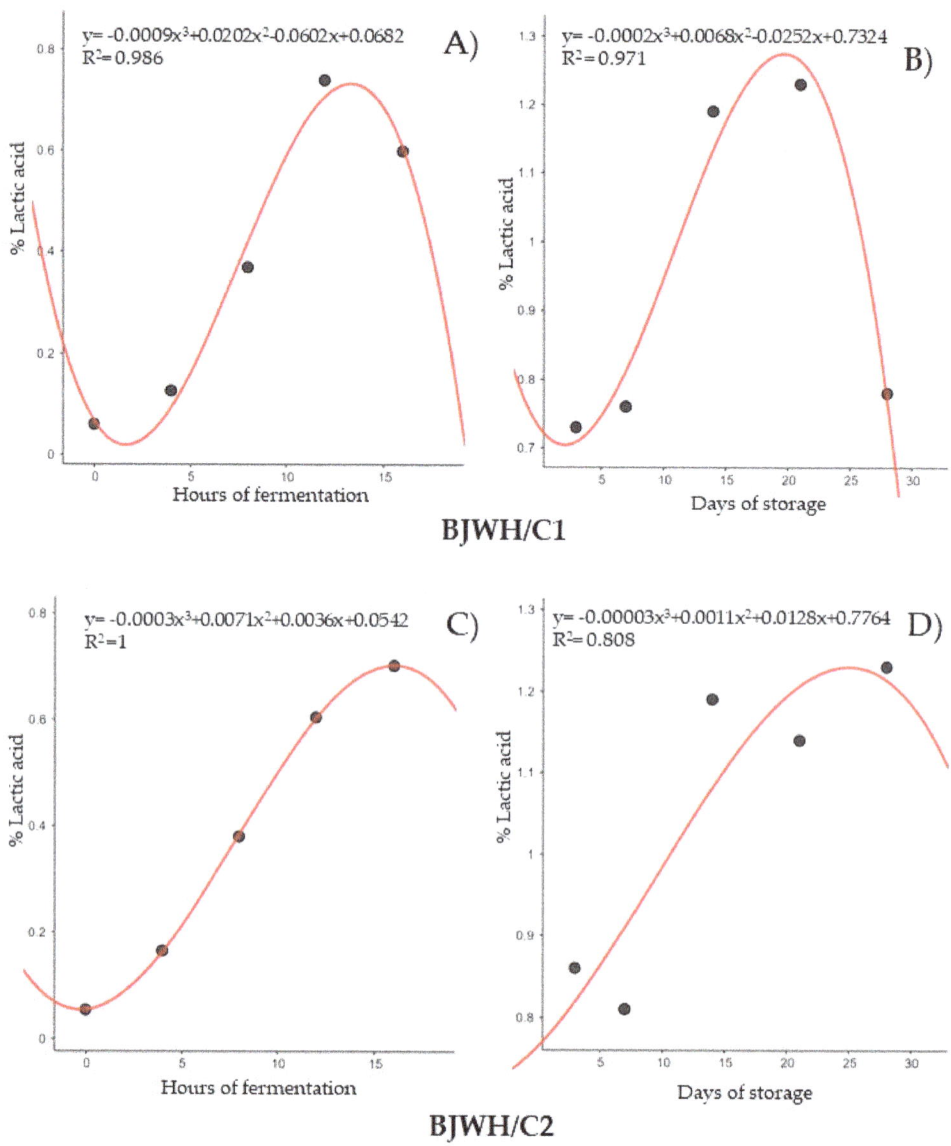

Figure 4. Non-linear regression analysis lactic acid values across the time of fermentation (16 h) and the time of storage (28 days) in blackberry juice + whey protein inoculated with consortium 1 of *Levilactobacillus brevis*, *Lactiplantibacillus plantarum*, and *Pediococcus acidilactici* (BJWH/C1) (**A**,**B**), consortium 2 of *Lacticaseibacillus casei* and *Lacticaseibacillus rhamnosus* (BJWH/C2) (**C**,**D**).

The total soluble solids were one °Brix lower in the LAB-inoculated systems (BJWH/C1 and BJWH/C2) than controls after the 8 h of fermentation and remained at 5 °Brix throughout storage. Reducing sugars decreased in BJWH/C1 and BJWH/C2 during fermentation and storage, with a more pronounced reduction in BJWH/C2. Overall, reducing sugars levels were significantly lower in BJWH/C1 and BJWH/C2 throughout fermentation and storage compared to the negative control, BJWH/C- (Table 3).

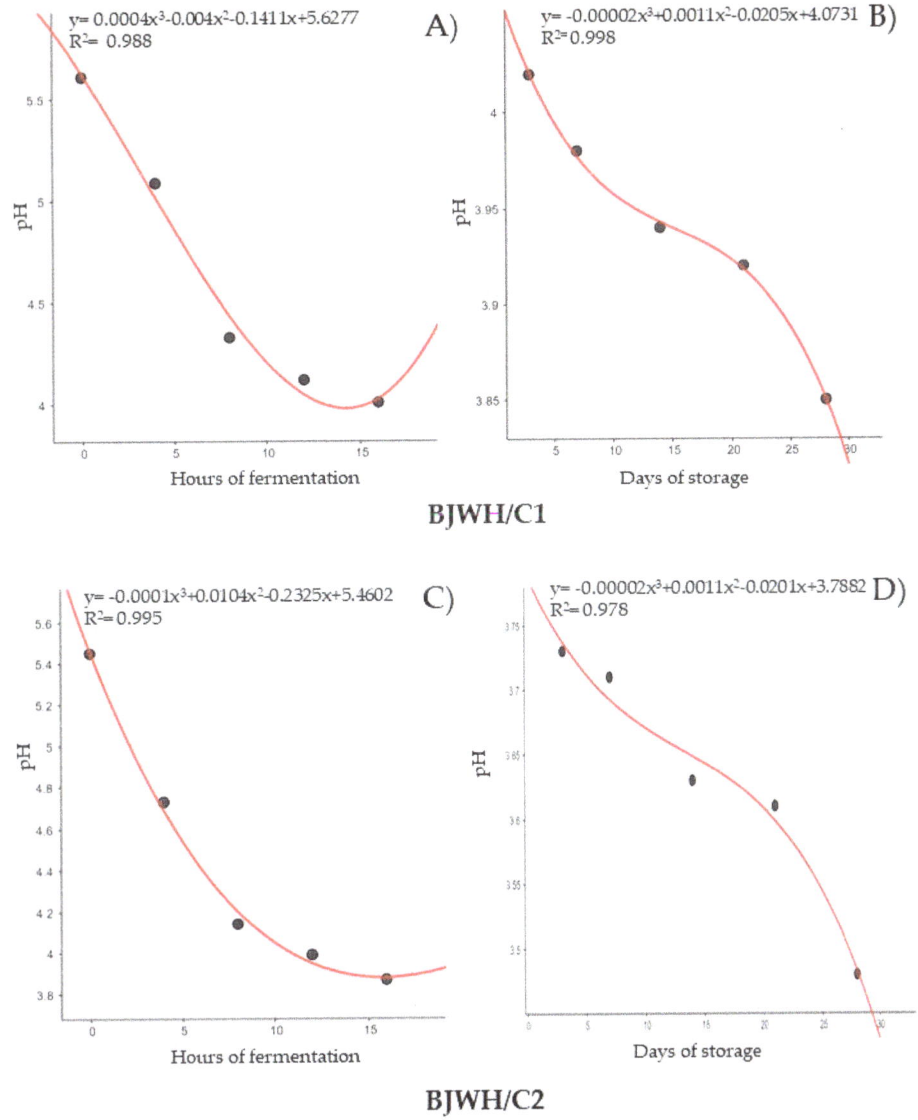

Figure 5. Non-linear regression analysis pH values across the time of fermentation (16 h) and the time of storage (28 days) in blackberry juice + whey protein inoculated with consortium 1 of *Levilactobacillus brevis*, *Lactiplantibacillus plantarum*, and *Pediococcus acidilactici* (BJWH/C1) (**A**,**B**), consortium 2 of *Lacticaseibacillus casei* and *Lacticaseibacillus rhamnosus* (BJWH/C2) (**C**,**D**).

The acceptability of beverages is influenced by their physical stability, viscosity, and color. Table 4 shows the stability (nonprecipitated solids) and viscosity (centipoise) values at the start and end of fermentation (16 h) and during storage. At the end of fermentation and throughout, BJWH inoculated with LAB exhibited greater stability than the negative control, with a mean difference of 35% compared to BJWH/C-. The stability remained consistent for both LAB-inoculated consortia during storage. Regarding viscosity, LAB-inoculated systems (BJWH/C1 and BJWH/C2) significantly decreased in cP values after fermentation, maintaining lower viscosity levels than the negative control throughout storage. There was no significant difference in viscosity between BJWH/C1 and BJWH/C2 (Table 4).

Table 3. Total soluble solids and reducing sugars at the time of fermentation (0–16 h) and storage (3–28 days) of blackberry juice supplemented with whey.

Time	Total Soluble Solids (°Brix) [1]			Reducing Sugars (g/100 mL)		
	BJWH/C1	BJWH/C2	BJWH/C-	BJWH/C1	BJWH/C2	BJWH/C-
0 h	6	6	6	3.4 ± 0.20 ^{AB,a}	2.9 ± 0.10 ^{B,a}	3.5 ± 0.20 ^{A,bcd}
4 h	6	6	6	2.1 ± 0.20 ^{B,d}	2.6 ± 0.10 ^{C,ab}	3.7 ± 0.40 ^{A,bcd}
8 h	5	5	6	2.6 ± 0.04 ^{B,b}	2.1 ± 0.30 ^{B,c}	3.3 ± 0.20 ^{A,cd}
12 h	5	5	6	2.2 ± 0.01 ^{B,cd}	2.2 ± 0.03 ^{B,c}	3.2 ± 0.02 ^{A,d}
16 h	5	5	6	2.1 ± 0.06 ^{B,d}	2.1 ± 0.08 ^{B,c}	3.4 ± 0.10 ^{A,bcd}
Day 3	5	5	6	2.5 ± 0.10 ^{B,bc}	2.0 ± 0.10 ^{C,c}	3.7 ± 0.07 ^{A,abd}
Day 7	5	5	6	2.2 ± 0.60 ^{B,bcd}	2.2 ± 0.04 ^{B,c}	3.4 ± 0.01 ^{A,bcd}
Day 14	5	5	6	2.1 ± 0.01 ^{C,d}	2.45 ± 0.06 ^{B,bc}	3.4 ± 0.06 ^{A,bcd}
Day 21	5	5	6	2.4 ± 0.08 ^{B,bcd}	2.1 ± 0.08 ^{C,c}	3.9 ± 0.05 ^{A,a}
Day 28	5	5	6	2.5 ± 0.10 ^{B,bc}	2.2 ± 0.03 ^{B,bc}	3.7 ± 0.10 ^{A,ab}

Blackberry juice + whey protein inoculated with consortium 1 (*Levilactobacillus brevis*, *Lactiplantibacillus plantarum*, and *Pediococcus acidilactici*, BJWH/C1), consortium 2 (*Lacticaseibacillus casei* and *Lacticaseibacillus rhamnosus*, BJWH/C2) and the negative control without LAB inoculum (BJWH/C-). Values represent three experiments' mean ± standard deviation (n = 3); [1] In these experiments, the SD was zero. Different capital letters indicate significant differences ($p < 0.05$) between the various strains or consortium for each fermentation time (0–48 h and 3–28 days). Different lowercase letters indicate a significant difference ($p < 0.05$) of each strain or consortium between their different fermentation and storage times.

Table 4. Stability and viscosity at the time of fermentation (0–16 h) and storage (3–28 days) of blackberry juice with whey.

Time	Stability (%)			Viscosity (cP)		
	BJWH/C1	BJWH/C2	BJWH/C-	BJWH/C1	BJWH/C2	BJWH/C-
0 h	73.7 ± 0.6 ^{B,b}	76.3 ± 0.5 ^{A,c}	71.3 ± 0.5 ^{C,ab}	64 ± 0.8 ^{A,a}	51.2 ± 0.01 ^{B,a}	63.7 ± 0.9 ^{A,ab}
16 h	84.6 ± 0.5 ^{B,a}	87.7 ± 0.5 ^{A,a}	73 ± 1 ^{C,a}	27.5 ± 1.7 ^{B,b}	24.9 ± 0.4 ^{B,b}	50.4 ± 1.5 ^{A,c}
Day 3	85.3 ± 0.5 ^{A,a}	86.6 ± 0.5 ^{B,ab}	67.4 ± 0.5 ^{C,cd}	22.4 ± 2.2 ^{B,b}	23.7 ± 2.5 ^{B,b}	57.6 ± 1.9 ^{A,bc}
Day 7	85.6 ± 0.5 ^{A,a}	85 ± 1 ^{A,b}	69.6 ± 0.5 ^{B,bc}	26.6 ± 5.2 ^{B,b}	23.3 ± 3 ^{B,b}	59.6 ± 2.6 ^{A,ab}
Day 14	85 ± 0.1 ^{B,a}	87 ± 0.07 ^{A,ab}	67.3 ± 0.5 ^{C,d}	29.7 ± 0.8 ^{B,b}	24.1 ± 2.1 ^{B,b}	62.5 ± 2.9 ^{A,ab}
Day 21	84.7 ± 1.1 ^{A,a}	86.4 ± 0.5 ^{A,ab}	66.5 ± 1.3 ^{B,d}	28.2 ± 0.6 ^{B,b}	25.1 ± 2.4 ^{B,b}	67.5 ± 5.9 ^{A,ab}
Day 28	86.3 ± 0.5 ^{A,a}	86 ± 1.7 ^{A,ab}	69.7 ± 0.6 ^{B,b}	27.8 ± 1 ^{B,b}	25.9 ± 0.3 ^{B,b}	70.5 ± 0.7 ^{A,a}

Blackberry juice + whey protein inoculated with consortium 1 of *Levilactobacillus brevis*, *Lactiplantibacillus plantarum*, and *Pediococcus acidilactici* (BJWH/C1), consortium 2 of *Lacticaseibacillus casei* and *Lacticaseibacillus rhamnosus* (BJWH/C2) and the negative control without LAB inoculum (BJWH/C-). Values represent three experiments with mean ± standard deviation (n = 3). Different capital letters indicate significant differences ($p < 0.05$) between the different strains or consortiums for each fermentation and storage time. Different lowercase letters indicate a significant difference ($p < 0.05$) of each strain or consortium between their different fermentation and storage times.

Regarding the colorimetric characteristics, the L parameter (luminosity) varied among samples with the LAB inoculum but remained consistent after fermentation and throughout storage (Table 5). The predominant color was red, indicated by increasing a* values post-fermentation, with higher values observed in BJWH/C2 samples during storage. The b* parameter, initially negative in LAB-inoculated samples (BJWH/C1 and BJWH/C2), indicating a blue hue, shifted to positive values in BJWH/C- during storage, reflecting a change to more yellow tones.

Table 5. Colorimetric characteristics of blackberry juice supplemented with whey and two consortia at the time of fermentation (16 h) and time of storage (28 days).

Parameter	Sample	Fermentation 0 h	16 h	Storage Day 3	Day 7	Day 14	Day 21	Day 28
L	BJWH/C1	33.6 ± 0.50 ^{B,b}	35.3 ± 0.40 ^{A,a}	36.1 ± 0.40 ^{A,a}	35.6 ± 0.20 ^{A,a}	34.8 ± 1.20 ^{A,ab}	35.2 ± 0.30 ^{A,ab}	35.7 ± 0.20 ^{A,a}
	BJWH/C2	35.4 ± 0.40 ^{A,a}	34.9 ± 0.60 ^{A,a}	36.3 ± 0.60 ^{A,a}	36.0 ± 0.10 ^{A,a}	35.6 ± 0.50 ^{A,a}	36.0 ± 0.10 ^{A,a}	35.4 ± 0.80 ^{A,a}
	BJWH/C-	34.5 ± 0.40 ^{AB,a}	33.5 ± 0.08 ^{B,a}	34.6 ± 0.60 ^{B,a}	34.4 ± 0.50 ^{B,a}	34.3 ± 0.50 ^{A,a}	34.0 ± 0.50 ^{B,a}	34.6 ± 10 ^{A,a}
a*	BJWH/C1	1.3 ± 0.10 ^{B,b}	7.8 ± 0.10 ^{A,a}	8.3 ± 0.20 ^{B,a}	7.9 ± 0.30 ^{B,a}	8.4 ± 0.20 ^{B,a}	8.1 ± 0.40 ^{B,a}	8.2 ± 0.10 ^{B,a}
	BJWH/C2	2.5 ± 0.30 ^{A,d}	8.1 ± 0.60 ^{A,c}	9.5 ± 0.50 ^{A,abc}	9.1 ± 0.10 ^{A,bc}	10 ± 0.70 ^{A,ab}	10.5 ± 0.10 ^{A,a}	9.9 ± 0.03 ^{A,ab}
	BJWH/C-	1.5 ± 0.20 ^{B,a}	1.3 ± 0.30 ^{B,a}	1.7 ± 0.30 ^{C,a}	1.2 ± 0.03 ^{C,a}	1.4 ± 0.10 ^{C,a}	1.1 ± 0.02 ^{C,a}	1.2 ± 0.04 ^{C,a}
b*	BJWH/C1	−0.8 ± 0.06 ^{B,a}	−1.3 ± 0.10 ^{B,a}	−1.3 ± 0.10 ^{B,a}	−1.2 ± 0.10 ^{B,a}	−1.2 ± 0.20 ^{C,a}	−1.0 ± 0.07 ^{C,a}	−1.0 ± 0.06 ^{C,a}
	BJWH/C2	−1.4 ± 0.02 ^{C,c}	−0.8 ± 0.10 ^{B,bc}	−1.0 ± 0.10 ^{B,bc}	−0.7 ± 0.10 ^{AB,ab}	−0.5 ± 0.03 ^{B,ab}	−0.4 ± 0.10 ^{B,a}	−0.5 ± 0.07 ^{B,ab}
	BJWH/C-	−0.6 ± 0.06 ^{A,c}	−0.1 ± 0.01 ^{A,bc}	−0.1 ± 0.30 ^{A,abc}	−0.1 ± 0.01 ^{A,abc}	0.04 ± 0.20 ^{A,abc}	0.3 ± 0.06 ^{A,ab}	0.3 ± 0.06 ^{A,a}
Chroma	BJWH/C1	1.7 ± 0.20 ^{B,b}	7.9 ± 0.10 ^{B,a}	8.4 ± 0.30 ^{B,a}	7.9 ± 0.30 ^{B,a}	8.4 ± 0.1 ^{B,a}	8.2 ± 0.40 ^{B,a}	8.2 ± 0.10 ^{B,a}
	BJWH/C2	2.8 ± 0.20 ^{A,d}	8.2 ± 0.60 ^{A,c}	9.6 ± 0.50 ^{A,abc}	9.1 ± 0.10 ^{A,bc}	10 ± 0.7 ^{A,ab}	10.5 ± 0.10 ^{A,a}	10 ± 0.03 ^{A,abc}
	BJWH/C-	1.7 ± 0.10 ^{B,a}	1.4 ± 0.30 ^{B,a}	1.7 ± 0.30 ^{C,a}	1.2 ± 0.03 ^{C,a}	1.4 ± 0.1 ^{C,a}	1.1 ± 0.01 ^{C,a}	1.2 ± 0.05 ^{C,a}
Hue	BJWH/C1	−27.4 ± 2.20 ^{A,b}	−8.5 ± 0.70 ^{A,a}	−9.2 ± 0.70 ^{A,a}	−8.5 ± 0.40 ^{A,a}	−7.5 ± 1.40 ^{A,a}	−7.3 ± 0.10 ^{B,a}	−6.7 ± 0.30 ^{C,a}
	BJWH/C2	−26.3 ± 1.80 ^{A,b}	−6.1 ± 0.60 ^{A,a}	−5.7 ± 0.60 ^{A,a}	−4.7 ± 0.90 ^{A,a}	−2.7 ± 0.20 ^{A,a}	−2.2 ± 0.60 ^{B,a}	−3.2 ± 0.40 ^{B,a}
	BJWH/C-	−20.7 ± 0.30 ^{A,c}	−7.4 ± 1.80 ^{A,cd}	−3.2 ± 0.20 ^{A,abc}	−7.3 ± 0.90 ^{A,c}	1.9 ± 0.10 ^{A,abc}	15.3 ± 2.60 ^{A,ab}	17 ± 1.60 ^{A,a}
Sistema color	BJWH/C1							
	BJWH/C2							
	BJWH/C-							

L: luminosity. Blackberry juice + whey inoculated with consortium 1 of *Levilactobacillus brevis*, *Lactiplantibacillus plantarum*, and *Pediococcus acidilactici* (BJWH/C1), consortium 2 of *Lacticaseibacillus casei* and *Lacticaseibacillus rhamnosus* (BJWH/C2), and the negative control without LAB inoculum (BJWH/C-) Values represent three experiments with mean ± standard deviation (n = 3). Different capital letters indicate significant differences ($p < 0.05$) between the different strains or consortiums for each fermentation and storage time. Different lowercase letters indicate a significant difference ($p < 0.05$) between fermentation and storage time of each strain or consortium.

Chroma measures color saturation, with values ranging from 0 to 100, reflecting the intensity of the sample's color. The results show that BJWH/C2 and BJWH/C1 are brighter than BJWH/C-. Additionally, negative hue values in BJWH/C1 and BJWH/C2 indicate a pinkish tone. A more detailed graphical representation of these color results is provided in Table 5.

Finally, the levels of phenolic compounds, anthocyanins, and the scavenging activity of 2,2-Azino-bis-3-ethylbenzothiazoline-6-sulfonic acid (ABTS) and 2,2-diphenyl-1-picrylhydrazyl (DPPH) were evaluated during fermentation and storage (Figure 6). The total phenolic content significantly decreased in all samples after 4 h of fermentation, averaging 30 mg GAE/100 mL ($p < 0.05$). This decline continued in BJWH/C1 through the end of storage, while BJWH/C2 showed a 19% increase after 3 days of storage, with higher levels observed in the control (BJWH/C-, Figure 6A). In contrast, the anthocyanin content was notably higher in BJWH/C1 after 4, 8, and 12 h of fermentation but showed an opposite trend at 16 h and 3 days. During storage, LAS-inoculated systems had a higher anthocyanin content compared to BJWH/C-, with BWH/C1 maintaining greater stability than BJWH/C2 by the end of storage (Figure 6B).

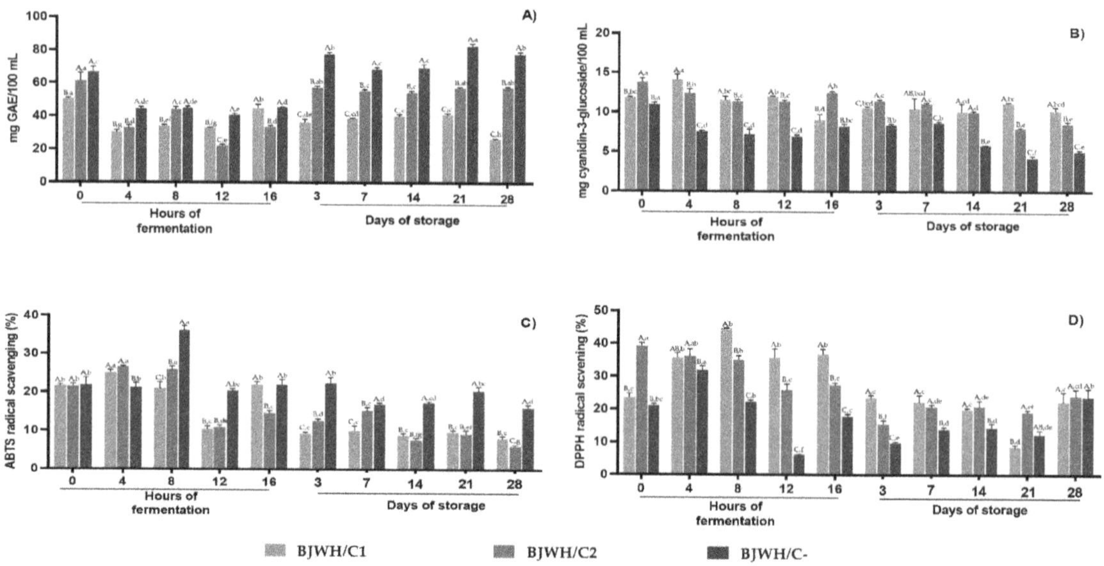

Figure 6. Content of total phenolic compounds (**A**), anthocyanins (**B**), ABTS scavenging (**C**), and DPPH scavenging (**D**). Values across the time of fermentation (16 h) and the time of storage (28 days) in Blackberry juice + whey inoculated with consortium 1 of *Levilactobacillus brevis*, *Lactiplantibacillus plantarum*, and *Pediococcus acidilactici* (BJWH/C1), consortium 2 of *Lacticaseibacillus casei* and *Lacticaseibacillus rhamnosus* (BJWH/C2) and the negative control without LAB inoculum (BJWH/C-). Values represent three experiments with mean ± standard deviation (n = 3). Different capital letters indicate significant differences ($p < 0.05$) between the different strains or consortiums for each fermentation and storage time. Different lowercase letters indicate a significant difference ($p < 0.05$) of each strain or consortium between their different fermentation and storage times.

For ABTS, which measures free radical scavenging activity, the BJWH/C1 fermentation system maintained its antioxidant activity at 0 h. However, during storage, the antioxidant capacity decreased by 57.7% in both consortia, with the non-inoculated sample exhibiting a significantly higher scavenging capacity ($p < 0.05$, Figure 6C). Despite this, DPPH scavenging activity was significantly higher at 8, 12, and 16 h of fermentation in BJWH/C1, averaging 38% activity. After 3 days of storage, the scavenging activity decreased by 36% in BJWH/C1, 43% in BJWH/C2, and 44.6% in BJWH/C- (Figure 6D).

3. Discussion

This study explored the effects of adding isolated whey protein to blackberry juice, using two consortia to optimize biomass growth at 24 and 48 h of fermentation. Additionally, we aimed to maintain stable levels of phytochemicals and antioxidants throughout storage. The growth of strains like *L. brevis* and *L. plantarum* was significantly higher at 24 h in the whey protein system. However, the blackberry juice (BJ) system was less favorable for all LAB strains. This outcome is similar to findings from fermenting elderberry juice with *L. plantarum*, *L. rhamnosus*, and *L. casei*. The authors noted that *L. casei* strains struggled to grow in the acidic environment, likely due to their origin in cheese, making them less adapted to acidic juices [24–26].

In the case of the limited growth observed in the BJ system, it could be attributed to the inherent chemical and physical properties of the fruit, such as its acidic environment, buffering capacity, high concentrations of indigestible nutrients (like fiber and fructooligosaccharides), and the presence of anti-nutritional compounds (such as phenols and tannins). Additionally, these factors may interact with the amino acids and proteins in whey [27].

Therefore, a protein source has been proposed to enhance biomass production and create a supportive environment for lactic acid bacteria (LAB). However, few studies have explored the impact of adding whey protein for fermentation systems to boost biomass in fruit beverages. Our research found that fermenting blackberry juice alone did not increase LAB counts. However, adding whey protein to blackberry juice significantly raised the LAB counts after 24 h (Figure 1C). Additionally, two different consortia were used, one consisting of *L. brevis*, *L. plantarum*, and *P. acidilactici* and the other of *L. casei* and *L. rhamnosus*, which proved to be more effective in promoting growth in the BJWH fermentation system, resulting in significantly higher LAB counts than other matrices (BJ and WH).

In addition, a similar pattern was observed in whey fermentation using *L. rhamnosus* and *S. thermophilus*, both in monoculture and co-culture. The authors found co-culturing these strains was more effective for biomass production [28]. The bacterial growth observed in BJWH/C1 and BJWH/C2 could be attributed to adding whey protein as a supplement, which enhances biomass production during fermentation and provides a valuable source for functional metabolites [3].

This study showed that combining blackberry juice with whey protein resulted in a more substantial increase in biomass than using the juice alone, where LAB counts decreased after 24 h. The synergy between these two food matrices (blackberries, which provide sucrose, glucose, and fructose as energy sources for LAB [29], and whey, which supplies proteins, lactose, fat, and mineral salts [30]) promotes enhanced biomass production. Phenolic compounds, especially ferulic and phenolic esters, play a crucial role in the growth and viability of probiotics in fruit-based food matrices [7]. Their presence has been shown to positively impact them, as they are susceptible to hydrolysis by *L. plantarum* esterases, including feruloyl esterase, tannase, and gallate decarboxylase [31]. Addotionally, blackthorn extract (up to 0.62 mg) has been found to exhibit prebiotic properties that support the growth of LAB strains, such as *L. plantarum*, *L. reuteri*, *L. rhamnosus*, and *S. thermophilus* [32].

Conversely, synergism within bacterial consortia establishes the essential conditions for strain growth. Compared to individual samples, the notable bacterial growth in BJWH/C1 and BJWH/C2 can be attributed to the LAB protocooperation system. In these systems, microorganisms release and exchange metabolites, promoting optimal development and impacting the growth and acidification of fermentation systems [3]. For example, in a protocooperative relationship, *L. rhamnosus* expresses the extracellular proteinase PrtR [33], releasing and accumulating peptides in the medium [34] by other *Lactobacillus* species. Furthermore, *L. plantarum* and *P. acidilactici* possess a biosynthetic pathway for lysine, which *L. brevis* lacks. Lysine is crucial as a cell wall constituent and an enzymatic substrate in metabolic pathways [35–37]. The combination of LAB in a fermentation medium could be pivotal for the development and viability of probiotics in producing symbiotic fermented foods.

During fermentation, LAB utilizes sugars (hexoses and pentoses) to produce lactic acid, decreasing pH [38–40]. Adjustments such as increasing pH are also necessary to enhance bacterial growth during fermentation [41]. Li et al. (2021) [42] previously reported a significant increase in LAB counts, exceeding 10 log CFU/mL, in blackberry juices using four *L. plantarum* strains and five *L. fermentum* strains, with a 35% increase in lactic acid production. In our study, a more significant reduction in pH was observed in the juice without whey and LAB consortium 1 (BJ/C1) at 24 and 48 h of fermentation, with an average reduction of up to 46%. This is associated with increased lactic acid at 24 h (75%) and 48 h (83%). However, the pH was lower in juice samples containing whey (BJWH) despite higher lactic acid production in BJWH/C1 and BJWH/C2 compared to the BJ and WH samples (Table 1). A similar relationship between pH changes and increased lactic acid production has been reported in whey fermentation with *L. acidophilus*, *L. delbrueckii* subsp. *bulgaricus* and *S. thermophilus* [30], wolfberry, and longan juice with *L. paracasei* and *L. lactis* subsp. *lactis* [42].

The BJWH/C1 and C2 samples showed a significant increase ($p < 0.05$) in bacterial count during fermentation, with exponential and continuous growth observed for up to 16 h (Figure 2A). This was accompanied by higher lactic acid production and a notable decrease in pH (Figure 3A). Similar results were reported for whey fermented with *L. casei* and *L. rhamnosus*, where exponential biomass growth occurred in the second fermentation stage, likely due to the consumption of lactose and galactose during the first 15 h of exponential growth [43]. Additionally, Wang et al. [44] noted that *L. acidophilus* JYLA-16 (La), *L. plantarum* JYLP-375 (Lp), and *L. rhamnosus* JYLR-005 (Lr) preferentially utilize glucose over fructose as a carbon source during fermentation. Therefore, adding milk whey protein to the fruit juice provides an additional substrate for fermentation, enhancing LAB growth. This effect was similarly observed in pomegranate juice supplemented with whey and fermented with *L. plantarum*, leading to increased LAB growth [45].

The superior performance of the BJWH/C2 system in kinetic parameters can be attributed to a combination of environmental factors and the complex interactions between microorganisms, particularly through protocooperation and synergism. Olvera-Rosales et al. (2023) [3] demonstrated that a co-culture system involving *Lactobacillus rhamnosus* GG and *Streptococcus thermophilus* in whey surpasses monocultures in both specific growth rate (μ) and generation time (g). This enhanced efficiency is reflected in higher lactic acid production and elevated levels of free amino groups [28]. Additionally, co-culture systems in milk fermentation are well documented to significantly enhance the production yields of lactic acid and secondary metabolites like peptides [34]. These findings highlight the crucial role of microbial interactions in optimizing fermentation processes and improving the overall yield of desired compounds.

During storage at 4 °C for 28 days, a significant decrease in LAB counts was observed during the first week, likely due to the adaptation of LAB in the BJWH/C1 and BJWH/C2. Notably, the LAB count remained above 9 Log CFU/mL by the end of the storage period, suggesting potential therapeutic benefits. Previous studies have shown that LAB mixtures in food matrices can exhibit heterogeneous growth behavior during storage at 4 °C [9,46]. However, by the end of 28 days, the strains survived at desirable and therapeutic levels (6–8 Log CFU/mL), depending on their specific adaptation conditions [46–48]. In consortia of LAB strains *L. acidophilus*, *L. casei*, and *L. plantarum*, added to ferment apple juice without additional nutrients, the maximum growth during fermentation reached 7.5 to 8.3 Log CFU/mL. However, during storage for 20 to 30 days, viable cell counts of *L. casei* rapidly decreased to 5.4 ± 0.04 Log CFU/mL, followed by *L. acidophilus* (5.6 ± 0.13 Log CFU/mL) and *L. plantarum* (5.7 ± 0.13 Log CFU/mL) [49].

During storage at 4 °C, lactic acid and physicochemical parameters (viscosity, color, soluble solids, reducing sugars) remained stable, with only slight variation observed in antioxidant capacity, total phenolic, and anthocyanin levels. The use of multiple LAB strains can enhance lactic acid production due to their homofermentative and heterofermentative characteristics, as well as the activation of various metabolic pathways (EMP, pentose phosphate, and phosphoketolase), depending on the strains used [50]. The increase in lactic acid production was particularly pronounced in BJWH/C1 and BJWH/C2.

Additionally, the total solids in a fermented product are influenced by differences in the activated enzyme systems and the microorganisms' ability to utilize matrix substrates [9]. As LAB consumes sugars, producing lactic acid and increased biomass can help maintain a balance in total solids [51]. Conversely, the reduction in soluble solids was more significant in the WH samples compared to BJ and BJWH at 24 and 48 h, as using protein substrates produces fewer soluble soils. The changes observed in BJ and BJWH are associated with a significant increase in lactic acid, over 45%, and a reduction of up to 48.6% in reducing sugars and bacterial growth.

During storage, the trends observed in lactic acid and pH remained consistent, while the content of reducing sugars decreased, and soluble solids stayed the same. This pattern mirrored the changes observed during fermentation. A similar decrease in pH during storage was reported for carrot juice supplemented with whey and fermented with kefir grains over 21 days [52]. In passion fruit juice fermented with *Lactobacillus plantarum*, the reduction in pH was attributed to the combined effects of the strains' post-acidification capacities and the buffering power of the substrate [53]. Additionally, LAB at low temperatures (4 °C) continue a slow fermentation process to meet their metabolic needs, resulting in ongoing production and consumption of metabolites during this period [54,55].

Lactic acid and pH levels correlated with fermentation or storage time, with correlation coefficients exceeding 0.97. This indicates a highly significant relationship between these parameters and time. The literature suggests that R^2 values close to 1 demonstrate a robust positive relationship between the variables, as seen in kimchi fermentation studies [56]. This strong correlation enables effective monitoring of the fermentation and storage process. Specifically, as the fermentation or storage time advances, lactic acid levels typically rise while pH decreases. These changes reflect the usual dynamics of fermentation, where increased lactic acid production lowers the medium's pH. This predictive relationship serves as a valuable tool for tracking process progress and can also be utilized to optimize and control the quality of the final product over time.

During fermentation, the physical stability of BJWH/C1 and BJWH/C2 improved, while viscosity decreased significantly by 57% for BJWH/C1 and 53% for BJWH/C2. Both properties remained stable throughout storage. The food matrix's structure influences the interplay between stability and viscosity. For instance, a thick and porous gel was observed in the fermentation of milk with strawberry juice using an initial kefir inoculum. However, a protein network with smaller pores enhanced stability and increased viscosity [57]. On the other hand, in a mixed juice (beet, carrot, apple, and strawberry) fermented with kefir grains, the viscosity initially increased within the first 12 h. Still, it later decreased due to the hydrolysis of exopolysaccharides by glycohydrolase [58]. This indicates that LAB-fermented products' physical stability and viscosity are largely determined by the strains' ability to integrate into the food matrix and their metabolic activities.

In the color analysis of BJWH/C1 and BJWH/C2 juices during fermentation, a shift from violet to red tones was observed at 16 h. During storage, the color parameters remained stable. This color change is linked to the stability of anthocyanins under varying pH levels, as anthocyanin cations remain stable in acidic conditions [20,53]. For BJWH/C2 samples, the color was red when the pH was below 7, violet between pH 7 and 8, and blue when the pH exceeded 11. Color variations in matrices containing antioxidants are also influenced by the concentration of glycosylated polyhydroxyl (violet) or polymethoxy (red) groups. Most anthocyanins have two, three, or more sugars attached as oligosaccharide side chains [53].

Lactic acid fermentation of antioxidant-rich foods affects these compounds, with changes dependent on the specific strain and the matrix (fiber, protein) in which they are present. During fermentation, a decrease in phenolic compounds was observed in all samples. This reduction can be attributed to thermal processing, which induces physical and chemical reactions, including the breakdown of bound phenolic compounds and their transformation into other forms [59]. LAB are known to metabolize phenolic compounds [60], possibly due to the biotransformation of these compounds through LAB reductases, decarboxylases, and glycosidases [49,61]. Additionally, LAB-produced hydrolases may hydrolyze phenolic complexes into simpler forms and disrupt cell walls, releasing antioxidant compounds, as seen in the co-fermentation of longan and goji juice [42]. In storage, however, sample BJWH/C2 (*L. casei* + *L. ramnhosus*) showed an increased total phenolic content. Phenolic acid decarboxylases can catalyze the non-oxidative decarboxylation of phenolic acids to generate P-vinyl derivatives [62].

The fermentation of a drink made from blackberry juice and whey demonstrated that phenolic compounds and anthocyanins became more stable after fermentation with *L. plantarum* and *L. casei*. Binding anthocyanins or phenolic hydroxyl groups stabilized NH, CN, and C [20]. In apple juice fermented with *L. acidophilus*, *L. casei*, and *L. plantarum*, the total phenolic content decreased after 72 h of fermentation and during the first 10 days of storage [49], with stability observed thereafter until day 20. Similarly, Wang et al. [63] reported on the fermentation of mango juice with *L. bulgaricus*, *S. thermophilus* 6063, and *L. plantarum*.

During fermentation, the anthocyanin content remained stable in the LAB-inoculated samples (BJWH/C1 and BJWH/C2). There was also a slight decrease in BJWH/C1 during storage and a more pronounced decrease in BJWH/C2 starting from day 21. In contrast, blackberry juice without an inoculum significantly reduced anthocyanins from day 14. In yogurt supplemented with 5% freeze-dried mulberry fruit juice (FDMJ), the anthocyanin content increased over 35 days (from 36.49 to 40.65 mg CGE cyanidin-3-glucoside/100 g) [64]. However, antioxidant activities decreased, likely due to the degradation of phenolic compounds or increased milk protein–polyphenol interactions [65]. LAB-produced-glucosidases contribute to reductions in anthocyanins during fermentation by breaking glycosidic bonds [27].

The radical scavenging capacity was measured using ABTS and DPPH assays to assess antioxidant activity. The BJWH/C1 and BJWH/C2 samples retained ABTS activity during the first eight hours of fermentation, followed by a decrease after 12 h. In contrast, DPPH assays showed increased antioxidant capacity during fermentation, especially in BJWH/C1, with activity maintained for the first 14 days of storage (Figure 6D). These antioxidant properties are attributed to proton donation and the samples' acidic and phenolic hydroxyl groups. Similar findings have been reported for fermented chokeberry [5] and mulberry pomace [66], which experienced a decrease in phenolic and anthocyanin contents during fermentation but showed increased ABTS and DPPH scavenging on the first day of fermentation.

One limitation of this study is the lack of analysis of the profiles of sugars and amino acids at the beginning and end of fermentation. Additionally, there is limited research on the effects of co-cultures or consortia, with most studies focusing on individual LAB and *Saccharomyces* strains. Future research should also explore how adding isolated whey to the medium might affect sensory preferences. Despite this, our study found that including whey enhances LAB growth conditions, resulting in good stability and antioxidant capacity during storage at low temperatures.

4. Materials and Methods

4.1. Materials

The blackberries were sourced from Atotonilco, Hidalgo, Mexico, and the milk whey (100% isolated whey protein) was purchased from HolixLab, Hilmar ingredients® (Guadalajara, Jalisco, Mexico). The autochthonous strains used (*Lactiplantibacillus plantarum* and *Pediococcus acidilactici* from aguamiel, *Levilactobacillus brevis* from pulque and *Lacticas-*

eibacillus rhamnosus and *Lacticaseibacillus casei* from fermented milk [23]) were isolated and provided by the Academic Area of Nutrition of Autonomous University of Hidalgo State, México.

Rogosa and Sharpe (MRS) agar and broth from MERCK (Rahway, NJ, USA) and Difco (Sparks, MD, USA), respectively, were utilized for culturing. The reagents used for principal assays included sodium hydroxide (NaOH Labessa, Ciudad De León, Guanajuato, Mexico), Folin–Ciocalteu reagent (Sigma-Aldrich, Louis, MO, USA), sodium carbonate (Meyer, Vallejo, CA, USA), gallic acid (Chem-Cruz, Dallas, TX, USA), potassium chloride (Meyer, Vallejo, CA, USA), sodium acetate (Meyer, Vallejo, CA, USA), hydrochloric acid (Meyer, Vallejo, CA, USA), ammonium 2,2′-azino-bis-(3-ethylbenzothiazoline-6-sulfonate (ABTS, Sigma-Aldrich, Louis, MO, USA), potassium persulfate (Sigma-Aldrich, Louis, MO, USA), ascorbic acid (Meyer, Vallejo, CA, USA), 2,2-Diphenyl-1-Picrylhydrazyl (DPPH, Sigma-Aldrich, Louis, MO, USA), absolute ethyl alcohol (Meyer, Vallejo, CA, USA), 6-hydroxy-2,5,7,8-tetramethylchroman-2-carboxylic acid (Trolox, Sigma-Aldrich, Louis, MO, USA).

4.2. Blackberry Juice and Whey Preparation

Blackberry fruit (*Rubus fruticosus*) was sourced locally from Atotonilco El Grande, Hidalgo, Mexico. The method outlined by Cervantes-Elizarrarás [67] was followed to ontain clarified juice. The fruits were washed, disinfected, ground, and filtered with a conventional strainer. The juice was then clarified through centrifugation (10,000 rpm for 30 min at 4 °C) and treated with thermoultrasound under optimal conditions (50 ± 1 °C, 17 ± 1 min at 80% amplituded). After processing, the juice was frozen at −35 °C until needed. The whey was prepared by dissolving at 9.3 g/L of water and sterilized before use (121 °C for 15 min).

4.3. Juice Fermentation

Blackberry juice (BJ) and whey (WH) were mixed in a 1:1 ratio to produce BJWH solution, and the pH of all samples (BJ, WH, and BJWH) was adjusted to 6 using 1 M NaOH. The samples were inoculated with LAB strains previously activated on MRS agar (24 h at 37 °C) and growth in 5 mL of MRS broth under the same conditions. The biomass was collected by centrifugation (10,000 rpm for 15 min), resuspended in the fermentation matrices, and adjusted to an approximate initial count of 9 Log CFU/mL. Inoculation was performed with the following consortia: consortium 1 (C1) containing *Lactiplantibacillus plantarum*, *Pediococcus acidilactici* and *Levilactobacillus brevis*; consortium 2 (C2) containing of *Lacticaseibacillus rhamnosus* and *Lacticaseibacillus casei*; and negative control (C-) without LAB. The samples were incubated at 37 °C for 48 h, with 20 mL aliquots taken 0, 24, and 48 h during the initial phase. Testing continued for 32 h in the subsequent phase, with aliquots collected every 4 h. For storage analysis, the fermented samples were kept in the dark at 4 °C, with aliquots taken on days 3, 7, 14, 21, and 28. All aliquots were centrifuged (10,000 rpm for 15 min at 4 °C) and the supernatants were stored at −32 °C to analyze reducing sugars and antioxidant properties.

4.4. Microbiological Analysis of Fermented Juice

Microbiological analysis was conducted using the micro drop method described by Strahsburger, Retamales, Estrada, and Seeger [68]. The fermented juice samples' serial dilutions (1:10) were prepared in a peptone water solution for microbial counting. Aliquots were then inoculated onto MRS agar and incubated at 37 °C for 36 h. The results were logarithms of ten colony-forming units per milliliter (Log10 CFU/mL).

4.5. Kinetic Parameters at the Time of Fermentation

The kinetic parameters were determined using Monod's mathematical model [69], which describes the growth of microorganisms during the exponential phase. The initial (N_0) and final (N_x) concentrations of growth were measured during the logarithmic phase, corresponding to the initial time (t_0) and final time (t_x). The increases in cell density (A), growth rate (μ), generation time (g) and growth constant (k) were calculated using the following equations:

$$A = \mathrm{LogN_x} - \mathrm{LogN_0}$$

$$\mu = [\mathrm{In}(N_x) - \mathrm{In}(N_0)]/(t_x - t_0)$$

$$g = \mathrm{In}(2)/\mu$$

$$k = 1/g$$

4.6. pH, Titratable Acidity and Soluble Solids of Fermented Juice

The pH was measured using a potentiometric method following standard procedures. Titratable acidity was determined by titrating a 1:10 dilution of the juice with deionized water using 0.01 N NaOH until a pH of 8.2 was reached. The titratable acidity was expressed as a percentage equivalent to lactic acid. Total soluble solids (°Brix) were measured with a manual ATC refractometer by placing a drop of the juice on the prism, calibrating, and then taking the reading. Non-linear regression analysis was performed on the lactic acid and pH variables for fermentation (48 h) and storage (28 days), using GeoGebra Classic (version 6.0.851.0, https://www.geogebra.org/classic?lang=es, 12 July 2024) to model polynomial behavior.

4.7. Stability and Viscosity of Fermented Juice

The stability percentage was determined using the following procedure: 10 g of sample (sW) was weighed and centrifuged at 3500 rpm (Hamilton, model V65000, Montvale, NJ, USA) for 30 min. The precipitate (SW) was decanted and weighed. Stability was calculated using the equation below and expressed as a percentage.

$$\text{Stability (\%)} = 100 - [(\mathrm{SW}/\mathrm{sW}) \times 100]$$

Viscosity was measured using the method described by Díaz-Jiménez, Sosa-Morales, and Vélez-Ruiz [70] with a Brookfield DV3T viscometer (Manassas, VA, USA), employing needle LV-4 at 60 rpm. The readings were taken in centipoises (cP). A 30 mL aliquot was used, and measurements were started after 30 s, at which point the cP values stabilized.

4.8. Measure of the Color of Fermented Samples

Color measurements were taken using a colorimeter (Minolta CM-80, 500 SM-508D, Osaka, Japan Co., Ltd.) with a D65 illuminant and a 10° observation angle. Then, 5 mL of fermented juice was placed in a container for analysis. The color parameters were recorded using the CIE—L* a* b* values, where L* indicates lightness (L = 0 for black and L = 100 for white), a* measures the chromaticity from green (−) to red (+), and b* measures the chromaticity from blue (−) to yellow (+). The colorimeter also provided the hue angle (h) and chroma (C) [71].

4.9. Antioxidant Properties

4.9.1. Total Phenolic Content

Total phenolic content (TPC) was measured following the method described by Stintzing et al. [72]. In brief, 100 μL of the sample was mixed with 500 μL of 1:10 diluted Folin–Ciocalteu reagent. After adding 400 μL of 7.5% sodium carbonate, the mixture was incubated at room temperature for 30 min. The absorbance was then measured at 765 nm using a microplate lector (Power Wave XS UV-Biotek Gen 5 software (version 3.16.10), KC Junior, Winooski, VT, USA). Gallic acid served as the reference standard, and the results were expressed as milligrams of gallic acid equivalents per 100 milliliters (mg GAE/100 mL).

4.9.2. Anthocyanins

The anthocyanin content was measured using the method outlined by Wallace and Giusti [73]. Two buffers were prepared: potassium chloride at 0.025 M and pH 1 (0.466 g in 250 mL of distilled water) and sodium acetate at 0.4 M and pH 4.5 (8.203 g in 250 mL of distilled water), with pH adjusted using 6 M HCl. A 4.5 mL aliquot of either buffer was mixed with 0.5 mL of the sample, and the mixture was allowed to stand for 15 min at room temperature and in complete darkness. Absorbance was then measured using a microplate lector (Power Wave XS UV-Biotek Gen 5 software (version 3.16.10) KC Junior, Winooski, VT, USA) at 510 and 700 nm wavelengths with the respective buffer. Results were expressed as milligrams of Cyanidin-3-glucoside per 100 milliliters of the sample (mg Cy-3-Gl/100 mL).

4.9.3. Antioxidant Activity by ABTS

Antiradical capacity was assessed following the method described by Kuskoski, Asuero, Troncoso, Mancini-Filho, and Fett [74]. An ABTS stock solution was prepared using 7 mmol/L ABTS and 2.45 mmol/L potassium persulfate and stored at a dark room temperature for 16 h. This stock solution was diluted with deionized water until an absorbance of 0.70 ± 0.10 at 754 nm was achieved. A 20 μL aliquot of the sample was added to 980 μL of the diluted ABTS solution, and absorbance readings were taken after 7 min incubation at room temperature. The absorbance was measured at 754 nm using a microplate reader (Power Wave XS UV-Biotek Gen 5 software (version 3.16.10), KC Junior, Winooski, VT, USA). Antioxidant capacity was expressed as milligrams of vitamin C equivalent antioxidant capacity per 100 milliliters (mg VCEAC/100 mL) of the fermented juice.

4.9.4. Antioxidant Activity by DPPH

Antiradical activity was assessed using DPPH radicals [75]. A stable DPPH radical solution was prepared in ethanol (7.4 mg/100 mL). In Eppendorf tubes, 100 μL of the extract was mixed with 500 μL of the DPPH solution and allowed to react for one hour at room temperature. Absorbance was then measured at 520 nm using a microplate reader (Power Wave XS UV-Biotek Gen 5 software (version 3.16.10), KC Junior, Winooski, VT, USA). The results are expressed as the fermented juice's micromoles of Trolox equivalents per 100 milliliters (μmol TE/100 mL).

4.10. Statistical Analysis

All results are presented as the mean and standard deviation (mean $\pm$ SD) of three independent experiments (n = 3) with a significance level of $p < 0.05$. Comparisons were made for the same strain or consortium at different fermentation times (0, 24, and 48 h) and storage periods (3, 7, 14, 21, and 28 days). Additionally, analyses were conducted between different strains or consortia across fermentation and storage time. Data were analyzed using one-way analysis of variance (ANOVA), with two-way ANOVA applied for bacterial growth during fermentation (48 h) and storage (28 days). Kinetic parameters were compared using Student's *t*-test and post hoc Tukey test, with statistical analyses performed using SPSS v. 19 software.

5. Conclusions

This study underscores the significant advantages of incorporating whey protein into fruit-based beverages, particularly highlighting its impact on enhancing the antioxidant content and maintaining probiotic viability during refrigerated storage. This approach provides a sustainable alternative for using blackberries and showcases the synergistic effects between probiotics, whey-derived peptides, and phenolic compounds, which act as potential prebiotics.

Statistical analysis reveals that fermenting blackberry juice with added whey protein and two or three strains of lactic acid bacteria (LAB) results in continuous and linear growth patterns, surpassing the fermentation outcomes of blackberry juice or whey alone. The regression analysis further confirms a strong correlation between pH and lactic acid production, dependent on the time, throughout fermentation and subsequent refrigerated storage. These findings indicate that the effectiveness of these fermentations depends on the specific proteolytic systems of the bacteria involved and their ability to adapt to the medium.

Moreover, the lactic acid fermentation process contributes to the stability of the beverage's physicochemical properties, notably enhancing its viscosity and color characteristics in both consortia (BJWH/C1 and BJWH/C2). Additionally, the fermentation of BJWH/C1 and BJWH/C2 during storage successfully preserves anthocyanins, ensuring sustained antioxidant potential over 28 days at low temperatures.

These advancements suggest that such formulations could improve consumer acceptability and offer potential health benefits. Future research is recommended to explore these potential benefits further and optimize the formulation for enhanced performance.

Author Contributions: L.L.-Z. investigation and methodology, L.L.-Z. and A.S.J.-O. conceptualization, formal analysis, writing—original draft and editing. L.D.-O., N.d.S.C.-C. and N.S.C.-J. investigation advisors and validation. L.L.-Z., A.S.J.-O. and L.D.-O. project administration. D.E.-L. manuscript review and edited. A.S.J.-O., N.d.S.C.-C. and L.D.-O. supervision and validation, L.G.G.-O. Data curation, formal analysis, writing—review and editing. All authors have read and agreed to the published version of the manuscript.

Funding: This research was funded by "Fondos Estatales Genéricos de la Universidad Autónoma del Estado de Hidalgo", Number 110201.

Institutional Review Board Statement: Not applicable.

Informed Consent Statement: Not applicable.

Data Availability Statement: The data presented in this study are available on request from the corresponding author.

Conflicts of Interest: The authors declare no conflicts of interest. The funders had no role in the design of the study; in the collection, analyses, or interpretation of data; in the writing of the manuscript; or in the decision to publish the results.

References

1. Rodzi, N.A.R.M.; Lee, L.K. Traditional Fermented Foods as Vehicle of Non-Dairy Probiotics: Perspectives in South East Asia Countries. *Food Res. Int.* **2021**, *150*, 110814. [CrossRef]
2. Rastogi, Y.R.; Thakur, R.; Thakur, P.; Mittal, A.; Chakrabarti, S.; Siwal, S.S.; Thakur, V.K.; Saini, R.V.; Saini, A.K. Food Fermentation—Significance to Public Health and Sustainability Challenges of Modern Diet and Food Systems. *Int. J. Food Microbiol.* **2022**, *371*, 109666. [CrossRef]
3. Olvera-Rosales, L.B.; Pérez-Escalante, E.; Castañeda-Ovando, A.; Contreras-López, E.; Cruz-Guerrero, A.E.; Regal-López, P.; Cardelle-Cobas, A.; González-Olivares, L.G. ACE-Inhibitory Activity of Whey Proteins Fractions Derived of Fermentation by *Lacticaseibacillus rhamnosus* GG and *Streptococcus thermophilus* SY-102. *Foods* **2023**, *12*, 2416. [CrossRef]
4. Duan, W.; Zhou, L.; Ren, Y.; Liu, F.; Xue, Y.; Wang, F.-Z.; Lu, R.; Zhang, X.-J.; Shi, J.-S.; Xu, Z.-H.; et al. Lactic Acid Fermentation of Goji Berries (*Lycium barbarum*) Prevents Acute Alcohol Liver Injury and Modulates Gut Microbiota and Metabolites in Mice. *Food Funct.* **2024**, *15*, 1612–1626. [CrossRef]
5. Wang, J.; Wei, B.; Xu, J.; Jiang, H.; Xu, Y.; Wang, C. Influence of Lactic Acid Fermentation on the Phenolic Profile, Antioxidant Activities, and Volatile Compounds of Black Chokeberry (*Aronia melanocarpa*) Juice. *J. Food Sci.* **2024**, *89*, 834–850. [CrossRef]
6. Li, Y.; Wang, W.; Deng, Y.; Gao, J.; Shi, J.; Cai, L. Antioxidant Properties and Changes in Vitro Digestion of the Fermented Kiwifruit Extract Prepared by Lactic Acid Bacteria and Yeasts. *Food Chem.* **2024**, *442*, 138416. [CrossRef]
7. Ruiz Rodríguez, L.G.; Zamora Gasga, V.M.; Pescuma, M.; Van Nieuwenhove, C.; Mozzi, F.; Sánchez Burgos, J.A. Fruits and Fruit By-Products as Sources of Bioactive Compounds. Benefits and Trends of Lactic Acid Fermentation in the Development of Novel Fruit-Based Functional Beverages. *Food Res. Int.* **2021**, *140*, 109854. [CrossRef]
8. Luo, Y.; Tang, R.; Qiu, H.; Song, A. Widely Targeted Metabolomics-Based Analysis of the Impact of *L. plantarum* and *L. paracasei* Fermentation on *rosa roxburghii* Tratt Juice. *Int. J. Food Microbiol.* **2024**, *417*, 110686. [CrossRef]
9. Yan, Y.; Zhang, F.; Chai, Z.; Liu, M.; Battino, M.; Meng, X. Mixed Fermentation of Blueberry Pomace with *L. rhamnosus* GG and *L. plantarum-1*: Enhance the Active Ingredient, Antioxidant Activity and Health-Promoting Benefits. *Food Chem. Toxicol.* **2019**, *131*, 110541. [CrossRef]
10. Mustafa, S.M.; Chua, L.S.; El-Enshasy, H.A.; Abd Majid, F.A.; Hanapi, S.Z.; Abdul Malik, R. Effect of Temperature and pH on the Probiotication of *Punica granatum* Juice Using *Lactobacillus* Species. *J. Food Biochem.* **2019**, *43*, e12805. [CrossRef]

11. Lima, W.D.L.; Monteiro, S.S.; Pasquali, M.A.d.B. Study of Fermentation Strategies by *Lactobacillus gasseri* for the Production of Probiotic Food Using Passion Fruit Juice Combined with Green Tea as Raw Material. *Foods* **2022**, *11*, 1471. [CrossRef]
12. Liu, M.; Zhang, L.; Li, J.; Xu, G.; Zong, W.; Wang, L. Effects of Lactic Acid Bacteria on Antioxidant Activity in Vitro and Aroma Component of *Eucommia ulmoides* Tea. *J. Food Sci. Technol.* **2024**, *61*, 169–177. [CrossRef]
13. Huynh, N.K.; Wilson, M.D.; Eyles, A.; Stanley, R.A. Recent Advances in Postharvest Technologies to Extend the Shelf Life of Blueberries (*Vaccinium* sp.), Raspberries (*Rubus idaeus* L.) and Blackberries (*Rubus* sp.). *J. Berry Res.* **2019**, *9*, 687–707. [CrossRef]
14. Aguilera, J.M. Berries as Foods: Processing, Products, and Health Implications. *Annu. Rev. Food Sci. Technol.* **2024**, *15*, 1–26. [CrossRef]
15. Khaire, R.A.; Gogate, P.R. Chapter 7—Whey Proteins. In *Proteins: Sustainable Source, Processing and Applications*; Galanakis, C.M., Ed.; Academic Press: Cambridge, MA, USA, 2019; pp. 193–223. ISBN 978-0-12-816695-6.
16. Pillai, A.T.; Morya, S.; Kasankala, L.M. Emerging Trends in Bioavailability and Pharma-Nutraceutical Potential of Whey Bioactives. *J. Nutr. Metab.* **2024**, *2024*, 8455666. [CrossRef] [PubMed]
17. Rosa, L.S.; Santos, M.L.; Abreu, J.P.; Rocha, R.S.; Esmerino, E.A.; Freitas, M.Q.; Mársico, E.T.; Campelo, P.H.; Pimentel, T.C.; Cristina Silva, M.; et al. Probiotic Fermented Whey-Milk Beverages: Effect of Different Probiotic Strains on the Physicochemical Characteristics, Biological Activity, and Bioactive Peptides. *Food Res. Int.* **2023**, *164*, 112396. [CrossRef]
18. Qian, Y.; Gui-Bing, L.; Wen-Qiong, W.; Cong-Cong, T.; Ruixia, G. The Mechanism of Whey Protein and Blueberry Juice Mixed System Fermented with *Lactobacillus* Inhibiting *Escherichia coli* during Storage. *Sci. Rep.* **2023**, *13*, 6614. [CrossRef]
19. Gordon, R.A.; Zumbro, E.L.; Castleberry, T.J.; Sokoloski, M.L.; Brisebois, M.F.; Irvine, C.J.; Duplanty, A.A.; Ben-Ezra, V. Whey Protein Improves Glycemia during an Oral Glucose Tolerance Test Compared to Vigorous-Intensity Aerobic Exercise in Young Adult Men. *BMC Sports Sci. Med. Rehabil.* **2022**, *14*, 147. [CrossRef] [PubMed]
20. Wen-Qiong, W.; Jie-Long, Z.; Qian, Y.; Ji-Yang, Z.; Mao-Lin, L.; Rui-Xia, G.; Yujun, H. Structural and Compositional Changes of Whey Protein and Blueberry Juice Fermented Using *Lactobacillus plantarum* or *Lactobacillus casei* during Fermentation. *RSC Adv.* **2021**, *11*, 26291–26302. [CrossRef]
21. Cai, Z.; Guo, Y.; Zheng, Q.; Liu, Z.; Zhong, G.; Zeng, L.; Huang, M.; Pan, D.; Wu, Z. Screening of a Potential Probiotic *Lactiplantibacillus plantarum* NUC08 and Its Synergistic Effects with Yogurt Starter. *J. Dairy Sci.* **2024**, *107*, 2760–2773. [CrossRef]
22. Karimkhani, M.M.; Jamshidi, A.; Nasrollahzadeh, M.; Armin, M.; Jafari, S.M.; Zeinali, T. Fermentation of *Rubus dolichocarpus* juice using *Lactobacillus gasseri* and *Lacticaseibacillus casei* and protecting phenolic compounds by Stevia extract during cold storage. *Sci. Rep.* **2024**, *14*, 5711. [CrossRef]
23. Cervantes-Elizarrarás, A.; Cruz-Cansino, N.D.S.; Ramírez-Moreno, E.; Vega-Sánchez, V.; Velázquez-Guadarrama, N.; Zafra-Rojas, Q.Y.; Piloni-Martini, J. In vitro probiotic potential of lactic acid bacteria isolated from aguamiel and pulque and antibacterial activity against pathogens. *Appl. Sci.* **2019**, *9*, 601. [CrossRef]
24. Filannino, P.; Cardinali, G.; Rizzello, C.G.; Buchin, S.; De Angelis, M.; Gobbetti, M.; Di Cagno, R. Metabolic Responses of *Lactobacillus plantarum* Strains during Fermentation and Storage of Vegetable and Fruit Juices. *Appl. Environ. Microbiol.* **2014**, *80*, 2206–2215. [CrossRef]
25. Ricci, A.; Cirlini, M.; Levante, A.; Dall'Asta, C.; Galaverna, G.; Lazzi, C. Volatile Profile of Elderberry Juice: Effect of Lactic Acid Fermentation Using *L. plantarum*, *L. rhamnosus* and *L. casei strains*. *Food Res. Int.* **2018**, *105*, 412–422. [CrossRef] [PubMed]
26. Ricci, A.; Cirlini, M.; Maoloni, A.; Del Rio, D.; Calani, L.; Bernini, V.; Galaverna, G.; Neviani, E.; Lazzi, C. Use of Dairy and Plant-Derived Lactobacilli as Starters for Cherry Juice Fermentation. *Nutrients* **2019**, *11*, 213. [CrossRef] [PubMed]
27. Di Cagno, R.; Coda, R.; De Angelis, M.; Gobbetti, M. Exploitation of vegetables and fruits through lactic acid fermentation. *Food Microbiol.* **2013**, *33*, 1–10. [CrossRef] [PubMed]
28. Olvera-Rosales, L.B.; Cruz-Guerrero, A.E.; Jaimez-Ordaz, J.; Pérez-Escalante, E.; Quintero-Lira, A.; Ramírez-Moreno, E.; Contreras-López, E.; González-Olivares, L.G. Differences in the Proteolytic System of Lactic Acid Bacteria Affect the Release of DPP-IV Inhibitory Peptides from Whey Proteins. *Dairy* **2023**, *4*, 515–526. [CrossRef]
29. Fan-Chiang, H.-J.; Wrolstad, R.E. Sugar and Nonvolatile Acid Composition of Blackberries. *J. AOAC Int.* **2010**, *93*, 956–965. [CrossRef] [PubMed]
30. Pescuma, M.; Hébert, E.M.; Mozzi, F.; de Valdez, G.F. Functional Fermented Whey-Based Beverage Using Lactic Acid Bacteria. *Int. J. Food Microbiol.* **2010**, *141*, 73–81. [CrossRef]
31. Surve, S.; Shinde, D.B.; Kulkarni, R. Isolation, characterization and comparative genomics of potentially probiotic *Lactiplantibacillus plantarum* strains from Indian foods. *Sci. Rep.* **2022**, *12*, 1940. [CrossRef]
32. Marčetić, M.; Samardžić, S.; Ilić, T.; Božić, D.D.; Vidović, B. Phenolic Composition, Antioxidant, Anti-Enzymatic, Antimicrobial and Prebiotic Properties of *Prunus spinosa* L. Fruits. *Foods* **2022**, *11*, 3289. [CrossRef]
33. Pastar, I.; Fira, D.; Strahinić, I.; Krstić, K.; Begović, J.; Topisirović, L.; Jovanović, G. Analysis of the presence of prtR proteinase gene in natural isolates of *Lactobacillus rhamnosus*. *Folia Microbiol.* **2006**, *51*, 535–540. [CrossRef] [PubMed]
34. Sebastián-Nicolas, J.L.; Contreras-López, E.; Ramírez-Godínez, J.; Cruz-Guerrero, A.E.; Rodríguez-Serrano, G.M.; Añorve-Morga, J.; González-Olivares, L.G. Milk fermentation by *Lacticaseibacillus rhamnosus* GG and *Streptococcus thermophilus* SY-102: Proteolytic profile and ace-inhibitory activity. *Fermentation* **2021**, *7*, 215. [CrossRef]
35. Hao, P.; Zheng, H.; Yu, Y.; Ding, G.; Gu, W.; Chen, S.; Zhao, G. Complete sequencing and pan-genomic analysis of *Lactobacillus delbrueckii* subsp. bulgaricus reveal its genetic basis for industrial yogurt production. *PLoS ONE* **2011**, *6*, e15964. [CrossRef] [PubMed]

36. Kobayashi, J. D-amino acids and lactic acid bacteria. *Microorganisms* **2019**, *7*, 690. [CrossRef]
37. Toe, C.J.; Foo, H.L.; Loh, T.C.; Mohamad, R.; Abdul Rahim, R.; Idrus, Z. Extracellular proteolytic activity and amino acid production by lactic acid bacteria isolated from Malaysian foods. *Int. J. Mol. Sci.* **2019**, *20*, 1777. [CrossRef]
38. Espirito-Santo, A.P.; Carlin, F.; Renard, C.M.G.C. Apple, Grape or Orange Juice: Which One Offers the Best Substrate for Lactobacilli Growth?—A Screening Study on Bacteria Viability, Superoxide Dismutase Activity, Folates Production and Hedonic Characteristics. *Food Res. Int.* **2015**, *78*, 352–360. [CrossRef]
39. Mukisa, I.M.; Byaruhanga, Y.B.; Muyanja, C.M.B.K.; Langsrud, T.; Narvhus, J.A. Production of Organic Flavor Compounds by Dominant Lactic Acid Bacteria and Yeasts from *Obushera*, a Traditional Sorghum Malt Fermented Beverage. *Food Sci. Nutr.* **2017**, *5*, 702–712. [CrossRef] [PubMed]
40. De Angelis, M.; Calasso, M.; Cavallo, N.; Di Cagno, R.; Gobbetti, M. Functional Proteomics within the Genus *Lactobacillus*. *Proteomics* **2016**, *16*, 946–962. [CrossRef]
41. Perjéssy, J.; Hegyi, F.; Nagy-Gasztonyi, M.; Zalán, Z. Effect of the Lactic Acid Fermentation by Probiotic Strains on the Sour Cherry Juice and Its Bioactive Compounds. *Food Sci. Technol. Int.* **2022**, *28*, 408–420. [CrossRef]
42. Li, S.; Tao, Y.; Li, D.; Wen, G.; Zhou, J.; Manickam, S.; Han, Y.; Chai, W.S. Fermentation of Blueberry Juices Using Autochthonous Lactic Acid Bacteria Isolated from Fruit Environment: Fermentation Characteristics and Evolution of Phenolic Profiles. *Chemosphere* **2021**, *276*, 130090. [CrossRef] [PubMed]
43. Alvarez, M.M.; Aguirre-Ezkauriatza, E.J.; Ramírez-Medrano, A.; Rodríguez-Sánchez, A. Kinetic Analysis and Mathematical Modeling of Growth and Lactic Acid Production of *Lactobacillus casei* Var. *rhamnosus* in Milk Whey. *J. Dairy Sci.* **2010**, *93*, 5552–5560. [CrossRef]
44. Wang, J.; Wei, B.-C.; Zhai, Y.-R.; Li, K.-X.; Wang, C.-Y. Non-Volatile and Volatile Compound Changes in Blueberry Juice Inoculated with Different Lactic Acid Bacteria Strains. *J. Sci. Food Agric.* **2024**, *104*, 2587–2596. [CrossRef] [PubMed]
45. Pontonio, E.; Montemurro, M.; Pinto, D.; Marzani, B.; Trani, A.; Ferrara, G.; Mazzeo, A.; Gobbetti, M.; Rizzello, C.G. Lactic Acid Fermentation of Pomegranate Juice as a Tool to Improve Antioxidant Activity. *Front. Microbiol.* **2019**, *10*, 1550. [CrossRef] [PubMed]
46. Fonseca, H.C.; Melo, D.d.S.; Ramos, C.L.; Menezes, A.G.T.; Dias, D.R.; Schwan, R.F. Sensory and Flavor-Aroma Profiles of Passion Fruit Juice Fermented by Potentially Probiotic *Lactiplantibacillus plantarum* CCMA 0743 Strain. *Food Res. Int.* **2022**, *152*, 110710. [CrossRef] [PubMed]
47. Maisto, M.; Annunziata, G.; Schiano, E.; Piccolo, V.; Iannuzzo, F.; Santangelo, R.; Ciampaglia, R.; Tenore, G.C.; Novellino, E.; Grieco, P. Potential Functional Snacks: Date Fruit Bars Supplemented by Different Species of *Lactobacillus* spp. *Foods* **2021**, *10*, 1760. [CrossRef]
48. Neffe-Skocińska, K.; Okoń, A.; Kołożyn-Krajewska, D.; Dolatowski, Z. Amino Acid Profile and Sensory Characteristics of Dry Fermented Pork Loins Produced with a Mixture of Probiotic Starter Cultures. *J. Sci. Food Agric.* **2017**, *97*, 2953–2960. [CrossRef] [PubMed]
49. Yang, J.; Sun, Y.; Gao, T.; Wu, Y.; Sun, H.; Zhu, Q.; Liu, C.; Zhou, C.; Han, Y.; Tao, Y. Fermentation and Storage Characteristics of "Fuji" Apple Juice Using *Lactobacillus acidophilus*, *Lactobacillus casei* and *Lactobacillus plantarum*: Microbial Growth, Metabolism of Bioactives and in Vitro Bioactivities. *Front. Nutr.* **2022**, *9*, 833906. [CrossRef]
50. Eiteman, M.A.; Ramalingam, S. Microbial Production of Lactic Acid. *Biotechnol. Lett.* **2015**, *37*, 955–972. [CrossRef]
51. Mojikon, F.D.; Kasimin, M.E.; Molujin, A.M.; Gansau, J.A.; Jawan, R. Probiotication of Nutritious Fruit and Vegetable Juices: An Alternative to Dairy-Based Probiotic Functional Products. *Nutrients* **2022**, *14*, 3457. [CrossRef]
52. Rejdlová, A.; Salek, R.N.; Míšková, Z.; Lorencová, E.; Kůrová, V.; Adámek, R.; Sumczynski, D. Physical Characterization of a Novel Carrot Juice Whey-Enriched Beverage Fermented with Milk or Water Kefir Starter Cultures. *Foods* **2023**, *12*, 3368. [CrossRef] [PubMed]
53. Wang, W.; Wang, Y.; Liu, X.; Yu, Q. The Characteristics of Whey Protein and Blueberry Juice Mixed Fermentation Gels Formed by Lactic Acid Bacteria. *Gels* **2023**, *9*, 565. [CrossRef]
54. Bontsidis, C.; Mallouchos, A.; Terpou, A.; Nikolaou, A.; Batra, G.; Mantzourani, I.; Alexopoulos, A.; Plessas, S. Microbiological and Chemical Properties of Chokeberry Juice Fermented by Novel Lactic Acid Bacteria with Potential Probiotic Properties during Fermentation at 4 °C for 4 Weeks. *Foods* **2021**, *10*, 768. [CrossRef]
55. Isas, A.S.; Mariotti Celis, M.S.; Pérez Correa, J.R.; Fuentes, E.; Rodríguez, L.; Palomo, I.; Mozzi, F.; Van Nieuwenhove, C. Functional Fermented Cherimoya (*Annona cherimola* Mill.) Juice Using Autochthonous Lactic Acid Bacteria. *Food Res. Int.* **2020**, *138*, 109729. [CrossRef]
56. Kim, S.J.; Lee, J.Y.; Yoon, S.R.; Lee, H.W.; Ha, J.H. Regression analysis for predicting the fermentation state of packaged Kimchi using a colorimetric indicator. *J. Food Eng.* **2019**, *240*, 65–72. [CrossRef]
57. Wang, H.; Wang, C.N.; Guo, M.R. Effects of Addition of Strawberry Juice Pre- or Postfermentation on Physiochemical and Sensory Properties of Fermented Goat Milk. *J. Dairy Sci.* **2019**, *102*, 4978–4988. [CrossRef]
58. Paredes, J.L.; Escudero-Gilete, M.L.; Vicario, I.M. A New Functional Kefir Fermented Beverage Obtained from Fruit and Vegetable Juice: Development and Characterization. *LWT* **2022**, *154*, 112728. [CrossRef]
59. Chen, W.; Xie, C.; He, Q.; Sun, J.; Bai, W. Improvement in color expression and antioxidant activity of strawberry juice fermented with lactic acid bacteria: A phenolic-based research. *Food Chem.* **2023**, *17*, 100535. [CrossRef]

60. Piekarska-Radzik, L.; Klewicka, E. Mutual influence of polyphenols and *Lactobacillus* spp. bacteria in food: A review. *Eur. Food Res. Technol.* **2021**, *247*, 9–24. [CrossRef]
61. Gaur, G.; Gänzle, M.G. Conversion of (Poly)Phenolic Compounds in Food Fermentations by Lactic Acid Bacteria: Novel Insights into Metabolic Pathways and Functional Metabolites. *Curr. Res. Food Sci.* **2023**, *6*, 100448. [CrossRef] [PubMed]
62. Rodríguez, H.; Curiel, J.A.; Landete, J.M.; de las Rivas, B.; López de Felipe, F.; Gómez-Cordovés, C.; Mancheño, J.M.; Muñoz, R. Food Phenolics and Lactic Acid Bacteria. *Int. J. Food Microbiol.* **2009**, *132*, 79–90. [CrossRef] [PubMed]
63. Wang, J.; Xie, B.; Sun, Z. Quality Parameters and Bioactive Compound Bioaccessibility Changes in Probiotics Fermented Mango Juice Using Ultraviolet-Assisted Ultrasonic Pre-Treatment during Cold Storage. *LWT* **2021**, *137*, 110438. [CrossRef]
64. Sung, J.M.; Kim, Y.B.; Kum, J.S.; Choi, Y.S.; Seo, D.H.; Choi, H.W.; Park, J.D. Effects of Freeze-dried Mulberry on Antioxidant Activities and Fermented Characteristics of Yogurt during Refrigerated Storage. *Korean J. Food Sci. Anim. Resour.* **2015**, *35*, 807–814. [CrossRef] [PubMed]
65. Yuksel, Z.; Avci, E.; Erdem, Y.K. Characterization of Binding Interactions between Green Tea Flavanoids and Milk Proteins. *Food Chem.* **2010**, *121*, 450–456. [CrossRef]
66. Tang, S.; Cheng, Y.; Wu, T.; Hu, F.; Pan, S.; Xu, X. Effect of Lactobacillus Plantarum-Fermented Mulberry Pomace on Antioxidant Properties and Fecal Microbial Community. *LWT* **2021**, *147*, 111651. [CrossRef]
67. Cervantes-Elizarrarás, A.; Piloni-Martini, J.; Ramírez-Moreno, E.; Alanís-García, E.; Güemes-Vera, N.; Gómez-Aldapa, C.A.; Zafra-Rojas, Q.Y.; Cruz-Cansino, N.D. Enzymatic inactivation and antioxidant properties of blackberry juice after thermoultrasound: Optimization using response surface methodology. *Ultrason. Sonochem.* **2017**, *34*, 371–379. [CrossRef]
68. Strahsburger, E.; Retamales, P.; Estrada, J.; Seeger, M. Microdot method: Used with chromogenic agar is a useful procedure for sanitary monitoring in aquaculture. *Lat. Am. J. Aquat. Res.* **2016**, *44*, 742–749. [CrossRef]
69. Annual Review of Microbiology, Growth of Bacterial Cultures. In *The Physiological Significance of the Growth Constant*; Annual Review, Inc.: San Mateo, CA, USA, 1949; Volume 3, pp. 378–393. Available online: https://archive.org/details/sim_annual-review-of-microbiology_1949_3/page/392/mode/2up (accessed on 15 May 2024).
70. Díaz Jiménez, B.; Sosa Morales, M.E.; Vélez Ruiz, J.F. Efecto de la adición de fibra y la disminución de grasa en las propiedades fisicoquímicas del yogurt. *Rev. Mex. Ing. Quím.* **2004**, *3*, 287–305.
71. Manzur-Valdespino, S.; Ramírez-Moreno, E.; Arias-Rico, J.; Jaramillo-Morales, O.A.; Calderón-Ramos, Z.G.; Delgado-Olivares, L.; Córdoba-Díaz, M.; Córdoba-Díaz, D.; Cruz-Cansino, N.D. *Opuntia ficus-indica* L. Mill Residues—Properties and Application Possibilities in Food Supplements. *Appl. Sci.* **2020**, *10*, 3260. [CrossRef]
72. Stintzing, F.C.; Herbach, K.M.; Mosshammer, M.R.; Carle, R.; Yi, W.; Sellappan, S.; Akoh, C.C.; Bunch, R.; Felker, P. Color, Betalain Pattern, and Antioxidant Properties of Cactus Pear (*Opuntia* spp.) Clones. *J. Agric. Food Chem.* **2005**, *53*, 442–451. [CrossRef]
73. Wallace, T.C.; Giusti, M.M. Determination of Color, Pigment, and Phenolic Stability in Yogurt Systems Colored with Nonacylated Anthocyanins from *Berberis boliviana* L. as Compared to Other Natural/Synthetic Colorants. *J. Food Sci.* **2008**, *73*, C241–C248. [CrossRef] [PubMed]
74. Kuskoski, E.M.; Asuero, A.G.; Troncoso, A.M.; Mancini-Filho, J.; Fett, R. Aplicación de Diversos Métodos Químicos Para Determinar Actividad Antioxidante En Pulpa de Frutos. *Food Sci. Technol.* **2005**, *25*, 726–732. [CrossRef]
75. Morales, F.J.; Jiménez-Pérez, S. Free radical scavenging capacity of Maillard reaction products as related to colour and fluorescence. *Food Chem.* **2001**, *72*, 119–125. [CrossRef]

International Journal of
Molecular Sciences

MDPI

Article

Effect of High-Dose Vitamin C on Tendon Cell Degeneration—An In Vitro Study

Shusuke Ueda [1,*,†], Toru Ichiseki [1,2,*], Miyako Shimasaki [3,†], Daisuke Soma [1], Masaru Sakurai [4], Ayumi Kaneuji [1] and Norio Kawahara [1]

1 Department of Orthopaedic Surgery, Kanazawa Medical University, Daigaku 1-1, Uchinada-machi, Kahoku-gun 920-0293, Japan
2 Division of Translational Research, Department of Life Science, Medical Research Institute, Kanazawa Medical University, Daigaku 1-1, Uchinada-machi, Kahoku-gun 920-0293, Japan
3 Department of Pathology 2, Kanazawa Medical University, Daigaku 1-1, Uchinada-machi, Kahoku-gun 920-0293, Japan; miya0807@kanazawa-med.ac.jp
4 Social and Environmental Medicine, Kanazawa Medical University, Daigaku 1-1, Uchinada-machi, Kahoku-gun 920-0293, Japan
* Correspondence: adeu221@kanazawa-med.ac.jp (S.U.); tsy-ichi@kanazawa-med.ac.jp (T.I.); Tel.: +81-76-286-2211 (ext. 23214) (S.U. & T.I.); Fax: +81-76-286-4406 (S.U. & T.I.)
† These authors contributed equally to this work.

Abstract: Tendinopathy is an aging-related disease, often caused by micro-scarring and degeneration due to overuse or trauma. Ascorbic acid (vitamin C) supplementation is reported to be a useful treatment for tendinopathy recovery. We compared the inhibitory effects of various ascorbic acid doses on tendon cell damage. H_2O_2 was added to human-derived tendon cells in vitro (Group H_2O_2, control), followed by incubation with 150 μM or 30 mM of ascorbic acid (Group C, Group HC). The oxidative injury degree was evaluated by determining reactive oxygen species levels. The cytoskeletal structure was examined via fluorescence immunostaining of actin filaments. Quantitative polymerase chain reaction (qPCR) was performed to analyze the expressions of mitochondria transcription factor A, adenosine triphosphate 5A, type I collagen, and p16. Cell death was reduced, and oxidative stress was inhibited in C and HC groups. The cytoskeleton was maintained in the HC group but not in the C group. qPCR analysis revealed that p16 expression was inhibited in both the C and HC groups compared to the H_2O_2 group; other markers had increased expression. The progression of cell death and cytoskeletal disruption was inhibited by the administration of high-dose vitamin C. Hence, high-dose vitamin C is a potential treatment for tendon cell degeneration.

Keywords: high-dose vitamin C; tendon cell degeneration; mitochondrial damage; oxidative injury; cellular senescence; actin filament

1. Introduction

In young individuals, tendon damage occurs as a result of severe trauma, but in older adults, tendon ruptures and damage can occur even with minor trauma or activities of daily living. One of the reasons for this may be the aging and degenerative changes of tendons and ligaments. Tendon rupture and tendon damage associated with tendon degeneration and aging can occur in any part of the body, including the Achilles tendon, patellar tendon, and rotator cuff. When tendon damage or rupture occurs, it can cause inflammation in the surrounding tissue, potentially resulting in serious disruption to daily living. In recent years, the involvement of oxidative stress in degenerative tendinopathy has attracted attention, and its inhibition is considered important [1,2].

Recent reports have shown that oxidative stress in the Achilles tendon is involved in tendonitis and rupture and that oxidative stress is also related to rotator cuff injury and degenerative rotator cuff tears in the shoulder [3–6]. Additionally, there are conservative and surgical treatments for tendon rupture, but both carry the risk of re-rupture, which is also related to tendon tissue degeneration and aging-related disorders [7]. Therefore, it is believed that inhibiting tendon tissue degeneration is important for tissue repair and prevention of tendon damage.

As mentioned previously, oxidative stress has been reported to play a significant role in the degeneration and damage of tendon tissue. Furthermore, the involvement of advanced glycation end products (AGEs) or apoptosis in tendon injury has also been reported [7–10]. The accumulation of the end products of oxidation, such as AGEs, that accompanies tissue peroxidation and associated pathological conditions, may be improved by drugs with antioxidant properties.

Ascorbic acid (vitamin C) is a nutrient with strong antioxidant properties, and its effectiveness has been demonstrated in various fields [11–13]. The effectiveness of ascorbic acid in treating tendon injuries has also been reported [14–16]. However, recent reports have shown that although administration of ascorbic acid alone can suppress oxidative stress and the generation of AGEs, it is insufficient to suppress mitochondrial-mediated apoptosis [17–19]. Since the caspase activity pathway (intrinsic factor pathway) is involved in tendon cell degeneration, there may be limitations to its treatment or prevention with ascorbic acid.

The therapeutic effectiveness of high-dose vitamin C has been reported as a recent treatment method in various applications [20–22]. Specifically, high-dose vitamin C therapy inhibited the progression of nonalcoholic fatty liver disease [23]. Further, it has attracted attention due to its anti-aging effect [24]. Previous research has shown the inhibitory effects of ordinary-dose ascorbic acid on tendon cell degeneration; however, these effects were considered weaker than those elicited by other antioxidant agents. Therefore, this study compared the inhibitory effects of ordinary-dose versus high-dose ascorbic acid on tendon cell degeneration.

2. Results

2.1. Vitamin C Inhibits Reactive Oxygen Species (ROS) Production Due to Oxidative Stress

Tendon cells were exposed to 2 mM H_2O_2 and then cultured for 24 h after adding 150 μM, 7.5 mM, 15 mM, or 30 mM ascorbic acid. The ROS of the cultured cells was measured. In tendon cells exposed to H_2O_2 only, a significant increase in ROS levels was observed ($p < 0.001$). In the groups to which ascorbic acid was added, ROS was reduced in a concentration-dependent manner. In addition, a significant decrease in ROS was observed at 30 mM compared to 150 μM ascorbic acid ($p < 0.05$).

Thus, in this study, 2 mM H_2O_2 + 150 μM ascorbic acid (Group C) or 2 mM H_2O_2 + 30 mM ascorbic acid (Group HC) was added to the media of tendon cell cultures. H_2O_2(-), ascorbic acid(-), FBS(-) culture medium (normoxia group), and the medium (Group H_2O_2) to which 2mM H_2O_2 was added served as controls. These media were incubated for 3 h to detect ROS.

In the H_2O_2 group, ROS was detected in many cells. ROS production was inhibited in the two groups treated with ascorbic acid. A noticeable inhibitory effect on ROS levels was found in the HC group (Figure 1).

2.2. Vitamin C Protects Against Cytoskeletal Damage in Tendon Cells Due to Oxidative Stress

After adding the respective treatments, the cell morphology was compared across the four groups. After 3 h of incubation, the H_2O_2 group showed a large number of round or shrunken cells that appeared to be dead [25,26]. In the C group, cells with round and shrunken nuclei were mixed with cells that maintained their original forms. In the HC group, the cell form was preserved to the same degree as that of the normoxia group (Figure 2).

Fluorescence immunostaining showed a definite disappearance of actin filaments in the H_2O_2 group. In the normoxia group, actin filaments were detected, and the appearance of the cytoskeleton was maintained. However, in the C group, the actin filaments had disappeared. In the HC group, the actin filaments and the cytoskeleton were maintained to the same degree as those of the normoxia group (Figure 3).

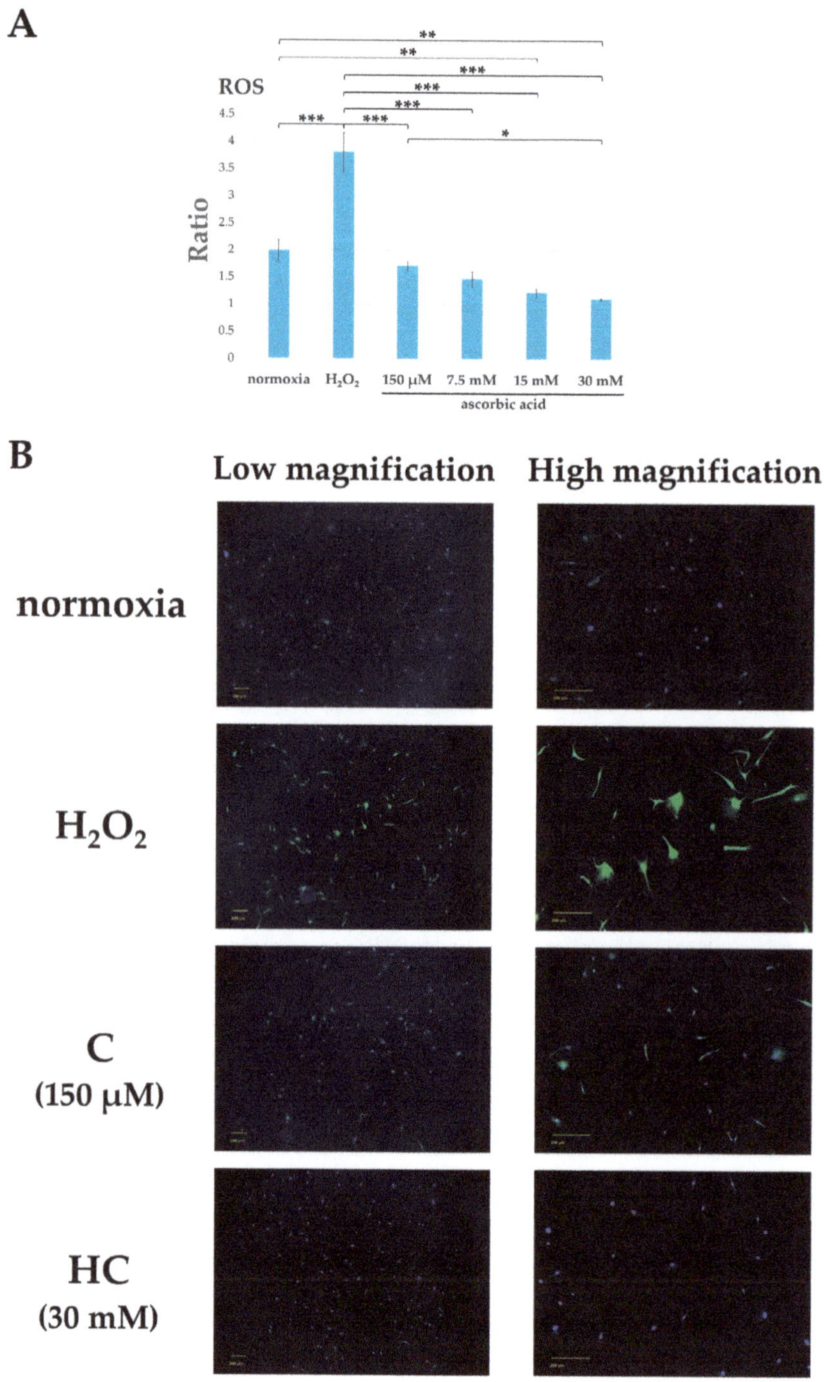

Figure 1. Reactive oxygen species (ROS) levels in tendon cells after H_2O_2 exposure. (**A**) ROS levels in tendon cells at 24 h after H_2O_2 exposure. Using a ROS Detection cell-based assay kit, ROS levels were measured in each group, as shown in the graphs (n = 5 each). Each graph indicates the ratio of

ROS under the indicated conditions. Columns and bars indicate means and standard deviations, respectively. ROS levels increased with H_2O_2 exposure, but in the group to which ascorbic acid was added, ROS decreased in a concentration-dependent manner. There was a significant decrease in ROS at 30 mM compared to 150 µM ascorbic acid (* $p < 0.05$, ** $p < 0.01$, and *** $p < 0.001$). (**B**) ROS levels in tendon cells at 3 h after H_2O_2 exposure. The figure shows the H_2O_2, 2 mM H_2O_2 + 150 µM ascorbic acid (C), and 2 mM H_2O_2 + 30 mM ascorbic acid (HC) groups. Low- and high-magnification images are shown on the left and right rows, respectively. Representative fluorescence images for DAPI (blue) and ROS (green) are also shown. In the H_2O_2 group, expression of ROS was seen. Compared with the H_2O_2 group, ROS levels were reduced in both the C and HC groups. ROS was minimally detected in the HC group. Scale bar: 200 µm. DAPI, 4′,6-diamidino-2-phenylindole.

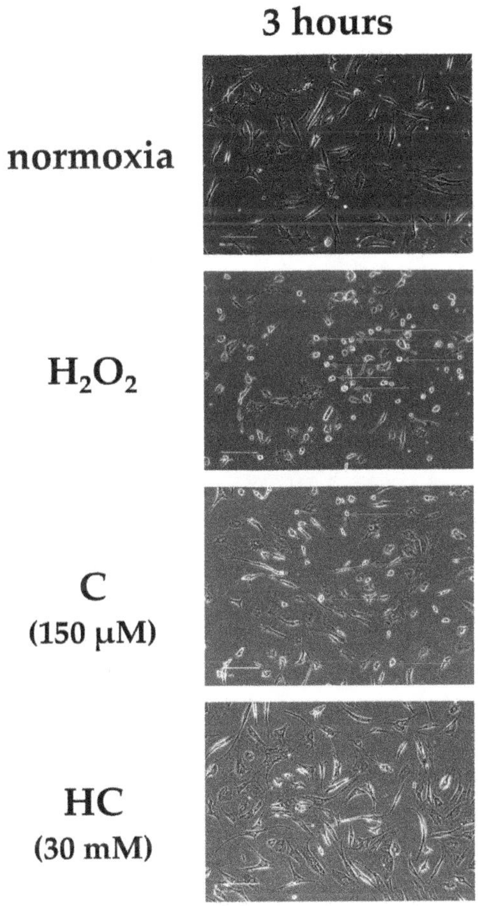

Figure 2. Phase-contrast images of tendon cells at 3 h after H_2O_2 exposure. The figures show representative images of various treatment groups. In the normoxia group, most of the cells were engrafted, and almost no detached cells were observed. The H_2O_2 group showed a large number of round or shrunken cells. The round and shrunken cells were released, and cell death was observed (red arrows). In the C group, cells with round and shrunken nuclei (red arrows) were mixed with cells that maintained their original forms. Compared with the H_2O_2 group, the number of dead cells was reduced in both the C and HC groups. In the HC group, the number of cell deaths was reduced compared to that in the C group. Scale bar: 200 µm.

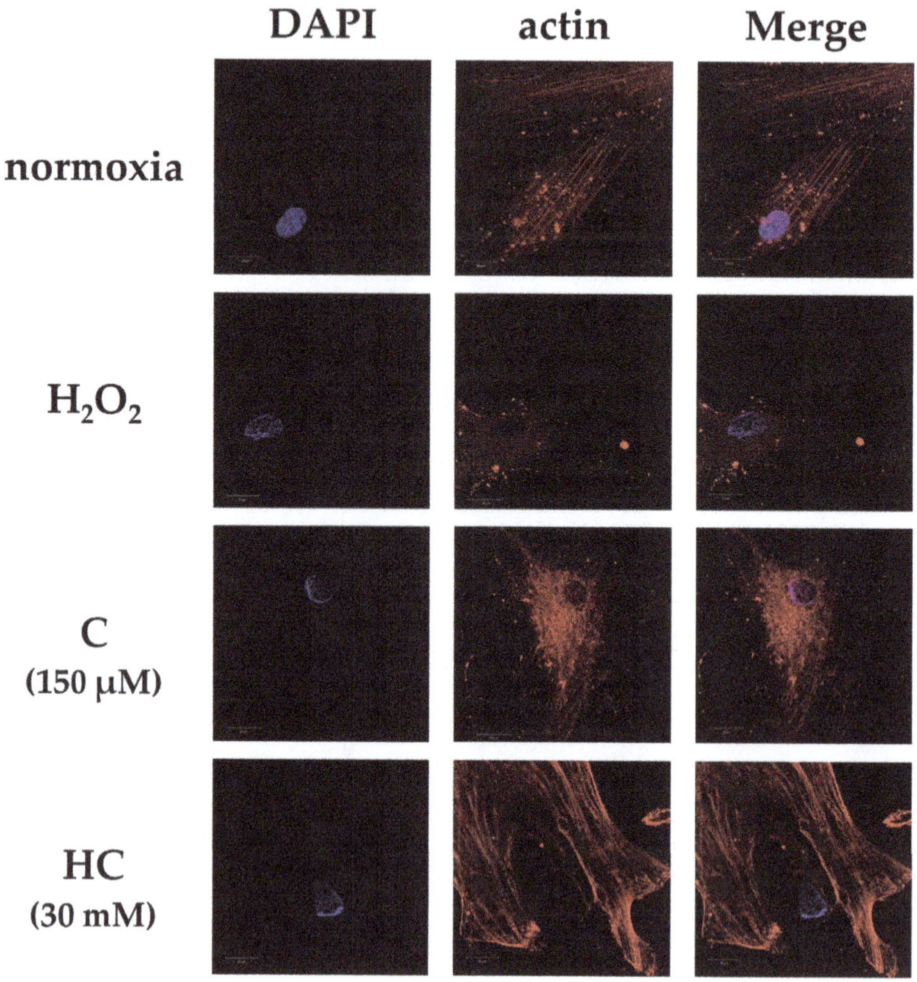

Figure 3. Fluorescence immunostaining of actin filaments in tendon cells at 3 h after H_2O_2 exposure. The figure shows representative fluorescence images for DAPI (blue) and/or actin (red) in the various treatment groups. Actin filaments had disappeared in the H_2O_2 and C groups. In the HC group, actin filaments were preserved to the same degree as those of the normoxia group. Scale bar: 20 μm.

2.3. *Vitamin C Maintains Mitochondrial Function, Energy Production, and the Extracellular Matrix from Cellular Damage Due to Oxidative Stress and Inhibits the Expression of Cellular Senescence-Related Genes*

Each group was cultured for 24 h, and adenosine triphosphate 5A (ATP5A) was compared using immunofluorescence staining. ATP5A expression was observed in all the groups, but the expression was decreased in the H_2O_2 group compared with the other groups. The HC group showed increased expression of ATP5A compared with the H_2O_2 and C groups. Shrinkage of the nucleus was observed in the H_2O_2 and C groups (Figure 4).

The expressions of the mitochondrial markers are as follows: mitochondria transcription factors A (TFAM) and ATP5A, the extracellular matrix synthetic protein type I collagen, and the senescence factor p16: CDKN2A (cyclin-dependent kinase inhibitor 2A). These were analyzed via quantitative polymerase chain reaction (qPCR). Each group was incubated for 3 h. The results showed that the expressions of TFAM, ATP5A, and type I collagen decreased in the H_2O_2 group compared to the normoxia group. In the C and HC groups, the expressions of these genes increased compared to the H_2O_2 group; gene expression was higher in the HC group than in the C group. The level of p16 expression was significantly higher in the H_2O_2 group than in the other groups. However, p16 expression was inhibited in both the C and HC groups; it was particularly noticeable in the HC group (* $p < 0.05$, ** $p < 0.01$, and *** $p < 0.001$) (Figure 5).

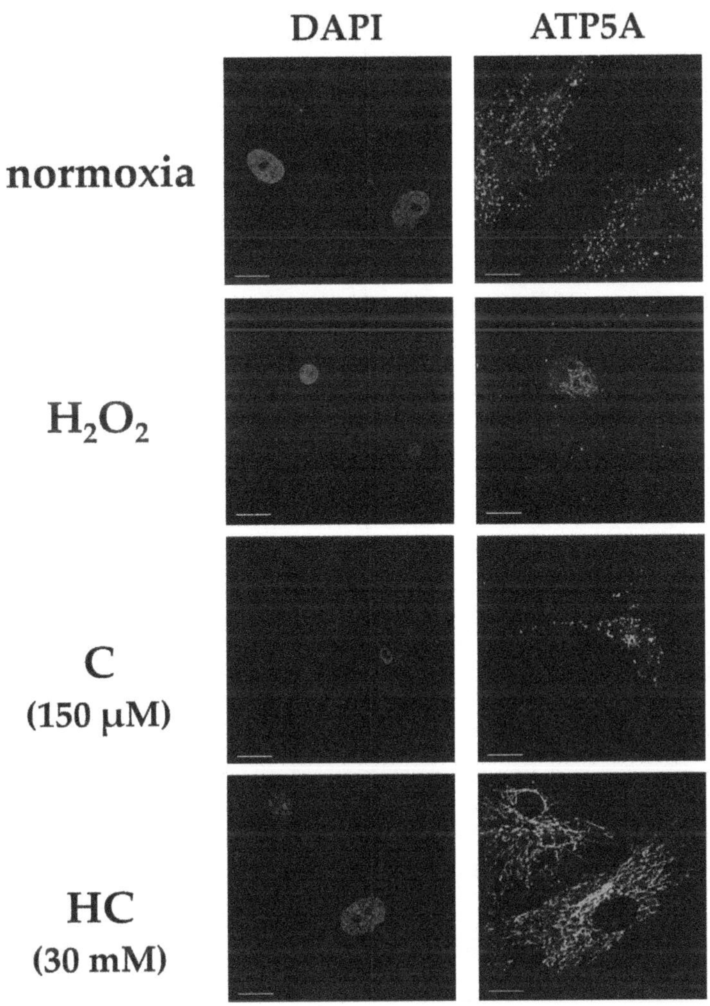

Figure 4. Fluorescence immunostaining of ATP5A in tendon cells at 24 h after H_2O_2 exposure. The figure shows representative fluorescence images for DAPI (blue) or ATP5A (green) in the various treatment groups. In the HC group, expression of ATP5A was seen. Compared with the HC group, ATP5A was reduced in both the H_2O_2 and C groups. Scale bar: 20 μm. ATP5A, adenosine triphosphate 5A.

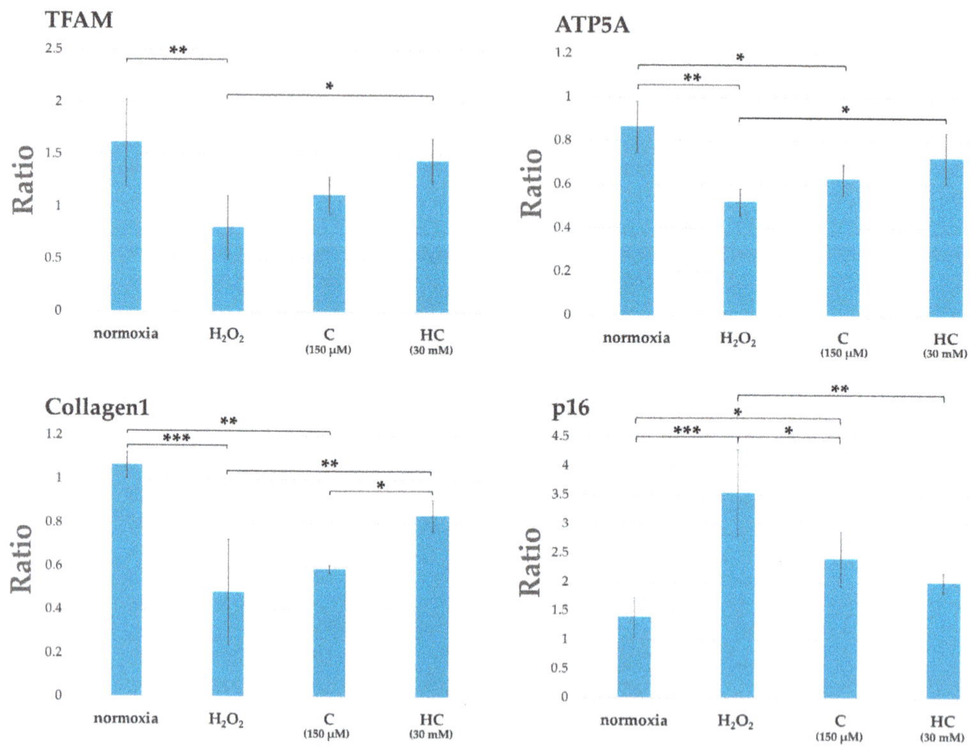

Figure 5. qPCR analysis of gene expression in tendon cells at 3 h after H_2O_2 exposure. Quantitative RT-PCR analysis of each gene expression level was conducted in each group, as shown in the graphs (n = 3 each). Each graph indicates the ratio of each gene expression level under the indicated conditions. Columns and bars indicate means and standard deviations, respectively. The H_2O_2 group showed a decrease in the expression levels of TFAM, ATP5A, and type I collagen and an increase in the p16 expression level. In the C and HC groups, an increase in the expression levels of TFAM, ATP5A, and type I collagen was seen as compared with the H_2O_2 group, although the expression level of p16 decreased. This was particularly noticeable in the HC group compared to the C group (* $p < 0.05$, ** $p < 0.01$, and *** $p < 0.001$). qPCR, quantitative polymerase chain reaction; RT-PCR, reverse transcription polymerase chain reaction.

2.4. High-Dose Vitamin C Maintains Cell Viability by Protecting the Cytoskeleton from Cell Damage in Tendon Cells Previously Exposed to Oxidative Stress

In this experiment, 150 μM or 30 mM ascorbic acid was added to the tendon cell culture medium that had been incubated for 1 h with H_2O_2 (C/after H_2O_2 group and HC/after H_2O_2 group). We then determined the cell morphology, viable cell count, and the presence of actin filaments after 2 h. One hour after exposure to H_2O_2 (H_2O_2 1 h group), round and shrunken cells were seen together with mature tendon cells. Two hours after adding a high dose of ascorbic acid, the cell form and presence of actin filaments were observed. The H_2O_2 group showed clear disruptions in nuclear and cellular morphology, a decrease in adherent cell count, and a marked loss of actin filaments. However, in the HC/after H_2O_2 group, the adherent cell count and cell form, actin filaments, and cytoskeleton were maintained. The C/after H_2O_2 group showed cell morphological destruction and a reduction in adherent cell number, as well as a loss of actin filaments, similar to the H_2O_2 group. The viable cell count was significantly maintained at 2 h after adding a high dose of ascorbic acid to the tendon cell culture medium that had been incubated with H_2O_2 for 1 h (HC/after H_2O_2 group), compared with the H_2O_2 group (* $p < 0.05$, ** $p < 0.01$, and *** $p < 0.001$) (Figures 6 and 7).

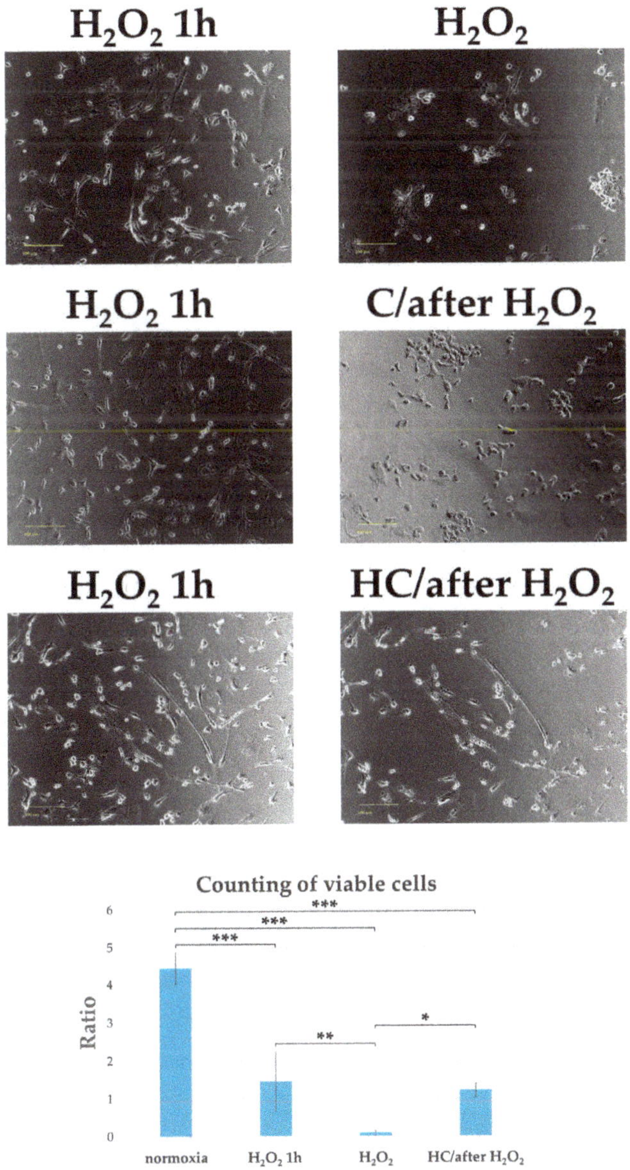

Figure 6. Phase-contrast image of tendon cells treated with 150 μM or 30 mM ascorbic acid added 1 h after H_2O_2 exposure. Shown are the representative phase-contrast images of the cells in each group at 1 h after H_2O_2 exposure and then at 2 h after adding 150 μM or 30 mM ascorbic acid (C/after H_2O_2 group and HC/after H_2O_2 group). The numbers of viable cells were measured in each group, as shown in the graphs ($n = 3$ each). Each graph indicates the ratio of viable cells under the indicated conditions. Columns and bars indicate means and standard deviations, respectively. One hour after exposure to H_2O_2 (H_2O_2 1 h group), round and shrunken cells were seen together with mature tendon cells. The number of adherent cells was reduced in the H_2O_2 group. Contrastingly, the number

of adherent cells in the HC/after H_2O_2 group was maintained. The number of adherent cells in the C/after H_2O_2 group was reduced to the same degree as in the H_2O_2 group. The number of viable cells was significantly maintained in the HC/after H_2O_2 group compared to the H_2O_2 group (* $p < 0.05$, ** $p < 0.01$, and *** $p < 0.001$). Scale bar: 200 μm.

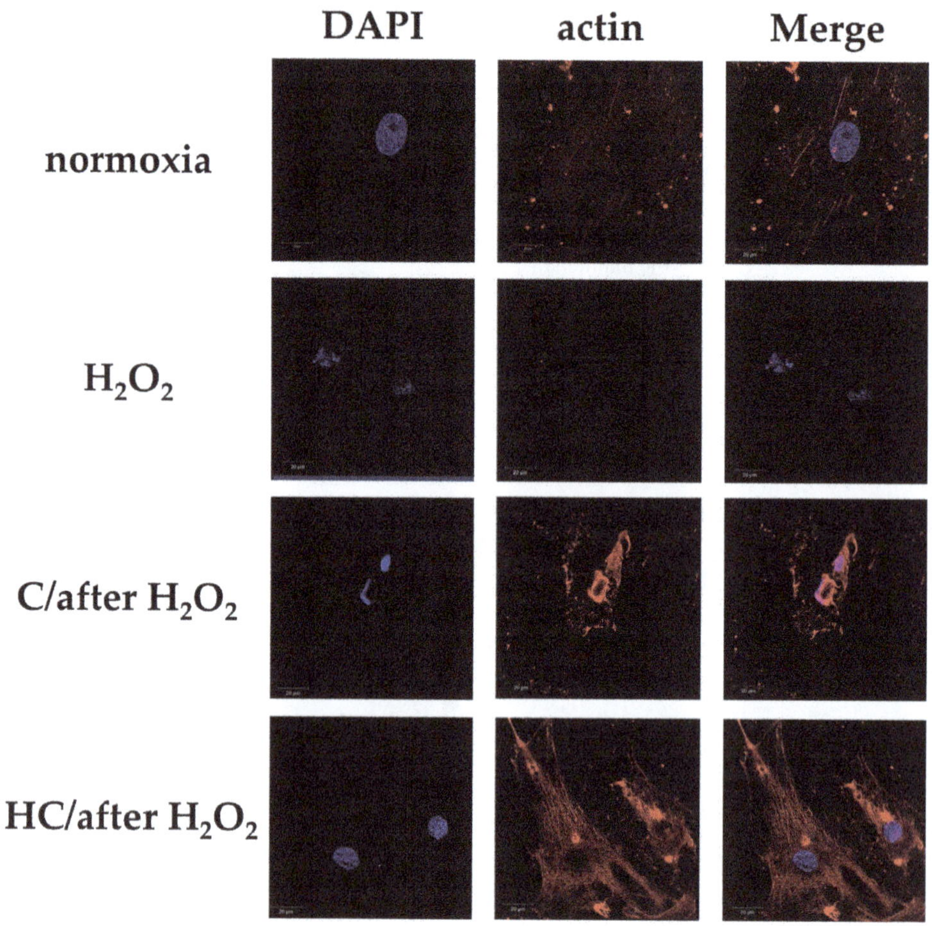

Figure 7. Fluorescent immunostaining of tendon cell actin filaments exposed to 150 μM or 30 mM ascorbic acid at 1 h after H_2O_2 exposure. Shown are representative fluorescence microscopy images for DAPI (blue) and/or actin (red) in the cells of all treatment groups. In the H_2O_2 group, a noticeable destruction in cell form and a loss of nuclear and actin filaments were observed. Contrastingly, the HC/after H_2O_2 group maintained the cell form, cytoskeleton, and actin filament expression. The C/after H_2O_2 group showed approximately the same degree of cell form destruction and actin filament loss as the H_2O_2 group. Scale bar: 20 μm.

3. Discussion

This study examined the prevention of tendon cell degeneration and its treatment. Inhibiting oxidative stress or mitochondrial damage at an early stage is important in suppressing the progression of tendon cell degeneration [19]. Thus, we studied the responses of the tendon cells in the early stages after H_2O_2 exposure. Our results showed that oxidative injury was significantly inhibited with high-dose vitamin C compared to ordinary-dose vitamin C, in addition to the maintenance of the tendon cells themselves.

The cyclin-dependent kinase (CDK) inhibitor and cellular senescence marker $p16^{INK4a}$ inhibit CDK4/6, reducing the phosphorylation of Rb protein, which is necessary for progression through the DNA replication phase of the cell cycle and for causing the cell cycle to stop [27,28]. Additionally, the removal of $p16^{INK4a}$-positive cells has been reported to improve various age-related pathologies [29]. Since $p16^{INK4a}$ also affects regeneration and repair in cardiac muscle cells, it regulates the promotion of cardiac muscle cell proliferation via CDK4/6 and ROS-related autophagy. Additionally, overexpression of $p16^{INK4a}$ overactivates autophagy, increasing myocardial damage and weakening myocardial repair [30]. In this study, an increase in $p16^{INK4a}$ expression was observed in tendon cells exposed to H_2O_2; however, treatment with high-dose vitamin C inhibited $p16^{INK4a}$ expression. This finding demonstrated that high-dose vitamin C administration is effective against muscle and tendon cell damage.

Mitochondria produce adenosine triphosphate (ATP), a molecule necessary for life activity, contributing to the maintenance of cell functions and reasonable stress responses. When mitochondrial dysfunction occurs, ATP production is reduced, and ROS production, a cause of DNA damage, is increased. This reduces cellular function, causes cellular transformation, and induces cellular senescence, cell death, or inflammation, finally leading to damage to body tissue or organ function [31–34]. Tendinopathy is an aging-related disease, and degeneration is said to be the cause. Degenerated tendons are a significant source of ROS. Recent studies have reported that mitochondrial dysfunction is involved in tendinopathy [35]. The presence of TFAM is responsible for maintaining and stabilizing mitochondrial DNA. Since mitochondrial DNA is easily subjected to oxidative injury, PCR-based gene expression analysis was performed to investigate the effects of oxidative injury on mitochondria in tendon cells where degeneration is induced. Both TFAM expression and mitochondrial function were reduced in degenerated tendon cells. However, treatment with high-dose vitamin C increased TFAM expression and maintained mitochondrial function. These results reveal that high-dose vitamin C protects mitochondria, inhibiting oxidative injury and tendon cell degeneration. Furthermore, oxidative stress causes the oxidation of actin, which promotes the formation of stress fibers and oxidizes actin regulatory proteins, resulting in the impairment of normal actin polymerization [36]. As mitochondria interact with the actin cytoskeleton, mutations in actin or actin-binding proteins affect the mitochondrial pathway and lead to cell death [37]. In this experiment, high-dose vitamin C suppressed actin oxidation by reducing the levels of ROS induced by H_2O_2. Moreover, it is believed that by protecting mitochondria, the actin cytoskeleton was maintained, and cell death was suppressed. These results also demonstrated that treating tendon cells with high-dose vitamin C maintains cytoskeletal morphology, raising the possibility that high-dose vitamin C is effective for inhibiting the progression of tendon cell degeneration. Collectively, the results of this study suggested that mitochondrial damage and senescence factors are involved in the development of tendon cell degeneration and injury. Therefore, mitochondria should be protected and the degeneration of tendon cells should be inhibited by suppressing cellular senescence. In this study, treatment with high-dose vitamin C maintained the cytoskeleton and increased the gene expression of type I collagen, suggesting a decrease in the risk of degenerated tendon injury.

Restoring a degenerated tendon tissue to a healthy tendon tissue is currently impossible. Only a few studies have addressed the therapeutic effectiveness or inhibitory effects for degenerative tendinopathy. It is important to inhibit mitochondrial damage and oxidative stress at an early stage to prevent mitochondria-mediated apoptosis and to inhibit tendon cell degeneration. Therefore, we investigated the therapeutic effect of vitamin C after H_2O_2 exposure and the occurrence of tendon cell degeneration; an increase in dead cells was observed 1 h after H_2O_2 exposure, followed by a high dose of vitamin C, which considerably suppressed cell death. Additionally, the cytoskeleton was maintained when high-dose vitamin C was added 1 h after H_2O_2 exposure. In other words, treatment of tendon cells with high-dose vitamin C showed promising results in inhibiting the progression of their degeneration. These results suggest that mitochondrial damage and senescent cells may be suppressed even when tendon cells are degenerated, and high-dose vitamin C may be expected to have a favorable postoperative effect even in cases where tendon tissue is degenerated and at risk of re-rupture.

Limitations of this study are as follows: (1) the study was conducted in vitro (not in vivo), and (2) the study was based on the occurrence of tendon cell death caused by H_2O_2; hence, it was conducted for a maximum of 24 h, and the effects of ascorbic acid on long-term tendon degeneration and tendon damage are not known. Although we believe it is necessary to conduct further studies in animal experiments, this study demonstrates the efficacy of high concentrations of ascorbic acid on tendon cells; these findings provide a valuable foundation for future studies. (3) In this study, we examined TFAM with a focus on mitochondrial function, but we did not examine Ca flux, so more detailed studies on the mechanisms of mitochondrial damage and cell death may be needed [38,39].

Ascorbic acid is an inexpensive and clinically used drug. Although the safety of high-dose administration of vitamin C has been reported [23,40], there are some caveats. Gastrointestinal discomfort and kidney stones are possible when large doses of vitamin C are administered. Caution should be applied when administering high doses of vitamin C, especially in patients suffering from kidney disease or hemochromatosis [40,41]. It is also reported that 40 mM ascorbic acid can be considered as the minimal concentration to induce significant cell changes, and thus the toxicological threshold and the concentration should be carefully considered when used [42]. Therefore, its use should be administered under the supervision of a health care provider or health care professional. Considering the results of this study and the safety concerns reported thus far, high-dose vitamin C therapy, which should be carefully monitored with careful monitoring of physical condition, may be an effective prevention and treatment for a class of injuries resulting from degeneration of tendons, including the Achilles tendon, patellar tendon, and shoulder rotator cuff.

4. Materials and Methods

4.1. Cell Culture

Human tendon cells (Zen-bio, Durham, NC, USA) were seeded into culture dishes and incubated for 24 h in a humidified incubator at 37 °C with 95% air and 5% CO_2 to allow for cell attachment and proliferation. Tendon cells were maintained as subconfluent monolayer cultures in a tenocyte growth medium (Zen-bio) supplemented with 10% fetal bovine serum (FBS; GIBCO-BRL, Grand Island, NY, USA) at 37 °C under 20% O_2 and 5% CO_2. When the cultures reached 70% confluence, 2 mM H_2O_2 (WAKO, Tokyo, Japan) + 150 μM ascorbic acid (WAKO) (Group C) or 2 mM H_2O_2 + 30 mM ascorbic acid (WAKO) (Group HC) was added to the medium, and the cells were incubated for 3 h. In the next experiment, 150 μM or 30 mM ascorbic acid was added to the tendon cell culture medium that had been incubated for 1 h with H_2O_2 (C/after H_2O_2 group, HC/after H_2O_2 group). The cells were incubated for a total of 3 h and then cultured under 20% O_2 in a medium without H_2O_2, ascorbic acid, or FBS (GIBCO-BRL; normoxia group) or in a medium containing H_2O_2 only (Group H_2O_2). Three independent experiments were performed for each group.

4.2. Analysis of Reactive Oxygen Species (ROS)

ROS was determined using a commercial DCFDA-cellular ROS detection assay kit following the manufacturer's instructions (Cayman Chemical Company, Ann Arbor, MI, USA). Briefly, tenocyte was plated on a 96-well plate (2.5×10^4 cells/well). After overnight attachment, the cells were treated with H_2O_2 at a designated concentration for 24 h, followed by staining with a cell permeant reagent, 2′,7′-dichlorofluorescein diacetate (DCFDA), for 30 min at 37 °C. Deacetylated DCFDA was fluorescently determined after being oxidized by ROS with Ex 495 nm/Em 529 nm by fluorescence microplates (Fluoroskan, Thermo Fisher Scientific, Waltham, MA, USA). In each group, 5 wells were assayed 3 times.

4.3. Detection of Intracellular ROS Generation

The experiments were conducted in accordance with the protocol provided with the ROS Assay Kit-Photo-oxidation Resistant DCFH-DA (Dojindo, Kumamoto, Japan). ROS production was analyzed using the ROS Assay Kit-Photo-oxidation Resistant DCFH-DA. Cells were seeded on dishes or plates and placed in an incubator at 37 °C with 95% air and 5% CO_2. The culture medium was then removed, and the cells were washed twice with Hank's basic salt solution (HBSS). An appropriate volume of photo-oxidation resistant DCFH-DA working solution was then added to the cells. The cells were then incubated as in step 1 for 30 min. The working solution was then discarded, and the cells were washed with HBSS. Next, a medium containing a ROS-inducing agent was then added, and the cells were incubated as in step 1 for an appropriate time. After removing the supernatant, the cells were washed twice with phosphate-buffered saline (PBS). The supernatant was then replaced with DAPI (1 μg/mL) in a blocking solution, and the cells were incubated at room temperature for 1 h. Finally, fluorescence microscopy (470 nm and 530 nm LED modules) was performed using a BZ-X700 fluorescence microscope (Keyence, Tokyo, Japan).

4.4. Quantitative RT-PCR Analysis

Total RNA was isolated according to an RNA extraction kit (Isogen, Nippon Gene, Tokyo, Japan), which included a DNase digestion step. cDNA was synthesized using a High-Capacity RNA-to-cDNA Kit (Thermo Fisher Scientific) in a mixture containing 1 μg of total RNA in 20 μL. Real-time qPCR analysis was performed on cDNA equivalent to 20 ng RNA in a 10 μL reaction volume using a QuantStudio 3 Real-Time PCR system (Thermo Fisher Scientific). Gene expression analysis in human tendon cells was measured using a TaqMan Fast Advanced Master Mix (Thermo Fisher Scientific) and one of the following TaqMan gene expression assays (Thermo Fisher Scientific): TFAM (Hs01082775_ml), ATP5A (Hs00900735_ml), CDKN2A (Hs00923894_ml), Collagen1 (Hs00164004_ml). It was then normalized to 18srRNA (Hs03003631_g1). The threshold cycle (Ct) was determined after setting the threshold in the linear amplification step of the PCR reaction. The ΔCt of a specific gene was defined as the Ct (target gene) − Ct (18srRNA). Data were analyzed using the relative quantification (ΔΔCt) method in QuantStudio Design & Analysis software v1.5 (Thermo Fisher Scientific). The data are shown as averages of three independent experiments.

4.5. Immunostaining for F-Actin and ATP Synthase (ATP5A)

The cultured cells were fixed in 4% paraformaldehyde, washed in PBS, and permeabilized with 0.3% Triton X-100 in PBS. Nonspecific binding was blocked by incubating the cells with 10% bovine serum albumin (Dako Cytomation, Santa Clara, CA, USA) in PBS for 15 min. They were then incubated with anti-actin antibodies (1:200; Sigma Aldrich, St. Louis, MO, USA) and anti-ATP5A antibodies (1:200; Proteintech, Rosemont, IL, USA) for 2 h. The cells were then incubated with Alexa 594-labeled secondary antibodies (Thermo Fisher Scientific, Waltham, MA, USA) for anti-actin and Alexa 488-labeled secondary antibodies (Thermo Fisher Scientific) for anti-ATP5A. The cells were then incubated with 4′,6-diamidino-2-phenylindole (DAPI) for 30 min. After washing, a Prolong diamond antifade mountant (Thermo Fisher Scientific) was added, and coverslips were mounted. Finally, fluorescence microscopy images were taken using a Zeiss-LSM710 microscope (Zeiss, Baden-Württemberg, Germany).

4.6. Determination of Cell Viability

The tendon cells were cultured at a density of 5.0×10^4 cells/well in a 24-well plate. One group comprised tendon cells exposed to 2 mM H_2O_2 for 1 h (H_2O_2 1 h group) before adding 30 mM ascorbic acid. The cells were then incubated in the tendon cell culture medium containing 30 mM ascorbic acid for 2 h (HC/after H_2O_2 group). The cells were incubated for a total of 3 h. Another group was exposed to 2 mM H_2O_2 only for 3 h (H_2O_2 group). After exposure to the respective treatments, the cells were stained using the Trypan blue (0.25%) dye exclusion method, the numbers of viable and nonviable cells were determined using a Countess 2 FL instrument (Thermo Fisher Scientific), and the survival rates were calculated. The control cells (normoxia group) were cultured in a medium without H_2O_2, ascorbic acid, or FBS (GIBCO-BRL). The tendon cells were cultured in three wells for all groups. In each group, the three wells were measured three times.

4.7. Statistical Analysis

All quantified results were expressed as the mean $\pm$ deviation (SD). Statistical significance was analyzed using a one-way analysis of variance followed by Fisher's protected least significant difference post-hoc test. Significance was defined at p-values < 0.05. Statistical analyses were performed using Stat View J-5.0 software (SAS Institute, Cary, NC, USA).

5. Conclusions

Although the tendon cell lines used in this study were in vitro studies derived from human Achilles and patellar tendons, the results of this study suggest that the addition of high-dose vitamin C, which has a strong antioxidant effect, may play a protective role in tendon cells.

This study may help in the prevention and treatment of tendon cells, although it is necessary to consider animal studies to see if similar effects can be obtained in different parts of tendons and ligaments (e.g., shoulder rotator cuff, Achilles tendon, finger tendons, etc.).

Author Contributions: S.U., T.I. and M.S. (Miyako Shimasaki) conceived and designed the experiments; S.U., T.I., M.S. (Miyako Shimasaki) and D.S. performed the experiments; S.U., T.I. and M.S. (Masaru Sakurai) analyzed the data; S.U., T.I., M.S. (Miyako Shimasaki), A.K. and N.K. interpreted the data and drafted the manuscript. All authors have read and agreed to the published version of the manuscript.

Funding: This study was partially supported by Grant-in-Aid for Young Scientists (No. 19K18511) to S.U. from The Ministry of Education, Culture, Sports, Science, and Technology of Japan.

Institutional Review Board Statement: Not applicable.

Informed Consent Statement: Not applicable.

Data Availability Statement: The datasets used and/or analyzed during the current study are available from the corresponding author upon reasonable request.

Acknowledgments: This work was supported by the grant for Promoted Research (S2018-1) to S.U.; the grant for Assist KAKEN (K2024-9) to S.U.; and the grant for Promoted Research (S2023-3) to M.S. (Miyako Shimasaki) from Kanazawa Medical University.

Conflicts of Interest: The authors declare no conflicts of interest.

References

1. Lui, P.P.Y.; Zhang, X.; Yao, S.; Sun, H.; Huang, C. Roles of oxidative stress in acute tendon injury and degenerative tendinopathy—A target for intervention. *Int. J. Mol. Sci.* **2022**, *23*, 3571. [CrossRef] [PubMed]
2. Shahid, H.; Morya, V.K.; Oh, J.U.; Kim, J.H.; Noh, K.C. Hypoxia-inducible factor and oxidative stress in tendon degeneration: A molecular perspective. *Antioxidants* **2024**, *13*, 86. [CrossRef] [PubMed]
3. Li, P.; Zhou, H.; Tu, T.; Lu, H. Dynamic exacerbation in inflammation and oxidative stress during the formation of peritendinous adhesion resulted from acute tendon injury. *J. Orthop. Surg. Res.* **2021**, *16*, 293. [CrossRef]
4. Lowes, D.A.; Wallace, C.; Murphy, M.P.; Webster, N.R.; Galley, H.F. The mitochondria targeted antioxidant MitoQ protects against fluoroquinolone-induced oxidative stress and mitochondrial membrane damage in human Achilles tendon cells. *Free Radic. Res.* **2009**, *43*, 323–328. [CrossRef] [PubMed]
5. Yuan, T.; Qian, H.; Yu, X.; Meng, J.; Lai, C.T.; Jiang, H.; Zhao, J.N.; Bao, N.R. Proteomic analysis reveals rotator cuff injury caused by oxidative stress. *Ther. Adv. Chronic. Dis.* **2021**, *12*, 2040622320987057. [CrossRef]
6. Yoshida, K.; Itoigawa, Y.; Wada, T.; Maruyama, Y.; Nojiri, H.; Kawasaki, T.; Kaneko, K. Association of superoxide-induced oxidative stress with rotator cuff tears in human patients. *J. Orthop. Res.* **2020**, *38*, 212–218. [CrossRef]
7. Yang, J.; He, J.; Yang, L. Advanced glycation end products impair the repair of injured tendon: A study in rats. *BMC Musculoskelet. Disord.* **2024**, *25*, 700. [CrossRef]

8. Skovgaard, D.; Svensson, R.B.; Scheijen, J.; Eliasson, P.; Mogensen, P.; Hag, A.M.; Kjær, M.; Schalkwijk, C.G.; Schjerling, P.; Magnusson, S.P.; et al. An advanced glycation endproduct (AGE)-rich diet promotes accumulation of AGEs in Achilles tendon. *Physiol. Rep.* **2017**, *5*, e13215. [CrossRef] [PubMed]
9. Mifune, Y.; Inui, A.; Muto, T.; Nishimoto, H.; Kataoka, T.; Kurosawa, T.; Yamaura, K.; Mukohara, S.; Niikura, T.; Kokubu, T.; et al. Influence of advanced glycation end products on rotator cuff. *J. Shoulder Elb. Surg.* **2019**, *28*, 1490–1496. [CrossRef] [PubMed]
10. Benson, R.T.; McDonnell, S.M.; Knowles, H.J.; Rees, J.L.; Carr, A.J.; Hulley, P.A. Tendinopathy and tears of the rotator cuff are associated with hypoxia and apoptosis. *J. Bone Jt. Surg. Br.* **2010**, *92*, 448–453. [CrossRef]
11. May, J.M.; Harrison, F.E. Role of vitamin C in the function of the vascular endothelium. *Antioxid. Redox Signal.* **2013**, *19*, 2068–2083. [CrossRef]
12. Carr, A.C.; Maggini, S. Vitamin C and immune function. *Nutrients* **2017**, *9*, 1211. [CrossRef] [PubMed]
13. Padayatty, S.J.; Katz, A.; Wang, Y.; Eck, P.; Kwon, O.; Lee, J.H.; Chen, S.; Corpe, C.; Dutta, A.; Dutta, S.K.; et al. Vitamin C as an antioxidant: Evaluation of its role in disease prevention. *J. Am. Coll. Nutr.* **2003**, *22*, 18–35. [CrossRef] [PubMed]
14. Morikawa, D.; Nojiri, H.; Itoigawa, Y.; Ozawa, Y.; Kaneko, K.; Shimizu, T. Antioxidant treatment with vitamin C attenuated rotator cuff degeneration caused by oxidative stress in Sod1-deficient mice. *JSES Open Access* **2018**, *2*, 91–96. [CrossRef] [PubMed]
15. Noriega-González, D.C.; Drobnic, F.; Caballero-García, A.; Roche, E.; Perez-Valdecantos, D.; Córdova, A. Effect of vitamin C on tendinopathy recovery: A scoping review. *Nutrients* **2022**, *14*, 2663. [CrossRef] [PubMed]
16. Uehara, H.; Itoigawa, Y.; Morikawa, D.; Koga, A.; Tsurukami, H.; Maruyama, Y.; Ishijima, M. The effect of vitamin C and N-acetylcysteine on tendon-to-bone healing in a rodent model of rotator cuff repair. *Am. J. Sports Med.* **2023**, *51*, 1596–1607. [CrossRef] [PubMed]
17. Yamanaka, S.; Tatsumi, T.; Shiraishi, J.; Mano, A.; Keira, N.; Matoba, S.; Asayama, J.; Fushiki, S.; Fliss, H.; Nakagawa, M. Amlodipine inhibits doxorubicin induced apoptosis in neonatal rat cardiac myocytes. *J. Am. Coll. Cardiol.* **2003**, *41*, 870–878. [CrossRef]
18. Leopold, J.A. Antioxidants and coronary artery disease: From pathophysiology to preventive therapy. *Coron. Artery Dis.* **2015**, *26*, 176–183. [CrossRef] [PubMed]
19. Ueda, S.; Ichiseki, T.; Shimasaki, M.; Hirata, H.; Kawahara, N.; Ueda, Y. Inhibitory effect of taurine on rotator cuff degeneration via mitochondrial protection. *Am. J. Transl. Res.* **2022**, *14*, 6286–6294.
20. Böttger, F.; Vallés-Martí, A.; Cahn, L.; Jimenez, C.R. High-dose intravenous vitamin C, a promising multi-targeting agent in the treatment of cancer. *J. Exp. Clin. Cancer Res.* **2021**, *40*, 343. [CrossRef] [PubMed]
21. Ko, S.H.; Jun, J.H.; Oh, J.E.; Shin, E.; Kwak, Y.L.; Shim, J.K. Effect of high-dose vitamin C on renal ischemia-reperfusion injury. *Biomed. Pharmacother.* **2024**, *173*, 116407. [CrossRef] [PubMed]
22. Cui, Y.N.; Tian, N.; Luo, Y.H.; Zhao, J.J.; Bi, C.F.; Gou, Y.; Liu, J.; Feng, K.; Zhang, J.F. High-dose Vitamin C injection ameliorates against sepsis-induced myocardial injury by anti-apoptosis, anti-inflammatory and pro-autophagy through regulating MAPK, NF-κB and PI3K/AKT/mTOR signaling pathways in rats. *Aging* **2024**, *16*, 6937–6953. [CrossRef]
23. Lee, S.W.; Lee, Y.J.; Baek, S.M.; Kang, K.K.; Kim, T.U.; Yim, J.H.; Kim, H.Y.; Han, S.H.; Choi, S.K.; Park, S.J.; et al. Mega-dose vitamin C ameliorates nonalcoholic fatty liver disease in a mouse fast-food diet model. *Nutrients* **2022**, *14*, 2195. [CrossRef]
24. Kim, S.M.; Lim, S.M.; Yoo, J.A.; Woo, M.J.; Cho, K.H. Consumption of high-dose vitamin C (1250 mg per day) enhances functional and structural properties of serum lipoprotein to improve anti-oxidant, anti-atherosclerotic, and anti-aging effects via regulation of anti-inflammatory microRNA. *Food Funct.* **2015**, *6*, 3604–3612. [CrossRef] [PubMed]
25. Rello, S.; Stockert, J.C.; Moreno, V.L.; Gamez, A.; Pacheco, M.; Juarranz, A.; Canete, M.; Villanueva, A. Morphological criteria to distinguish cell death induced by apoptotic and necrotic treatments. *Apoptosis* **2005**, *10*, 201–208. [CrossRef]
26. Wang, T.; Lei, H.; Zhou, L.; Tang, M.; Liu, Q.; Long, F.; Li, Q.; Su, J. Effect of Fumonisin B1 on Proliferation and Apoptosis of Intestinal Porcine Epithelial Cells. *Toxins* **2022**, *14*, 471. [CrossRef] [PubMed]
27. Lukas, J.; Parry, D.; Aagaard, L.; Mann, D.J.; Bartkova, J.; Strauss, M.; Peters, G.; Bartek, J. Retinoblastoma-protein-dependent cell-cycle inhibition by the tumour suppressor p16. *Nature* **1995**, *375*, 503–506. [CrossRef]
28. Stott, F.J.; Bates, S.; James, M.C.; McConnell, B.B.; Starborg, M.; Brookes, S.; Palmero, I.; Ryan, K.; Hara, E.; Vousden, K.H.; et al. The alternative product from the human CDKN2A locus, p14(ARF), participates in a regulatory feedback loop with p53 and MDM2. *EMBO J.* **1998**, *17*, 5001–5014. [CrossRef]
29. Baker, D.J.; Wijshake, T.; Tchkonia, T.; LeBrasseur, N.K.; Childs, B.G.; van de Sluis, B.; Kirkland, J.L.; van Deursen, J.M. Clearance of p16Ink4a-positive senescent cells delays ageing-associated disorders. *Nature* **2011**, *479*, 232–236. [CrossRef]
30. Sun, J.; Zhou, L.; Yang, T.; Deng, B.; Bao, Y.; Gu, L.; Wang, H.; Wang, L. P16INK4a regulates ROS-related autophagy and CDK4/6-mediated proliferation: A new target of myocardial regeneration therapy. *Oxid. Med. Cell. Longev.* **2023**, *2023*, 1696190. [CrossRef] [PubMed]
31. Sun, N.; Youle, R.J.; Finkel, T. The mitochondrial basis of aging. *Mol. Cell* **2016**, *61*, 654–666. [CrossRef] [PubMed]
32. López-Otín, C.; Blasco, M.A.; Partridge, L.; Serrano, M.; Kroemer, G. The hallmarks of aging. *Cell* **2013**, *153*, 1194–1217. [CrossRef] [PubMed]
33. Lima, T.; Li, T.Y.; Mottis, A.; Auwerx, J. Pleiotropic effects of mitochondria in aging. *Nat. Aging* **2022**, *2*, 199–213. [CrossRef] [PubMed]
34. Cui, H.; Kong, Y.; Zhang, H. Oxidative stress, mitochondrial dysfunction, and aging. *J. Signal Transduct.* **2012**, *2012*, 646354. [CrossRef] [PubMed]

35. Cheng, L.; Zheng, Q.; Qiu, K.; Elmer Ker, D.F.; Chen, X.; Yin, Z. Mitochondrial destabilization in tendinopathy and potential therapeutic strategies. *J. Orthop. Translat.* **2024**, *49*, 49–61. [CrossRef] [PubMed]
36. Patel, V.S.; Sampat, V.; Espey, M.G.; Sitapara, R.; Wang, H.; Yang, X.; Ashby, C.R.; Thomas, D.D.; Mantell, L.L. Ascorbic acid attenuates hyperoxia-compromised host defense against pulmonary bacterial infection. *Am. J. Respir. Cell Mol. Biol.* **2016**, *55*, 511–520. [CrossRef]
37. Boldogh, I.R.; Pon, L.A. Interactions of mitochondria with the actin cytoskeleton. *Biochim. Biophys. Acta* **2006**, *1763*, 450–462. [CrossRef] [PubMed]
38. Koh, J.H.; Kim, Y.W.; Seo, D.Y.; Sohn, T.S. Mitochondrial TFAM as a Signaling Regulator between Cellular Organelles: A Perspective on Metabolic Diseases. *Diabetes Metab. J.* **2021**, *45*, 853–865. [CrossRef]
39. Kaur, S.; Sehrawat, A.; Mastana, S.S.; Kandimalla, R.; Sharma, P.K.; Bhatti, G.K.; Bhatti, J.S. Targeting calcium homeostasis and impaired inter-organelle crosstalk as a potential therapeutic approach in Parkinson's disease. *Life Sci.* **2023**, *330*, 121995. [CrossRef]
40. Padayatty, S.J.; Sun, A.Y.; Chen, Q.; Espey, M.G.; Drisko, J.; Levine, M. Vitamin C: Intravenous use by complementary and alternative medicine practitioners and adverse effects. *PLoS ONE* **2010**, *5*, e11414. [CrossRef]
41. Gurm, H.; Sheta, M.A.; Nivera, N.; Tunkel, A. Vitamin C-induced oxalate nephropathy: A case report. *J. Community Hosp. Intern. Med. Perspect.* **2012**, *2*, 17718. [CrossRef] [PubMed]
42. Melin, A.M.; Perromat, A.; Déléris, G. Pharmacologic application of FTIR spectroscopy: Effect of ascorbic acid-induced free radicals on Deinococcus radiodurans. *Biospectroscopy* **1999**, *5*, 229–236. [CrossRef]

MDPI AG
Grosspeteranlage 5
4052 Basel
Switzerland
Tel.: +41 61 683 77 34

International Journal of Molecular Sciences Editorial Office
E-mail: ijms@mdpi.com
www.mdpi.com/journal/ijms

www.ingramcontent.com/pod-product-compliance
Lightning Source LLC
Chambersburg PA
CBHW041947260926
51850CB00018B/1956
9783725876631